T0135673

Development of New Synthetic Approaches towards the Anthraquinone-Xanthone Heterodimeric Structure of Beticolin 0

Zur Erlangung des akademischen Grades eines

DOKTORS DER NATURWISSENSCHAFTEN

(Dr. rer. nat.)

von der KIT-Fakultät für Chemie und Biowissenschaften

des Karlsruher Instituts für Technologie (KIT)

genehmigte

DISSERTATION

von

M. Sc. Janina Beck

aus Mannheim

Dekan: Prof. Dr. Manfred Wilhelm

Referent: Prof. Dr. Stefan Bräse

Korreferent: Prof. Dr. Joachim Podlech

Tag der mündlichen Prüfung: 21.04.2020

Band 91
Beiträge zur organischen Synthese
Hrsg.: Stefan Bräse

Prof. Dr. Stefan Bräse
Institut für Organische Chemie
Karlsruher Institut für Technologie (KIT)
Fritz-Haber-Weg 6
D-76131 Karlsruhe

Bildquelle: https://www.flickr.com/photos/scotnelson/15332755807

Bibliografische Information der Deutschen Bibliothek

Die Deutsche Nationalbibliothek verzeichnet diese Publikation in der
Deutschen Nationalbibliografie; detaillierte bibliografische Daten sind
im Internet über http://dnb.d-nb.de abrufbar.

ISBN 978-3-8325-5146-9
ISSN 1862-5681

Logos Verlag Berlin GmbH
Georg-Knorr-Str. 4, Geb. 10
12681 Berlin
Tel.: +49 030 42 85 10 90
Fax: +49 030 42 85 10 92
INTERNET: http://www.logos-verlag.de

Für Patricia.

Die vorliegende Arbeit wurde von November 2016 bis April 2020 am Institut für Organische Chemie des Karlsruher Instituts für Technologie (KIT) unter der Leitung von Prof. Dr. Stefan Bräse durchgeführt.

Die Arbeit wurde durch ein Promotionsstipendium der Carl-Zeiss-Stiftung gefördert.

Hiermit erkläre ich, die vorliegende Arbeit selbständig verfasst und keine anderen als die angegebenen Quellen und Hilfsmittel verwendet, sowie Zitate kenntlich gemacht zu haben. Die Dissertation wurde bisher an keiner anderen Hochschule oder Universität eingereicht.

Janina Beck   Karlsruhe, April 2020

Table of Contents

1 Abstract 1

2 Kurzzusammenfassung – Abstract in German 3

3 Introduction 5

   3.1   Natural Product Synthesis over Time 5

   3.2   Mycotoxins 9

      3.2.1   Mycotoxins Derived from Tetrahydroxanthones 9

      3.2.2   Anthraquinone-Xanthone Heterodimers 10

   3.3   Beticolins 13

      3.3.1   Structure and Nomenclature 13

      3.3.2   Biological Activities 15

      3.3.3   The Biosynthesis of Beticolins 18

   3.4   Synthetic Studies on Beticolin Subunits 19

      3.4.1   Tetrahydroxanthones 19

      3.4.2   Anthraquinone Derivatives 22

4 Objective 25

5 Results and Discussion 27

   5.1   Retrosynthetic Analysis 27

   5.2   Naphthoquinone Derivatives 30

   5.3   Synthesis of Functionalized Dienes 33

   5.4   Synthesis of Anthraquinone Derivatives 35

      5.4.1   The Diels-Alder Reaction 35

      5.4.2   Anthraquinone Model System 37

      5.4.3   Diels-Alder Reaction with Cyclic Dienes 38

      5.4.4   Scope of the Cycloaddition 39

      5.4.5   Studies on the Control of the Selectivity in Diels-Alder Reactions 44

   5.5   Intramolecular Cyclization of Anthraquinone Derivatives 48

5.5.1   Radical Cyclization 48

5.5.2   First Studies on Intramolecular Heck Reactions 49

5.6   Further Studies on Diels-Alder Reactions 60

5.6.1   Towards Highly Functionalized Anthraquinone Derivatives 60

5.7   Intramolecular Heck Reactions with Functionalized Anthraquinone Derivatives 65

5.8   Modifications of Anthraquinones 68

5.8.1   Silyl Ether Deprotection 68

5.9   Synthesis of a Xanthone Dienophile 70

5.10   Alternative Coupling Strategies: Suzuki-Miyaura Coupling 76

5.10.1   Halogenated Dienes 76

5.10.2   Borylated Dienophiles 78

5.10.3   Diels-Alder Reactions with Halogenated Dienes 79

6   Summary 83

7   Experimental Part 89

7.1   General Remarks 89

7.1.1   Preparative Work 89

7.1.2   Solvents and Reagents 89

7.1.3   Analytics and Equipment 90

7.2   Syntheses and Characterizations 93

7.2.1   General Procedures (GPs) 93

7.2.2   Precursor Synthesis 95

7.2.3   Synthesis of Naphthoquinone Derivatives 101

7.2.4   Syntheses of Dienes 106

7.2.5   Synthesis of Anthraquinone Derivatives 112

7.2.6   Synthesis of the Heck Products 143

7.2.7   Products of the Modifications 156

7.2.8   Synthesis of a Xanthone Dienophile 159

7.3   Crystallographic Data... 165

    7.3.1   Crystallographic Data Solved by Dr. Martin Nieger................................. 165

    7.3.2   Crystallographic Data Solved by Dr. Olaf Fuhr...................................... 181

8   List of Abbreviations.. 211

9   References 213

10   Appendix 219

1 Abstract

Mycotoxins are fungal secondary metabolites exhibiting adverse effects on humans, animals as well as crops, resulting in diseases and economic loss. Beticolins are mycotoxins produced by the fungus *Cercospora beticola* which is responsible for cercosporiosis, commonly known as leaf spot disease, causing heavy damages to crops worldwide. With total synthesis approaches access to such biologically active compounds as well as novel derivatives thereof, expected to exhibit similar activities, can be facilitated in order to study their mechanism of action.

Beticolins consist of a chlorinated tetrahydroxanthone linked to an anthraquinone subunit *via* a unique bicyclo[3.2.2]nonane ring system. As this bridged core structure is the most interesting and at the same time most challenging part of the molecule, its contruction represents the key step of a total synthesis approach. Since beticolin 0 (**1**) can easily be converted to its related compounds, this thesis was focused on synthetic studies towards this parent compound. First, a simple three-step route towards diverse naphthoquinone derivatives was established. Subsequent Diels-Alder cycloadditions between these naphthoquinone dienophiles and functionalized dienes afforded the highly functionalized anthraquinone subunit of beticolin 0 (**1**) containing a sterically congested quaternary center. The regio- and stereoselectivity of the reactions was analyzed as well as the reactivity of the compounds. For many of the synthesized anthraquinone derivatives single crystals were obtained and the structures could be confirmed by X-ray diffraction.

In order to pave the way for the implementation of the tetrahydroxanthone subunit of beticolin 0 (**1**) into the pursued synthetic route, important steps towards the synthesis of a naphthoquinone bearing a halogenated xanthone subunit were accomplished.

For the construction of the bicyclo[3.2.2]nonane ring system of beticolin 0 (**1**), various intramolecular couplings were performed with the anthraquinone derivatives. The formation of the desired scaffold turned out to be challenging, however a variety of novel bicyclo[3.3.1]systems was obtained, representing interesting scaffolds whose biological properties remain to be investigated.

During a research stay at the University of Copenhagen, the functionalization of β-peptoid foldamers was examined. Peptidomimetics adopting helical structures with well-defined display of functional groups while being resistant to proteolysis, are of interest for the development of foldamers with a desired function, such as catalysis or as a scaffold enabling polyvalent display. Therefore, azide groups were successfully introduced into robustly folded β-peptoid helices, by synthesis and incorporation of novel chiral building blocks. The obtained hexamers were further functionalized with polar side chains as well as metal chelating ligands *via* click chemistry.

2 Kurzzusammenfassung – Abstract in German

Mykotoxine sind Sekundärmetabolite aus Pilzen, die schädliche Effekte auf Menschen, Tiere und Anbaukulturen ausüben und somit in Krankheiten und wirtschaftlichem Schaden resultieren. Zu diesen Toxinen zählen Beticoline, die aus dem Pilz *Cercospora beticola* isoliert werden, der für erhebliche Ernteschäden weltweit, verursacht durch die Blattfleckenkrankheit von Zuckerrüben, verantwortlich gemacht wird. Um die Wirkmechanismen solch biologisch aktiver Substanzen genauer untersuchen zu können und um Leitstrukturen zu identifizieren, werden Totalsynthesen dieser Verbindungen und entsprechender Derivate entwickelt.

Beticoline bestehen aus einem chlorierten Tetrahydroxanthon, das über ein strukturell einzigartiges Bicyclo[3.2.2]nonan-Ringsystem mit einer Anthrachinon-Untereinheit verbunden ist. Da die überbrückte Kernstruktur den synthetisch interessantesten und zugleich herausforderndsten Teil des Moleküls darstellt, repräsentiert dessen Konstruktion den Schlüsselschritt einer Totalsynthese. Da Beticolin 0 (**1**) mittels einfacher Transformationen in verwandte Strukturen überführt werden kann, lag der Fokus dieser Arbeit auf der Entwicklung eines synthetischen Zugangs zu dieser Stammverbindung. Dafür wurde zunächst eine simple dreistufige Route zur Synthese verschiedenster Naphthochinon-Derivate ausgearbeitet. Diels-Alder Cycloadditionen zwischen Letzteren und substituierten Dienen lieferten hoch funktionalisierte Derivate der Anthrachinon-Untereinheit von Beticolin 0 (**1**) mit einem sterisch anspruchsvollen, quartären Zentrum. Die Regio- und Stereoselektivität der Reaktionen wurde untersucht, sowie die Reaktivität der beteiligten Verbindungen. Zudem konnte die dreidimensionale Anordnung vieler Anthrachinone mittels Röntgenstrukturanalyse aufgeklärt werden.

Um den Weg für die Integration der Tetrahydroxanthon-Untereinheit von Beticolin 0 (**1**) in die verfolgte Syntheseroute zu ebnen, wurden wichtige Schritte zur Synthese eines Naphthochinon-Dienophils, welches ein funktionalisiertes Xanthon trägt, verwirklicht.

Zur Konstruktion des Bicyclo[3.2.2]nonan-Ringsystems von Beticolin 0 (**1**) wurden verschiedene intramolekulare Kupplungen mit Anthrachinonen durchgeführt. Die Synthese der gewünschten Struktur stellte eine Herausforderung dar, jedoch lieferten die Reaktionen verschiedene, neuartige Bicyclo[3.3.1]nonan-Ringsysteme, die interessante Gerüststrukturen darstellen und deren biologische Eigenschaften untersucht werden sollen.

Während eines Forschungsaufenthaltes an der Universität Kopenhagen wurde die Funktionalisierung von helikalen β-Peptoiden untersucht. Proteolysestabile Peptidomimetika, die Tertiärstrukturen mit definierter Präsentation von funktionellen Gruppen einnehmen, sind interessant für die Entwicklung von Foldameren mit spezifischen Funktionen wie Katalyse oder als Gerüststruktur für die polyvalente Präsentation von Funktionalitäten. Durch die Synthese und den Einbau neuartiger chiraler Bausteine, wurden Azid-Seitengruppen in stabil gefaltete Hexamere eingebaut. Die so erhaltenen Peptoid-Helices wurden mittels Click-Chemie weiter funktionalisiert, um beispielsweise polare Seitenketten, sowie Liganden zur Komplexierung von Metallen einzuführen.

3   Introduction

Nature represents an almost unlimited source for organic molecules and thanks to evolutionary processes a structural optimization that led to highly specific and potent structures took place. As a matter of fact, a lot of these compounds have unique three-dimensional shapes equipped with functionalities that result in a precise configuration. With their high potential of being pharmacologically or biologically active components, natural products are of great importance for basic scientific research, as well as the pharmaceutical and agricultural industry.

Mycotoxins represent a huge compound class of naturally occurring toxins produced by certain fungi.[1-2] Presumably those compounds exist for as long as crops are grown but their chemical nature was only started to be investigated after the discovery of aflatoxins in the early 1960s.[3] By colonizing and thus infecting crops, the fungi give rise to massive harvest losses as well as severe diseases in both humans and animals that feed from the plants.[4-6] The understanding of biochemical pathways and the mode of action of the diseases is therefore inevitable for the discovery and development of lead structures for drug design. As nature provides only small amounts of these toxins, total synthesis approaches are essential in order to investigate such diseases.

3.1   Natural Product Synthesis over Time

The term natural product describes organic compounds found in nature and produced by living organisms such as plants, bacteria or fungi. The finding that the natural products' biological activities are not limited to the species they originate from, makes them a source of inspiration for lead compounds. Moreover, with their mostly complex structures, natural products represent challenging tasks for synthetic organic chemists and therefore played a central role in the development of organic chemistry.[7]

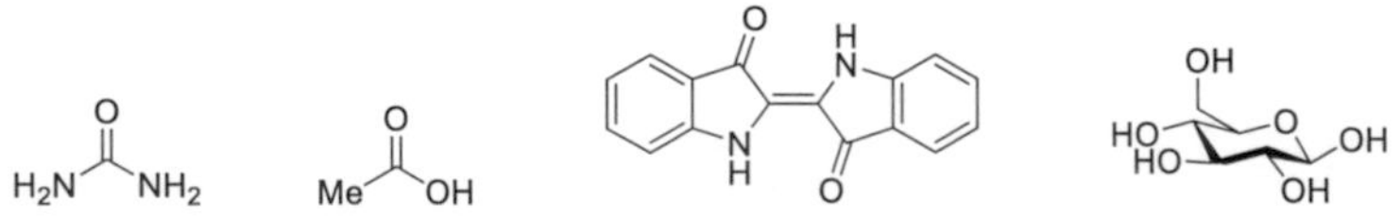

Figure 1. A collection of natural products related to the development of total synthesis in organic chemistry.

In the early nineteenth century when Wöhler successfully synthesized urea (2) from an inorganic salt, total synthesis slowly started to emerge (Figure 1).[8-9] Wöhler's finding demonstrated that despite being derived from natural organisms, natural products can be synthetically prepared in a chemical laboratory. Subsequently, achievements like the preparation of acetic acid (3) from elemental carbon by Kolbe or the synthesis of indigo (4) by Baeyer in 1882 represented examples for the targeted synthesis of natural products (Figure 1).[9-11] Since in the early times knowledge on the constitution of compounds was based on the structural theories of Kekulé and Butlerov, it was important to be able to securely determine structures.[12-14] As a consequence, only very well established reactions were applied to simple, structurally defined building blocks in order to synthesize natural products and in particular to confirm their proposed structures. An outstanding achievement by Fischer in 1890 was the first stereocontrolled synthesis of a chiral molecule, namely D-glucose (5).[15]

In the twentieth century important tools for analytical chemistry such as ultraviolet–visible (UV/Vis) and infrared (IR) spectroscopy as well as mass spectrometry were developed consecutively. A major advance followed in the 1950s when nuclear magnetic resonance (NMR) spectroscopy was discovered and applied for the structure elucidation of compounds. Since then, natural product synthesis provided a challenge to design and develop innovative reactions and reactivity patterns which attracted brilliant organic chemists and enabled a boom in the field. The liberty in the development of new synthetic methods as well as the better understanding of reaction mechanisms, paved the way for the strategic design of molecules with increased molecular complexity. Methods to stereochemically control reactions and the usage of protecting groups were for instance developed by Woodward who among many other total syntheses described the synthetic route towards quinine (6) in 1944 (Figure 2).[16]

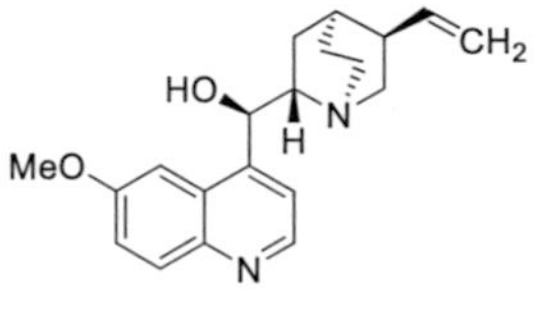

quinine (6) 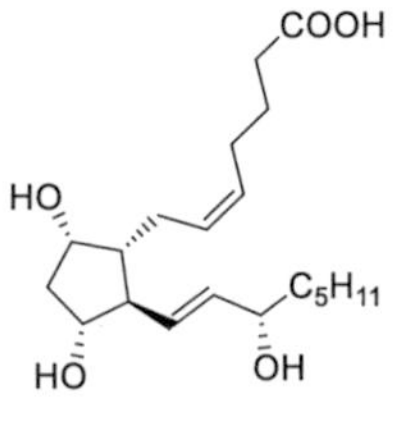   prostaglandin PGF$_2$a (7)

Figure 2. Natural products quinine (6) and prostaglandin PGF$_2\alpha$ (7).

Corey's synthesis of prostaglandin PGF$_2\alpha$ (7) proceeded with control of the stereogenic centers, as well as the planned formation of intermediates that serve as precursors for the synthesis of other compounds of the family.[17] In a publication about strategies towards complex molecules Corey stated: "*Such an effort is surely more than an intriguing theoretical exercise; it is a prerequisite to a deeper comprehension of synthesis and the methodologies which are fundamental to it, and it is likely to be a keystone in the rational development of synthesis to still higher forms.*"[18]

Techniques like the retrosynthetic analysis of structures introduced by Corey and applied by ingenious chemists together with the improvement of analytical methods allowed for total synthesis to evolve rapidly.[19]

The years between 1980 and 2000 were the most prosperous years in natural product synthesis. As soon as a new class of compounds was isolated and described, the investigation of a synthetic approach followed. The success rate was high, despite the outstanding complexity of the target molecules which is illustrated with examples shown in Figure 3: eleutherobin (**8**) (Nicolaou, 1997; Danishefsky, 1998) and ionomycin (**9**) (Hanessian, Evans, 1990).[20-25] With the development of organometallic reagents, the formation of C–C and C–N bonds was facilitated *via* methods like Stille, Negishi and Suzuki-Miyaura cross-coupling, olefin metathesis and the Heck reaction. These breakthrough innovations significantly extended the toolbox of highly effective synthetic methods by enabling C–C bond formation with high functional group tolerance.

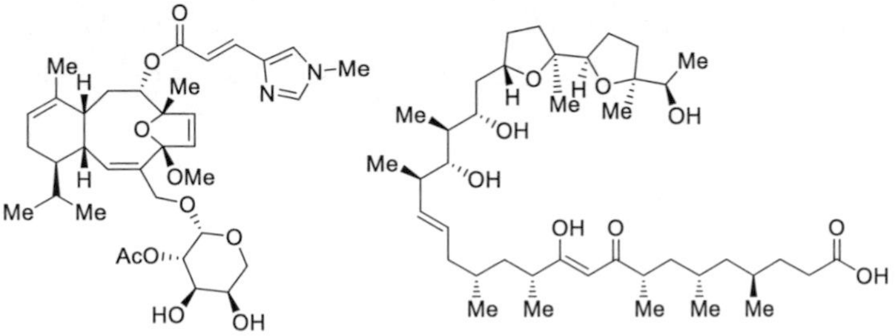

eleutherobin (**8**)   ionomycin (**9**)

Figure 3. Structures of eleutherobin (**8**) and ionomycin (**9**).

Research considerably changed in the field of natural product synthesis in recent years. The focus shifted from the synthesis of a specific natural product towards the development of new strategies and methods for the synthesis of a broad family of compounds. Moreover, by modifying the original structure of natural products and therefore tuning the properties of the active compounds, pharmacological properties can be optimized while unwanted side effects can be avoided. For the implementation of validated biological studies with natural products and their analogues, the final compounds are usually required in gram scale. As a consequence new problems arise: Nowadays the challenge of the total synthesis of natural products is not only the development of a synthetic route but also the consideration of factors like green chemistry, low cost of reagents, high yield and scalability.[26-29] Despite the countless improvements of the past decades, total synthesis remains a very challenging discipline for organic chemists.

Modern advances like flow chemistry can be used to enable complex multi-step syntheses in microfluidic systems.[30] Continuous flow conditions provide chemists with advantages including safety, scalability, and reproducibility which can enhance reactivity, pave the way for new reactions and even involve economic benefits. Flow chemistry methods are a rapidly growing area of research which provides new opportunities for natural product synthesis.

In addition, today computational approaches assist in the design of new total syntheses.[31] These methods facilitate the modeling of molecular structures and make predictions on the likely outcome of synthetic strategies.[32] The combination of experimental and theoretical achievements enables better understanding of the reactivity of complex molecular architectures. Due to the great amount of information contained within databases provided by research, artificial intelligence can be trained to interpret efficiently and utilize its potential.[33-34]

To conclude, modern natural product synthesis is an interdisciplinary research field that combines chemical, biological and pharmacological aspects. It is an exciting and dynamic area of research that allows for the invention of powerful new methods and strategies and therefore access to functionally useful molecules.

3.2   Mycotoxins

Natural products with diverse biological activities can be found among secondary metabolites which are produced by plants, bacteria or fungi and, in contrast to primary metabolites such as amino acids, play no crucial role in growth, development or reproduction.[35-37]

Fungi represent widespread organisms that metabolize during their lifespan producing a wide variety of organic compounds, including mycotoxins. The latter are secondary metabolic products which are mostly toxic and believed to be necessary to eliminate competing microorganisms.[38] Well-known representatives of this compound class are aflatoxin B$_1$ (**10**) and ergotamine (**11**) causing human as well as animal diseases (Figure 4).[39-40]

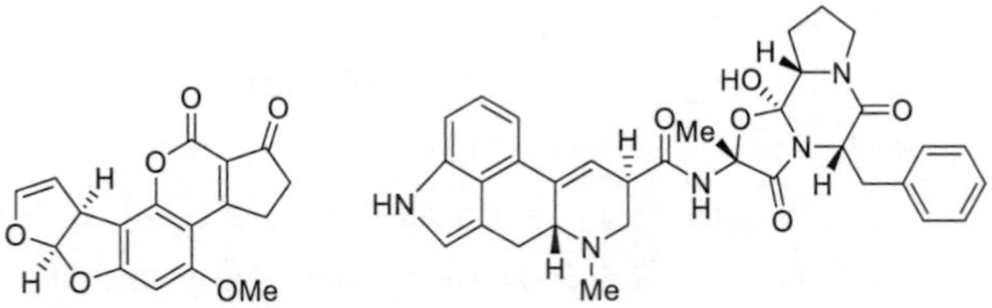

aflatoxin B$_1$ (**10**)   ergotamine (**11**)

Figure 4. Structures of aflatoxin B$_1$ (**10**) and ergotamine (**11**).

3.2.1   Mycotoxins Derived from Tetrahydroxanthones

In the design and discovery of new biologically active compounds, mycotoxins play an important role. The numbers of mycotoxins possessing tetrahydroxanthone scaffolds increased over the last years including complex structures like parnafungins **12** and secalonic acids **13** (Figure 5). Comprehensive studies towards their total syntheses were conducted in several research groups, including the group of Stefan Bräse.[41-45] Parnafungins such as **12** can be extracted from the the fungus *Fusarium larvarum* and are known to exhibit antifungal activity.[46-47] The natural products contain a tetrahydroxanthone framework linked to an isoxazolidinone and comprise two stereogenic centers. Secalonic acids **13** are 2,2'-connected, symmetrical or unsymmetrical dimers of tetrahydroxanthones.[48-50] The seven different members of this compound family differ in the configuration of the three stereogenic centers of the monomeric building blocks and show various biological activities such as cytostatic and antibacterial effects.[51-53]

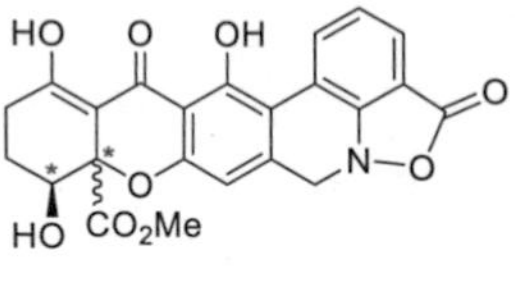

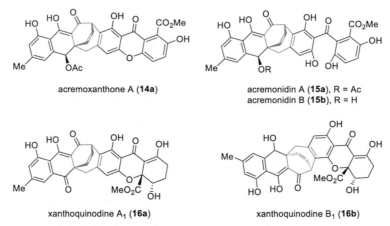

parnafungin A (**12a**) 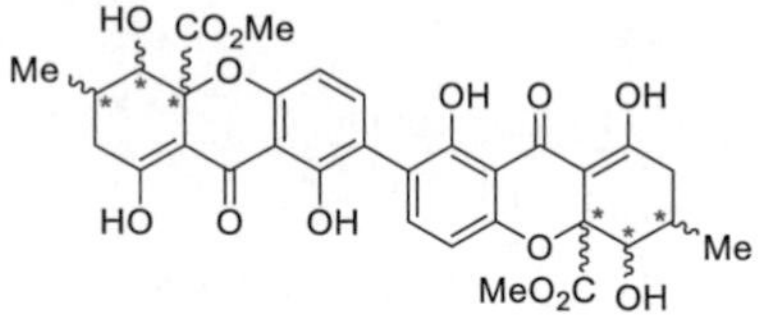   secalonic acids A – G (**13**)

Figure 5. Tetrahydroxanthone containing mycotoxins parnafungin A (**12a**) and secalonic acids A – G (**13**).

3.2.2 Anthraquinone-Xanthone Heterodimers

Anthraquinone-xanthone heterodimers belong to a large group of related compounds including acremoxanthones A – D, acremonidins A – C, as well as xanthoquinodins $A_1 – A_6$ and $B_1 – B_5$ (Figure 6).[54-58] These fungal metabolites stem from aromatic polyketides and exhibit diverse biological activities including the antibacterial activity of acremonidines and engyodontochones as well as the anticoccidial activity (coccidiosis: parasitic disease caused by coccidian protozoa) of xanthoquinodine.[57, 59-61] The structures contain a rare bicyclo[3.2.2]nonane fragment surrounded by a polycyclic system in various oxidation states. This synthetically challenging core structure attracted the attention of several research groups.[62-64]

acremoxanthone A (**14a**)   acremonidin A (**15a**), R = Ac
                                    acremonidin B (**15b**), R = H

xanthoquinodine A$_1$ (**16a**)   xanthoquinodine B$_1$ (**16b**)

Figure 6. Structures of selected anthraquinone-xanthone heterodimers **14a**, **15a/b** and **16a/b**.

In 2017 Suzuki *et al.* published the synthesis of an advanced carbocyclic scaffold relevant to acremoxanthone A (**14a**).[64] Thereby, initial attention was paid to the construction of the bridged bicyclic structure of the natural product. In a first attempt radical cyclization of enone **17** through application of tributyltin hydride (*n*Bu$_3$SnH) and azobisisobutyronitrile (AIBN) failed. However conjugate addition of **17** catalyzed by palladium acetate (Pd(OAc)$_2$) and with tris(*o*-tolyl)phosphine (POT) being the ligand gave products **18** and **19** in a 1 : 1 mixture

(Scheme 1). The yield of the desired product **18** was significantly improved by using triphenylphosphine (PPh₃) and *N,N*-diisopropylethylamine (*i*Pr₂NEt). In further experiments ketone **18** containing the bicyclo[3.2.2]nonane skeleton was extended to a pentacyclic system.[64]

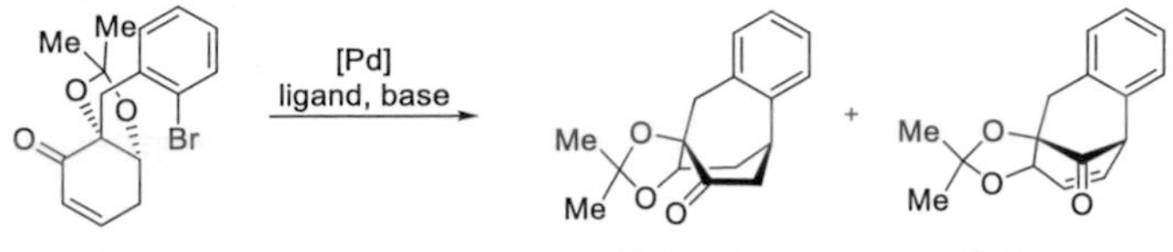

Scheme 1. Palladium-catalyzed intramolecular addition of enone **17** towards bicyclo[3.2.2]nonane skeleton **18** performed by Suzuki *et al.*[64]

Pronin and Holmbo later developed a five-step synthetic route towards the pentacyclic core of anthraquinone-xanthone heterodimers (Scheme 2) which was finalized in the synthesis of acremoxanthone A (**14a**) in ten steps.[62]

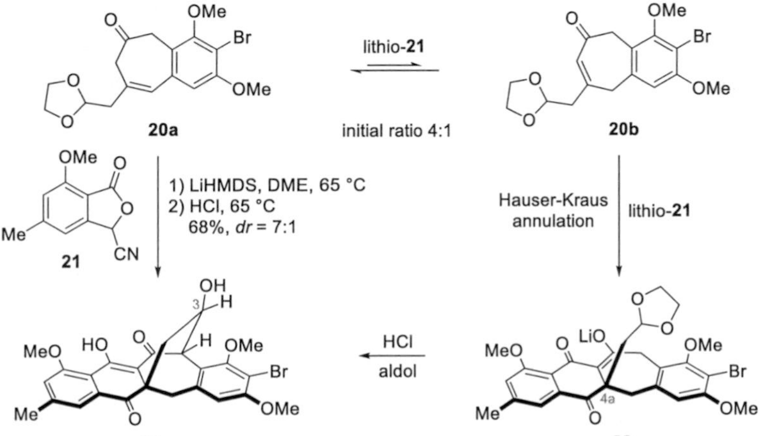

Scheme 2. Assembly of the bridged polycyclic skeleton **23** developed by Pronin and Holmbo.[62] LiHMDS = lithium bis(trimethylsilyl)amide; DME = dimethoxyethane; *dr* = diastereomeric ratio.

To begin with, the 4 : 1 mixture of ketones **20a** and **20b** was synthesized in three steps from commercially available 2-(2-bromoethyl)-1,3-dioxolane *via* an oxidative arylation protocol. The addition of a lithio derivative of cyanophthalide **21** to benzocycloheptenones **20a** and **20b** resulted in an isomerization of **20a** to **20b** and by Hauser-Kraus annulation of ketone **20b** with lithio-**21**, intermediate **22** containing a quaternary center at C4a was formed with good diastereoselectivity. Treatment of **22** with a strong acid at elevated temperatures gave the corresponding enol and deprotection of the dioxolane moiety which triggered an intramolecular aldol reaction of the formed aldehyde with the 1,3-dicarbonyl motif. The resulting characteristic bicyclo[3.2.2]nonene motif **23** was obtained with good stereoselectivity of the hydroxy functionality at C3.

Among anthraquinone-xanthone heterodimers, beticolins which will be discussed in detail in the following chapter, as well represent privileged structures due to their broad array of bioactivities.

3.3   Beticolins

Beticolins are anthraquinone-xanthone heterodimers isolated from mycelial extracts of the fungal phytopathogen *Cercospora beticola*. Numerous studies on these mycotoxins show interesting biological as well as physicochemical properties including a broad cytotoxic profile as well as antibiotic activity.[65-68] Since in nature beticolins are only produced in small amounts that require elaborate isolation processes, a synthetic access to the molecules would be a great advantage for further investigations on their specific properties.

3.3.1   Structure and Nomenclature

So far twenty closely related compounds (named beticolin 0 – 19) sharing a characteristic polycyclic skeleton were isolated. Beticolins consist of an anthraquinone derivative that is connected to a chlorinated tetrahydroxanthone *via* a bicyclo[3.2.2]nonane ring system. The latter is a rare scaffold in natural products which in beticolins is bridged by either a methylated alkene or an epoxide. Eight members of the beticolin family so far have been assigned by X-ray crystallography and NMR analysis (Figure 7).[69-70]

Beticolins can be divided into subgroups according to the ring closure of the heterocycle occurring either with the oxygen O1 in *ortho*-position of the chlorine atom at C13 or in *para* position. Furthermore, the subgroups are divided by the relative configuration of the methyl carboxylate at C2 and the hydroxy group at C3. *Ortho*-beticolins have *trans*-configuration while the substituents are *cis* in *epi-ortho* beticolins. *Epi-para* beticolins to date have not been observed.

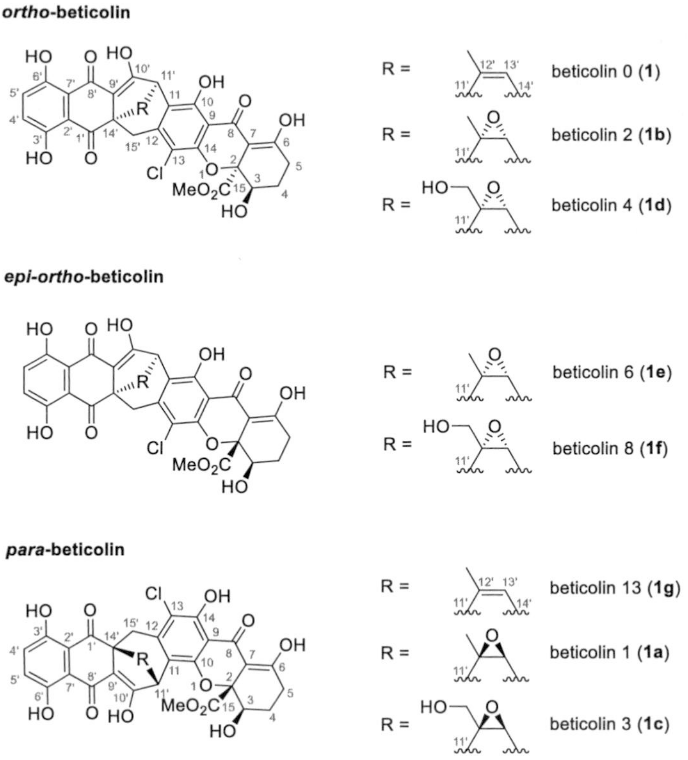

ortho-beticolin

epi-ortho-beticolin

para-beticolin

Figure 7. Structures of beticolins verified by X-ray crystallography.[67, 69]

As beticolin 0 (**1**) can be transformed into its related structures rather easily, it represents a synthetically useful building block. Under basic conditions, *o*-beticolins (e.g. **1**, **1b**, **1d**) can be converted to *epi-ortho* beticolins (e.g. **1e**, **1f**) as well as the thermodynamically more stable *p*-beticolins (e.g. **1g**, **1a**, **1c**) (Scheme 3).[71] A retro-oxa-Michael reaction that leads to the opening of the heterocycle can occur, and by rotation and recyclization the described transformations take place.

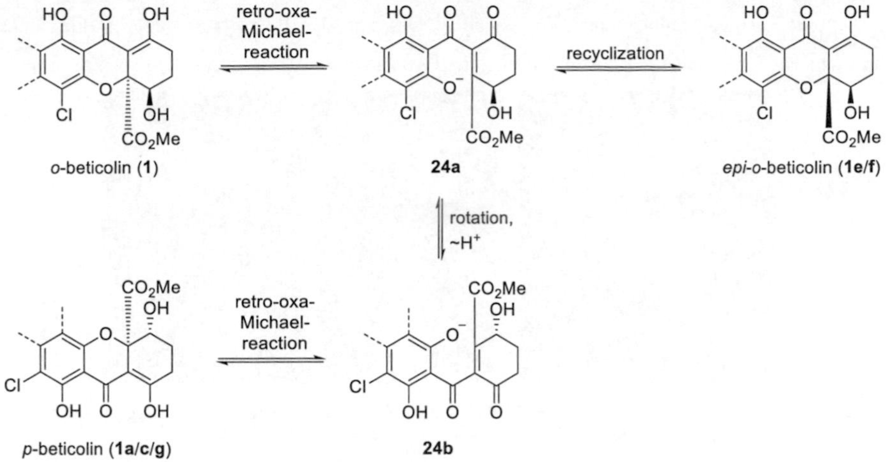

Scheme 3. Base catalyzed chemical interconversion of *ortho*-beticolin (**1**) to its related structures.[71]

Moreover, it was shown that beticolin 0 (**1**) can selectively be transformed into beticolin 2 (**1b**) by epoxidation of the alkene on the bridged bicycle with *meta*-chloroperoxybenzoic acid (*m*-CPBA) (Scheme 4).[72]

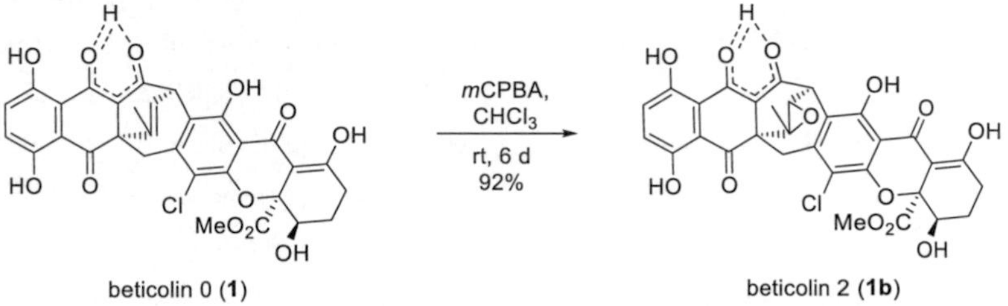

Scheme 4. Regio- and stereoselective epoxidation of beticolin 0 (**1**) to beticolin 2 (**1b**) with *m*-CPBA.[72]

As soon as there is an access to the structure of *ortho* beticolin 0 (**1**), other beticolins can be generated *via* simple chemical transformations. Hence, this thesis is focused on synthetic approaches towards the structure this parental compound.

3.3.2 Biological Activities

Beticolins have been biologically examined long before their structures were elucidated, for instance Schlösser *et al.* discovered an antibiotic activity of beticolins already in 1962.[68] An infection of sugar beet root with the phytopathogen *Cercospora beticola*, which is producing beticolins, causes the destructive leaf spot disease (cercosporiosis) which results in heavy damages to crops worldwide (Figure 8).[6, 65, 73] Sugarbeet supplies around one third of sucrose which is an important dietary supplement worldwide.[74] Despite significant research in the field, cercosporiosis remains a serious impediment to the production of sugarbeet.

Figure 8. Sugar beet plants infected with *Cercospora beticola*. Initially, round brown spots occur on the leaves and as the disease progresses the leaves turn brown and eventually the plants die.[75-76]

The effects caused by the non-host-specific toxin correspond to a broad cytotoxic profile. It was found that in plants, beticolins induce a dramatic loss of amino acids and β-cyanin from root tissues.[65, 77] A study on the effects of beticolin 1 (**1a**) on the plasma membrane H^+-ATPase of corn root showed that the hydrolysis activity of the proton pump decreases after incubation with the toxin.[78] H^+-ATPases are enzymes which consume ATP (adenosine triphosphate), create an electrochemical gradient in the plasma membrane and catalyze transmembrane movement of substances. By *in vitro* inhibiting this system, secondary active transport processes across the membrane responsible for intake and storage of metabolites were disrupted.[66, 79-80] This effect was also shown by intoxicating tobacco cells with beticolin 1 (**1a**) and 2 (**1b**).[80] Blein *et al.* showed that the inhibition of the plasmalemma ATPase of beticolin is stronger at high pH values and competitive with ATP-magnesium, indicating that beticolin chelating magnesium is the active inhibitory form.[66] Indeed, studies showed that in the presence of Mg^{2+} and under physiological pH range (pH 6 – 8) the dimeric neutral complex of beticolin 1 (**1a**) or 2 (**1b**) with magnesium is the major form present. In the early stage of research on beticolins, the dimeric form of beticolin 1 (**1a**) chelating two magnesium ions was isolated and characterized by Jalal *et al.* (Figure 9).[67, 81] A correlation between inhibited H^+-ATPase and the concentration of the beticolin-magnesium-complex indicating the activity against the enzyme was determined.[82] Also, the beticolin 1 (**1a**) complex with magnesium has a 40-fold higher affinity to lipid membranes compared to the monomer.[83]

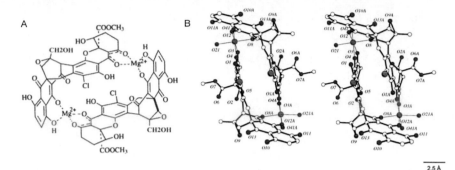

Figure 9. The pore-like structure formed by dimeric beticolin 3 (**1c**). A) Molecular structure of the dimer in complexation with Mg^{2+}[67, 84]; reprinted with permission from Elsevier. B) X-ray study of magnesium chelating beticolin 3 (**1c**) microcrystals[84-85]; reprinted with permission. Permission conveyed through Copyright Clearance Center, Inc. The section of the central lumen assumed to path the way for ions is 10×5 Å which means that 3–4 stacked dimers can form a tunnel long enough to cross the lipid bilayer (30 Å).

One of the most common killing mechanisms of cytotoxic molecules is induced permeabilization. It is assumed that the above described deleterious effects of beticolins originate from the ability to build pores. To examine the *in vitro* behavior of beticolins in the presence of cations, Ducrot exposed *p*-beticolins to magnesium carbonate in acetone under sonication to access stable dimeric complexes.[86] By assembling itself into a multimeric structure, beticolin 0 (**1**) was shown to form ion channels into planar lipid bilayers at cytotoxic concentration.[87] Thereby the membrane conductance of plant and animal cells or planar lipid bilayers was strongly increased with the formation of beticolin 0 (**1**) channels being dependent on the availability of free Mg^{2+}. Further studies described the formation of voltage-independent and barely selective ion channels by beticolin 3 (**1c**).[84] The formed ion channels have different properties such as permeability, conductance values and ion selectivity depending on the beticolin they are made of.[88] With a pore size of 7.5 Å solutes such as glucose can permeate through these channels, which is consistent with the findings describing a depolarized transmembrane potential and loss of amino acids and electrolytes, induced by beticolins.[77, 80]

Other biological studies made use of peroxidase-mediated luminol chemoluminescence to reveal *in vitro* $O_2^{-\cdot}$ scavenging activity of beticolin 1 (**1a**).[89] In comparison with typical radical scavengers like vitamin E and tiron, beticolin 1 (**1a**) showed an anti-radical effect without inhibiting peroxidase activity, already at low concentrations. This activity is assigned to the phenolic nature of the mycotoxin. A study on the effect of beticolin 1 (**1a**) and 2 (**1b**) on mammal cells showed an inhibition of cell proliferation at submicromolar concentrations by modulation of a step in the steroid metabolism.[90]

3.3.3 The Biosynthesis of Beticolins

Parts of the biosynthetic pathway towards beticolin toxins were elucidated by Nasini *et al.* in 1993.[81] Experiments on growing *Cercospora beticola* cultures in the presence of either $^{13}CH_3CO_2Na$ or $^{13}CH_3^{13}CO_2Na$ demonstrated, that all carbon atoms in beticolins derive from acetate building blocks **25**. The latter form octaketide **26** from which both the anthraquinone **27** and the xanthone subunit **28** are made (Scheme 5, ^{13}C marked with red dots).[57] Submission of the labelled compounds to ^{13}C NMR experiments enabled the identification of these structural fragments.

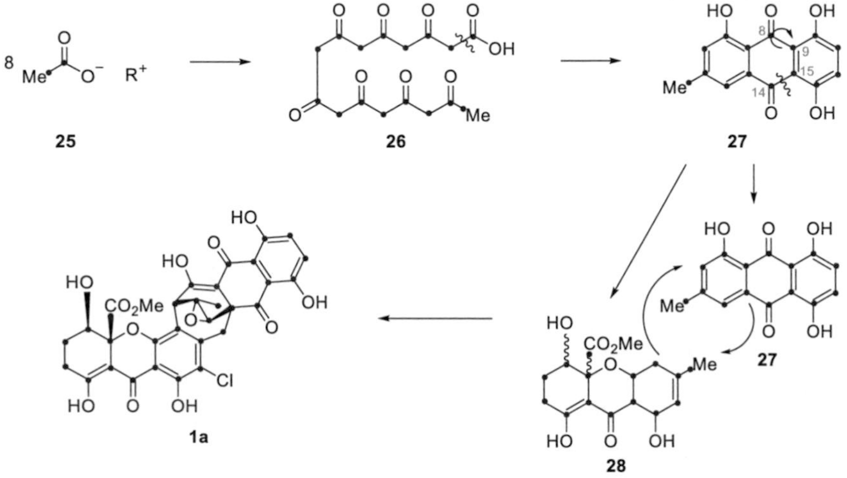

Scheme 5. A potential biosynthetic route towards beticolin 1 (**1a**) starting from acetates **25**.[81]

The fungus *Cercospora beticola* first synthesizes anthraquinone **27** on the basis of octaketide **26** by decarboxylation and multiple cyclizations. By oxidative cleavage of the C14 – C15 bond and rotation around the C8 – C9 bond, followed by an addition-elimination recyclization, xanthone **28** is build. This clevage is similar to the one occurring in other fungal tetrahydroxanthones like diversonolic esters.[91] After this rearrangement, chlorine is probably inserted. Coupling of reduced anthraquinone **27** and reduced xanthone **28** in a head-to-head fashion gives beticolin 1 (**1a**).

3.4 Synthetic Studies on Beticolin Subunits

Beticolin 0 (**1**) consists of two well-known natural product subunits: tetrahydroxanthones (blue) and anthraquinones (green) (Figure 10). The two building blocks are linked *via* a bicyclo[3.2.2]nonane ring system (orange).

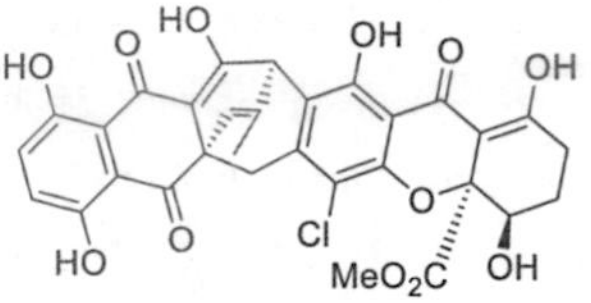

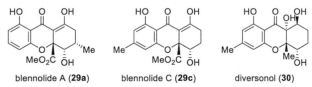

beticolin 0 (**1**)

Figure 10. Beticolin 0 (**1**) and the subunits it is comprised of.[72]

Tetrahydroxanthones as well as anthraquinones are part of many more natural products and have a wide range of applications. Hence, synthetic access towards their structures has already been extensively investigated.

3.4.1 Tetrahydroxanthones

Tetrahydroxanthones are a class of secondary metabolites of fungi. Among these, blennolides A – C (cf. **29a/c**) and diversonol (**30**) are structurally very similar, monomeric tetrahydroxanthones (Figure 11). Blennolides A – G (cf. **29a/c**) were isolated from the fungus *Blennoria sp.* in 2008[92] and are reported to inhibit algae growth and show antifungal as well as antibacterial activity.[93] Diversonol (**30**) which exhibits antibacterial activity, was isolated for the first time from *Penicillium diversum* by Turner *et al.* in 1978.[94]

blennolide A (**29a**)   blennolide C (**29c**)   diversonol (**30**)

Figure 11. Structures of tetrahydroxanthone natural products blennolide A (**29a**) and C (**29c**) and diversonol (**30**).

3.4.1.1 Syntheses of Tetrahydroxanthones

With regard to a synthetic access towards beticolin 0 (**1**), this chapter is focused on the synthesis of suitable tetrahydroxanthone building blocks. In fact, the structure of blennolides very much resembles the subunit of the toxin **1**.

In the context of the racemic total synthesis of diversonol (**30**), Bräse *et al.* developed a simple access towards functionalized tetrahydroxanthones in 2006 (Scheme 6).[44] A key step was the domino *oxa*-Michael aldol condensation of salicylic aldehyde **31** with cyclohexenone **32** to access the tetrahydroxanthone scaffold **33** of the target compound. Further functionalization was achieved *via* stereoselective formation of bromohydrin **34**. After elimination of HBr and oxidation under Ley conditions, diketone **35** was obtained. In a successive publication by Bräse *et al.* a synthetic route towards racemic (±)-blennolide C (**29c-rac**) was described.[95] Conjugate addition of tris(methylthio)methyllithium to α,β-unsaturated dienone **35** and removal of the bromine gave *trans* configured diketone **36**. Hydrolysis of the *ortho* thio unit of **36** into an ester and subsequent deprotection resulted in the desired (±)-blennolide C (**29c-rac**). The synthetic access to this functionalized tetrahydroxanthone facilitates the preparation of many other compounds as this structural unit is part of various natural products.

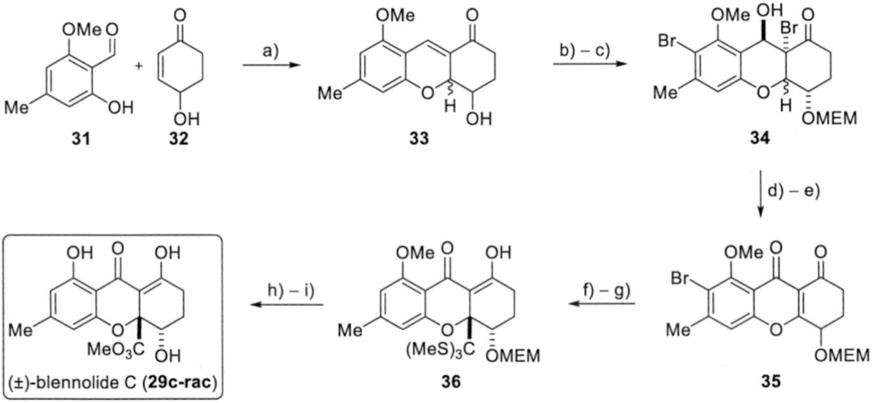

Scheme 6. Racemic synthesis of tetrahydroxanthone (±)-blennolide C (**29c-rac**) developed by Bräse *et al.*[44, 95] Reaction conditions: a) imidazole; b) MEMCl, *i*-Pr$_2$NEt; c) *n*-Bu$_4$NBr$_3$; d) DABCO; e) TPAP, NMO; f) HC(SMe)$_3$, *n*-BuLi; g) *t*-BuLi; h) HgO, HgCl$_2$; i) BBr$_3$. MEMCl = (2-methoxyethoxy)methyl chloride, DABCO = 1,4-diazabicyclo[2.2.2]octane, TPAP = tetrapropylammonium perruthenate, NMO = 4-methylmorpholine *N*-oxide.

A few years later, Tietze and his working group published the first enantioselective total synthesis of (−)-blennolide C (**29c**).[96] Starting from commercially available orcinol, six reaction steps including methylation, formylation, installation of the side chain by Wittig transformation, hydrogenation, incorporation of the terminal alkene with another Wittig transformation and chemoselective methylester cleavage provided **38** in 62% overall yield (Scheme 7). A significant step was the domino-Wacker/carbonylation/methoxylation using chiral BOX ligand **44** to obtain chromane **39** with a new stereocenter at C4a. After reduction and subsequent elimination, a stereocenter at C4 was established *via* a diastereoselective Sharpless dihydroxylation of vinyl chromane **40**. Aldehyde **41** was accessed *via* silylation of both hydroxy functionalities, selective deprotection of the primary alcohol and oxidation with Dess–Martin periodinane (DMP). Wittig-Horner reaction with trimethyl phosphonoacetate enabled chain elongation of **41**, subsequent reduction of the double bond and cleavage of the benzyl ether provided primary alcohol **42**. A stepwise oxidation sequence of the latter, subsequent dehydrogenation to generate a double bond, its hydroxylation and oxidation afforded chromanone **43**. Treatment with trichlorotitanium isopropoxide and triethylamine provided the tetrahydroxanthenone scaffold. Silyl deprotection and cyclization gave (–)-blennolide C (**29c**).

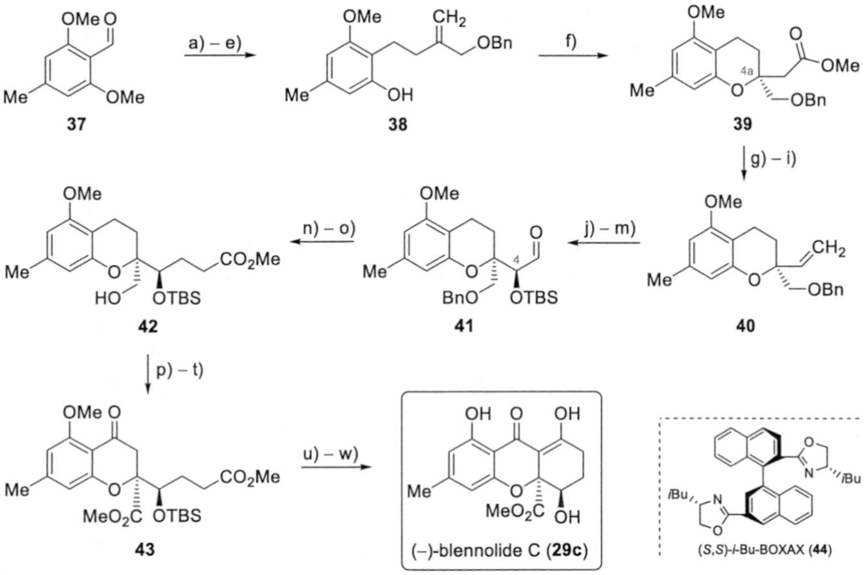

Scheme 7. First enantioselective total synthesis of (–)-blennolide C (**29c**) developed by Tietze *et al.*[96] Reaction conditions: a) Ph$_3$PCHCOCH$_2$OBn; b) H$_2$, PtO$_2$; c) IBX; d) Ph$_3$PCH$_3$Br, *n*-BuLi; e) NaSEt; f) Pd(OTFA)$_2$, (*S,S*)-*i*-Bu-BOXAX, *p*-benzoquinone, CO, 99% ee; g) LiAlH$_4$; h) *n*-Bu$_3$P, *o*-NO$_2$-C$_6$H$_4$SeCN; i) *m*-CPBA, Na$_2$HPO$_4$·2H$_2$O, *i*-Pr$_2$NH; j) K$_2$OsO$_2$(OH)$_4$, (DHQ)$_2$AQN, K$_2$CO$_3$, K$_3$[Fe(CN)$_6$], MeSO$_2$NH$_2$; k) TBSOTf, 2,6-lutidine; l) HF·pyridine; m) DMP; n) NaH, (MeO)$_2$P(O)CH$_2$CO$_2$Me; o) Pd/C, H$_2$; p) DMP; q) KOH, I$_2$; r) DDQ; s) [Mn(dpm)$_3$], PhSiH$_3$, O$_2$; t) TPAP, NMO; u) Ti(O*i*-Pr)Cl$_3$, Et$_3$N; v) H$_2$SiF$_{6(aq)}$; w) BBr$_3$. IBX = 2-iodoxybenzoic acid; Pd(OTFA)$_2$ = palladium(II) trifluoroacetate; (DHQ)$_2$AQN = hydroquinine anthraquinone-1,4-diyl diether; DMP = Dess-Martin periodinane; DDQ = 2,3-dichloro-5,6-dicyano-1,4-benzoquinone.

3.4.2 Anthraquinone Derivatives

Many plants including rhubarb and aloe as well as bacteria, fungi, and insects are rich sources of anthraquinones which represent an important class of natural products with versatile biological properties and diverse applications. [97-100] Besides their utilization as colorants as well as for medical applications, they make valuable tool compounds for biochemical and pharmacological studies.[101-102] In 1840, Laurent synthesized 9,10-anthraquinone (**45**) through oxidation of anthracene with nitric acid (Figure 12).[103] An example for an anthraquinone-based natural product is doxorubicin (**46**) which acts as an anti-cancer drug and is based on a tetracyclic ring system linked to an amino-sugar.[104-105] The bisanthraquinone flavoskyrin (**47**) is a naturally occurring toxin from *Penicillium islandicum Sopp* capable of inhibiting the mitochondrial respiration.[102, 106]

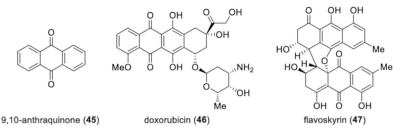

| 9,10-anthraquinone (**45**) | doxorubicin (**46**) | flavoskyrin (**47**) |

Figure 12. Structure of 9,10-anthraquinone (**45**) and anthraquinone containing natural products doxorubicin (**46**) and flavoskyrin (**47**).

3.4.2.1 Syntheses of Anthraquinones

As anthraquinones play an important role as building blocks in a large number of drugs and natural products, researchers developed several synthetic pathways towards these structures.[107-108] Synthetically useful and highly reactive molecules in Diels-Alder cycloaddition reactions are naphthoquinones which represent one of the most important dienophiles for total synthesis applications.[109] The typical products of these reactions are anthraquinone derivatives which are known to be substantial carbocyclic scaffolds.

In 1997, Brimble *et al.* performed Diels-Alder reactions with 2-substituted naphthoquinone derivatives **48** to generate the corresponding adducts **50** (Scheme 8).[110] These revealed that naphthoquinones bearing electron-withdrawing residues give the desired tricycles in good yields whereas a free carboxylic acid afforded aromatization towards 9,10-anthracenedione (**45**).

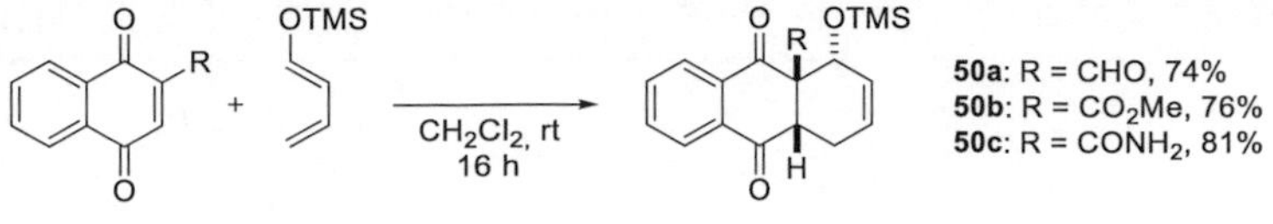

Scheme 8. Cycloaddition between substituted 1,4-naphthoquinones **48** and TMS diene **49** investigated by Brimble et al.[110]

In 2014, Bräse et al. described the synthesis of functionalized anthraquinones via Diels-Alder reactions of naphthoquinone monoketal dienophiles.[63] The study commenced with the reduction of naphthol **51**, subsequent silylation and oxidative ketalization with phenyliodine bis(trifluoroacetate) (PIFA) to give ketal **54a** representing the dienophile (Scheme 9).

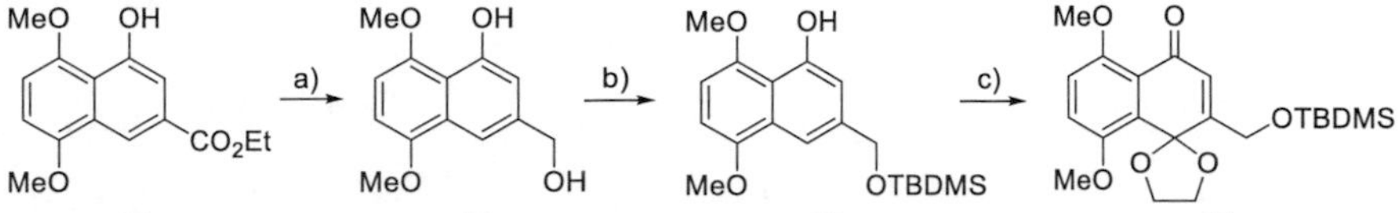

Scheme 9. Three-step synthesis of ketal **54a** developed by Bräse et al.[63] Reaction conditions: a) LiAlH₄; b) TBDMSCl, imidazole; c) PIFA, ethylene glycol.

Via a Diels-Alder cycloaddition of ketal **54a** with TBDMS diene **55a**, tricycle **56a** was accessed as a single diastereomer containing a congested quaternary center (Table 1, entry 1). With R¹ being an ester functionality, a mixture of regioisomers **56a/b** and **56c** was obtained from the reaction with TBDMS diene **55a** (entry 2). More activated dienophile **54c** bearing a phenyl residue yielded anthraquinone derivative **56c/d** as an undesired mixture of diastereomers (entry 3) despite the expected similar activity of both carbonyl groups in benzylic positions. Ketal **54d** decomposed upon heating through de-iodination (entry 4). In conclusion, it was found that the electronic rather than the steric nature of substituent R¹ had the strongest influence on the regioselective outcome of the reaction. However, when aldehyde **54e** reacted with **55b** under microwave conditions, steric repulsion lead to an unexpected mixture of regioisomers **56b/c**.

Table 1. Scope and regioselectivities of Diels-Alder reactions performed by Bräse *et al.*[63]

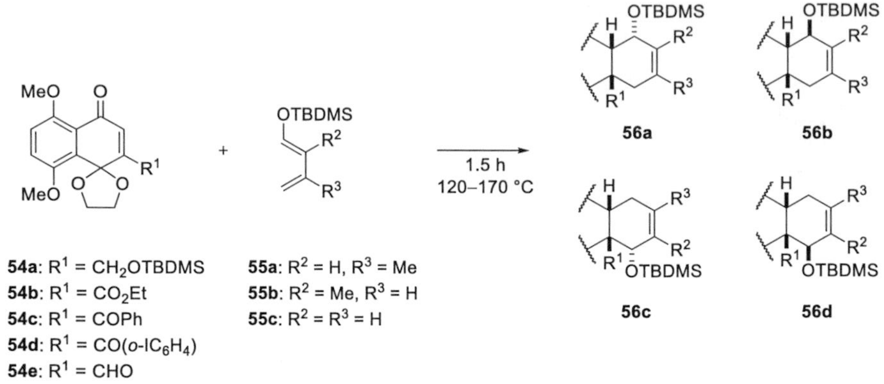

54a: R[1] = CH$_2$OTBDMS   55a: R^2 = H, R^3 = Me
54b: R[1] = CO$_2$Et   55b: R^2 = Me, R^3 = H
54c: R[1] = COPh   55c: R^2 = R^3 = H
54d: R[1] = CO(*o*-IC$_6$H$_4$)
54e: R[1] = CHO

Entry	Dienophile	R^1	Diene	R^2	R^3	Product, yield, *dr*
1	54a	CH$_2$OTBDMS	55a	H	Me	56a, 91%
2	54b	CO$_2$Et	55a	H	Me	56a/b, 81%, 4:1; 56c, 16%
3	54c	COPh	55a	H	Me	56c/d, 43%, 2:1
4	54d	CO(*o*-IC$_6$H$_4$)	55c	H	H	degradation
5	54e	CHO	55b	Me	H	56b/c, 60%, 1:1.1

dr = diastereomeric ratio.

Due to the lack of selectivity of the reaction and to incorporate a halogenated aromatic ring, 1,4-naphthoquinone precursor **57** was synthesized which is known to exhibit higher reactivity relative to its monoketal analogs due to electronic and steric reasons.[63] Diels-Alder cycloaddition of iodo dienophile **57** with TBDMS diene **55c** provided one regioisomer in a mixture of *exo* **58a** and *endo* **58b** products which could be separated by HPLC (Scheme 10). Switching to this system required an elaborate synthetic route towards precursor **57** (eight steps from naphthol **51**) but facilitated mild reaction conditions and a good yield.

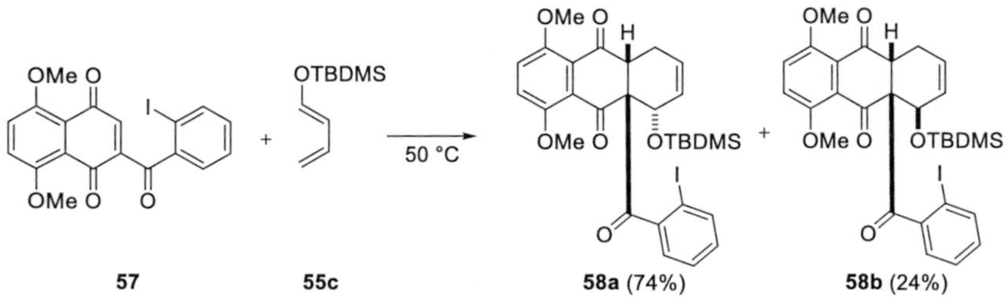

Scheme 10. Diels-Alder reaction of **57** with **55c** towards anthraquinone derivatives **58a** and **58b** developed by Bräse *et al.*[63]

4 Objective

Beticolins are mycotoxins associated with the destructive leaf spot disease of sugar beet root and thus are of great significance for the agricultural industry. Since the mechanistic details of the disease are not yet clear, a synthetic access to the toxin is required in order to elucidate its mode of action and fight the cause. The natural products furthermore show a broad cytotoxic profile making them potential lead structures, however, despite interesting biological activities to date no total synthesis was reported.

Therefore, the aim of this thesis was the development of a synthetic route towards the anthraquinone-xanthone heterodimeric structure of beticolin 0 (**1**). Considering the great number of publications on the synthesis of tetrahydroxanthones, this investigation was focused on the anthraquinone subunit of the natural product (green) with the specific challenge being represented by the characteristic central bicyclo[3.2.2]nonane ring system (orange) (Figure 13).

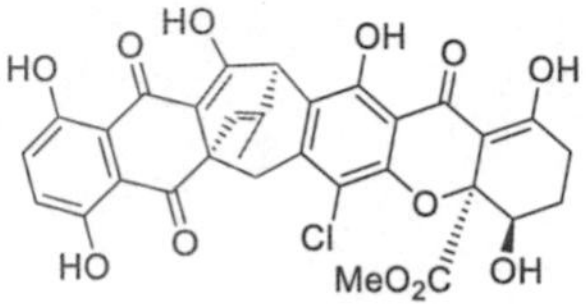

beticolin 0 (**1**)

Figure 13. Structure of beticolin 0 (**1**).

It was envisioned to develop a straightforward synthetic route towards naphthoquinone derivatives, which should be functionalized *via* a Diels-Alder approach, in order to construct the anthraquinone subunit of beticolin 0 (**1**) with its sterically congested all-carbon quaternary stereocenter. Bearing a halogenated benzene ring attached *via* a short alkyl chain, the anthraquinone derivatives were supposed to serve as precursors for intramolecular coupling reactions to build the [3.2.2]bicyclic system. By applying an anthraquinone model system in various intramolecular couplings, reaction conditions suitable for the construction of the characteristic skeleton of beticolin 0 (**1**) should be investigated. The incorporation of functionalities on the benzene ring attached to the anthraquinones was envisaged in order to facilitate the formation of the tetrahydroxanthone moiety through a domino oxa-Michael–aldol condensation with a cyclohexene derivative.

5   Results and Discussion

5.1   Retrosynthetic Analysis

In order to plan the synthesis of a natural product, the first step is a retrosynthetic analysis of the target molecule. Therefore, the molecule is divided into building blocks of lower complexity which in turn are transformed into simpler precursor structures step by step until basic materials are reached. Since there is no general approach for the fragmentation of a molecule, individual and creative solutions are required for each retrosynthetic challenge.

The polycyclic skeleton of beticolin 0 (**1**) can be considered as the combination of three subunits (Figure 14). A chlorinated tetrahydroxanthone (rings E/F/G) and an anthraquinone subunit (rings A/B/C) are connected *via* a unique bicyclo[3.2.2]nonane ring system (rings C/D). Tetrahydroxanthones as well as anthraquinones represent well-known organic structures and different strategies for their synthesis have already been investigated.[44, 95-96, 109, 111-114] The most interesting and synthetically challenging part of beticolin 0 (**1**) as a consequence is represented by the rather lipophilic core (C/D).

Thus, the design of the below described retrosynthesis is focused on the preparation of this subunit of the molecule. For the synthetic approach towards beticolin 0 (**1**), it is envisioned to enable a simple access to functionalized anthraquinone derivatives which serve as a starting material for the key step represented by an intramolecular coupling to build the bicyclo[3.2.2]nonane ring system.

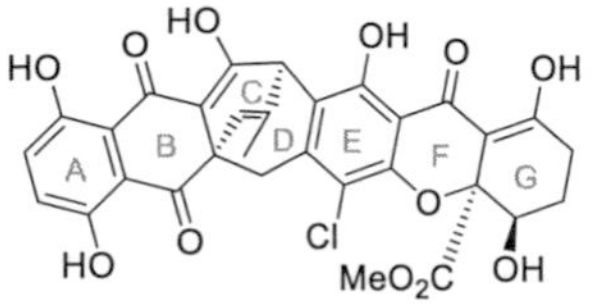

beticolin 0 (**1**)

Figure 14. The structure of beticolin 0 (**1**). The seven cycles are labelled with letters A – G.

In the envisaged route towards beticolin 0 (**1**), the final steps are planned to be various functionalization and deprotection reactions on heptacycle **59** (Scheme 11). Protecting groups of the hydroxy functionalities need to be removed, a chlorine atom has to be established at C13 and the methyl ester is considered to be attached *via* a Michael addition at the enone (C2). Since a selective chlorination on C13 is difficult to realize, an incorporation of a benzoic acid **66** containing a chlorine atom into the naphthoquinone derivative **65** through the acylation of dimethoxynaphthalene derivative **67** is considered instead. The construction of the tetrahydroxanthone part (E/F/G) of the natural product is envisioned to proceed *via* a domino oxa-Michael–aldol condensation reaction with a suitable cyclohexenone **60**. For this purpose, pentacycle **61** needs to be functionalized with an aldehyde and a free hydroxy group on ring E. For detailed procedures concerning the functionalization and the stereochemistry of the tetrahydroxanthone subunit, numerous publications on the synthesis of tetrahydroxanthones can be found in the literature.[44, 95-96]

The ring closure to pentacycle **61** with the highly complex, bridged seven-membered ring represents the key step of the synthetic route and is thought to be accomplished with an intramolecular, palladium catalyzed Heck reaction of halogenated compound **62**. Anthraquinone derivative **62** contains the required halogen residue and double bond in the right positions. Such anthraquinones can easily be prepared from naphthoquinone derivatives by Diels-Alder cycloadditions.[109] In this route a suitably functionalized benzene moiety is attached to the naphthoquinone derivative **64** *via* a keto group which is required for the synthesis of the tetrahydroxanthone unit. Diene **63** bears a methyl as well as a protected hydroxy group that can later be found in the core structure of beticolin 0 (**1**).

By oxidative demethylation with ceric ammonium nitrate (CAN), naphthoquinone derivative **64** can be generated from compound **65**. The latter is envisioned to be synthesized *via* acylation of naphthalene derivative **67** with a functionalized, *ortho*-halogenated benzoic acid **66**. Methylation provides dimethoxynaphthalene derivative **67** from commercially available dihydroxynaphthalene derivative **68** which represents the starting point of the designed synthetic route.

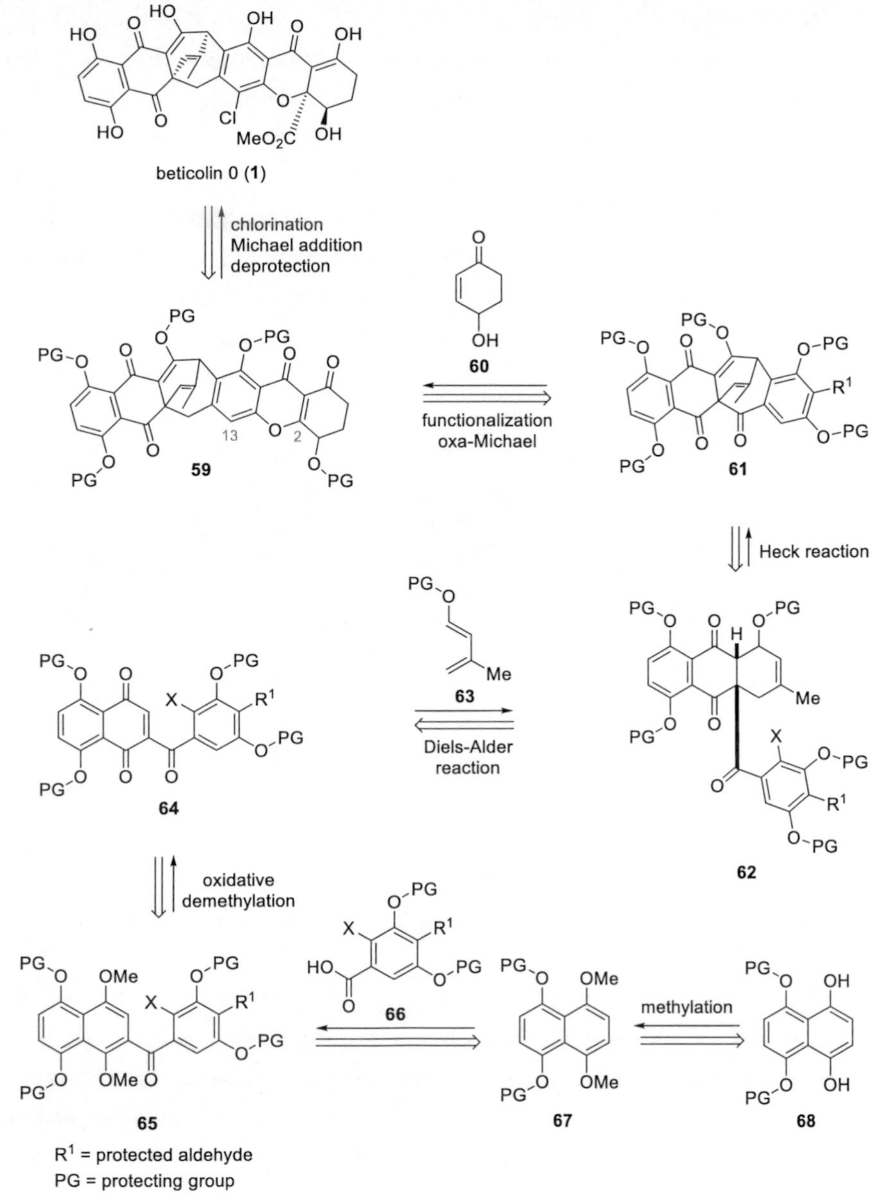

Scheme 11. Retrosynthetic analysis of beticolin 0 (**1**) starting from dihydroxynaphthalene derivative **68**.

5.2 Naphthoquinone Derivatives

Based on the findings of Kramer *et al.*[63] it was envisioned to find a straightforward and modular access to highly activated naphthoquinone dienophiles. Thus, a reported synthetic route towards naphthoquinone derivative **73a** starting from commercially available 1,4-dihydroxynaphthalene (**69**) was elaborated. As a first step methylation of the hydroxy groups of dihydroxynaphthalene **69** was accomplished using methyl iodide (MeI) by following a procedure published by Sakamoto *et al.* (Scheme 12).[115] The reaction proceeded smoothly and resulted in up to 90% yield. Hereafter 1,4-dimethoxynaphthalene (**70**) was acylated with 2-iodobenzoic acid (**71a**) in trifluoroacetic anhydride (TFAA) to give arylketone **72a** in 67% yield, according to a literature known procedure.[116] The following oxidative demethylation with ceric ammonium nitrate (CAN) provided quinone **73a** in quantitative yield.

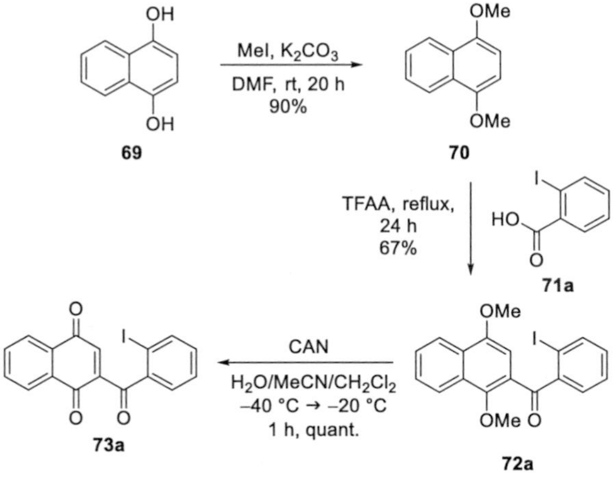

Scheme 12.   Three-step   synthesis   of   naphthoquinone   derivative   **73a**   starting   from 1,4-dihydroxynaphthalene (**69**).[116]

To conclude, the three-step synthesis provided naphthoquinone derivative **73a** which represents a bench-stable precursor for the anthraquinone subunit of beticolin 0 (**1**) in an overall yield of 60%. Naphthoquinone **73a** not only represents a highly activated dienophile bearing electron-withdrawing groups in conjugation with the alkene, but also contains an iodine atom in an aromatic ring which facilitates subsequent intramolecular coupling reactions. Since naphthoquinone derivative **73a** is a strong dienophile that can easily be accessed *via* a literature known procedure, it was used as a model system for the subsequent studies on the cycloaddition reactions. The investigation of more complex structures will be described in the following section.

With regard to the structure of beticolin 0 (**1**) and its tetrahydroxanthone subunit, the incorporation of functionalities into the halogenated aromatic ring of naphthoquinone derivatives **73a–g** was investigated. In this way the polarity of the molecules can be modified, different halogens can be integrated and the foundation for the establishment of a tetrahydroxanthone moiety can be laid. To implement these plans, different benzoic acids **71** were used in the acylation step of the dienophile synthesis. By following the previously described procedure, several derivatives of brominated benzoic acids containing methoxy groups, hydroxy groups, an *N*-acetyl or an *N*-trifluoroacetyl residue were incorporated (see Scheme 13).

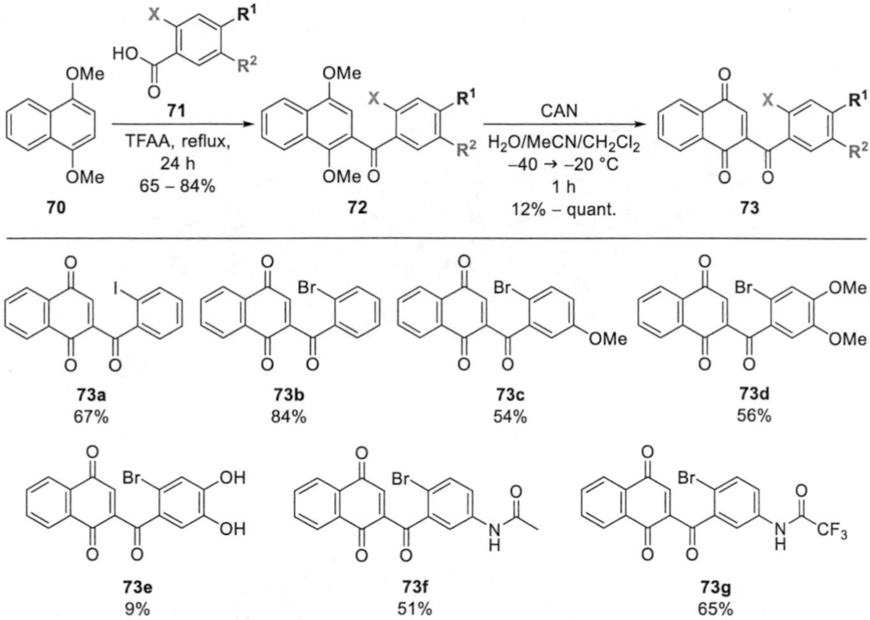

Scheme 13. Synthetic route towards naphthoquinone derivatives **73a–g** and an overview of the obtained products with the corresponding overall yields.

As shown for the seven examples, the two reaction steps tolerated different functional groups. Various 2-substituted 1,4-naphthoquinones **73a–g** were obtained in up to quantitative yield after crystallization from cyclohexane. For the derivatives bearing substituents additional to the halogen on the benzene ring, a minor decrease in yields was observed. Naphthoquinone derivative **73e** was isolated with only 9% overall yield since the ether cleavage did not proceed satisfactory, and the increased polarity of product **73e** hampered the purification process. Thus, for further studies the synthetic route towards **73e** would require optimization.

The overall benefit of the two-step route by Buccini *et al.* is its applicability for various derivatives and the scalability up to the multi-gram range. In subsequent experiments the differently functionalized naphthoquinone derivatives **73a–g** were supposed to function as dienophiles in Diels-Alder reactions as described in detail in chapter 5.4.

For some of the synthesized compounds single crystals could be obtained and the molecular structures in the solid state were confirmed by X-ray diffraction. A selection of these structures is shown in Figure 15 and Figure 16 whereas further structures can be found in the appendix (chapter 7.3). In all of the verified structures the 1,4-naphthalenedione part of the molecule is planar since it represents a conjugated system, while the halogenated benzene ring is twisted towards that plane.

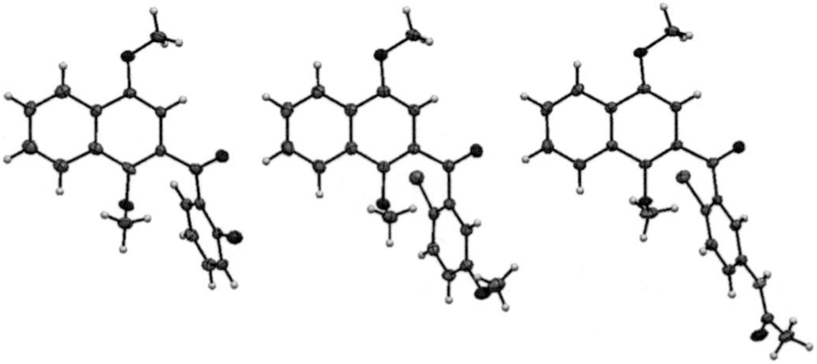

Figure 15. Molecular structures of **72a, 72c, 72f** (from left to right) determined by single-crystal X-ray diffraction. Displacement parameters are drawn at 50% probability level. Crystallographic data can be found in chapter 7.3.

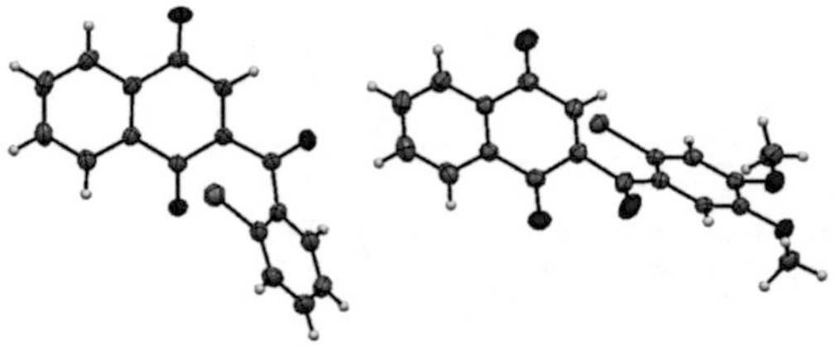

Figure 16. Molecular structures of **73b** (left) and **73d** (right) determined by single-crystal X-ray diffraction. Displacement parameters are drawn at 50% probability level. Crystallographic data can be found in chapter 7.3.

With these seven different naphthoquinone derivatives (**73a–g**) in hand, syntheses of diverse dienes as counterparts for the Diels-Alder reactions, were started to be developed (see following chapter).

5.3   Synthesis of Functionalized Dienes

For Diels-Alder cycloadditions with the just described naphthoquinone derivatives **73a–g**, activated dienes bearing EDGs (electron-donating groups) as well as suitable functionalities were required. The following figure depicts the primarily chosen series of commercially available molecules subjected to cycloaddition reactions: 1,3-butadiene (**74a**), isoprene (**74b**), 2,3-dimethyl-1,3-butadiene (**74c**) and (trimethylsiloxy)-1,3-butadiene (**74d**) (Figure 17). 1,3-Butadiene (**74a**) was generated *in situ* from 3-sulfolene (**75**) which will be explained in detail in chapter 5.4.2.

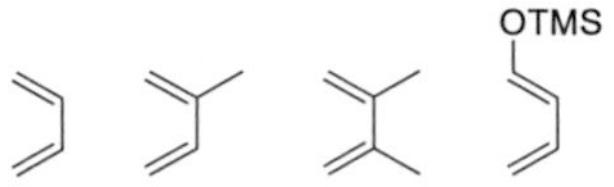

Figure 17. Commercially available dienes **74a–d** that were used for Diels-Alder reactions.

By taking the existence of a hydroxy group in ring C of beticolin 0 (**1**) into account, further dienes with protected alcohol functionalities were synthesized. Following a procedure published by Kalesse *et al.*, silylated dienes **74e** and **74f** were prepared (Scheme 14).[117] By condensation of *trans*-2-methyl-2-butenal (**76**) with trimethylsilyl (TMS) chloride in the presence of triethylamine and zinc chloride as a lewis acid, TMS diene **74e** was obtained in 23% yield. By following a method described by Duhamel *et al.* in 1993, TMS ether **74e** was converted to *tert*-butyldimethylsilyl (TBDMS) diene **74f** with 62% yield. The yields obtained for the two dienes are rather low due to the instability of the molecules and the required purification *via* distillation.

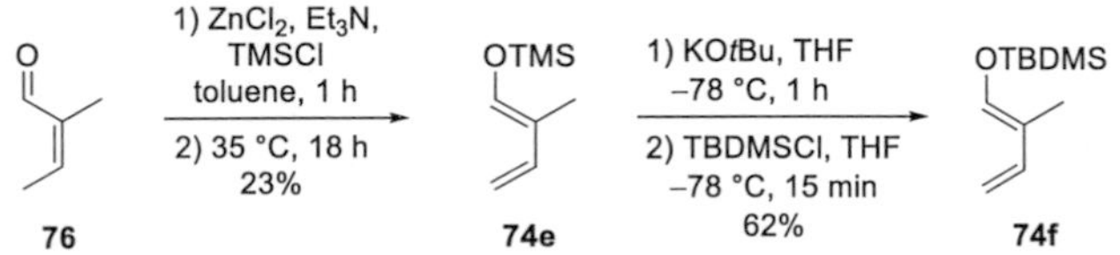

Scheme 14. A synthetic route yielding silylated dienes **74e** and **74f** bearing a protected alcohol functionality.[117-118]

Additionally, a triisopropylsilyl (TIPS) protected diene **74g** was synthesized starting from but-3-en-2-one (**77**) which was reacted with triisopropylsilyl trifluoromethanesulfonate (TIPSOTf) to give diene **74g** in 27% yield after distillation (Scheme 15).[119] Compound **74g** as well as TBDMS diene **74f** bear a space demanding silyl protecting group, which might have an interesting effect on the outcome of the coupling reaction that is going to be performed with the anthraquinone derivative accessed *via* Diels-Alder cycloaddition (see chapter 5.7).

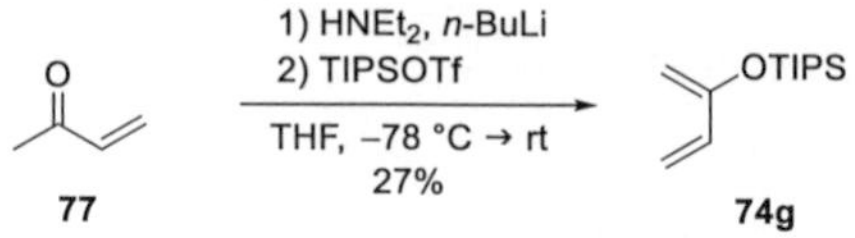

Scheme 15. Synthesis of diene **74g** with a TIPS protected alcohol functionality at C2.[119]

Following a literature known procedure *tert*-butyldiphenylsilyl (TBDPS) diene **74h** was synthesized from but-3-en-2-one (**77**) (Scheme 16).[120] Product **74h** was isolated with 33% yield as estimated from ^{1}H NMR analysis.

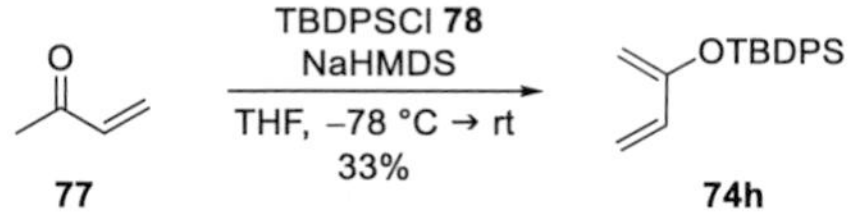

Scheme 16. Synthesis of TBDPS protected diene **74h** from but-3-en-2-one (**77**).[120]

By combining the seven different naphthoquinone dienophiles (chapter 5.2) with the above described eight dienes, cycloaddition reactions can be performed to generate highly functionalized anthraquinones.

5.4   Synthesis of Anthraquinone Derivatives

In order to build the anthraquinone subunit of beticolin 0 (**1**), Diels-Alder reactions between the previously described dienes and dienophiles were performed. Considering the high reactivity of 1,4-naphthoquinones in cycloaddition reactions, the effect of bulky substituents at the 2-position of the naphthoquinone derivatives on the outcome of the pericyclic reactions was investigated. It is worth mentioning that the products of these Diels-Alder reactions represent valuable precursors for intramolecular transition-metal catalyzed C–C coupling reactions, which are supposed to build the anthraquinone subunit with the bicyclo[3.2.2]nonane ring system of the natural product.

5.4.1   The Diels-Alder Reaction

Nearly one century after its discovery, the Diels-Alder reaction still is one of the most fundamental reactions in organic synthesis, applied for the introduction of complexity into chemical structures.[121-122] It is a powerful and versatile transformation that in an atom-economical process unites two fragments while generating up to four stereocenters. Two σ-bonds are formed stereoselectively in a single step to afford six-membered rings in a concerted way.[123-124]

Quinones as dienophiles are the very first example investigated by Diels and Alder in 1928 (Scheme 17).[121] The reaction between 1,4-benzoquinone (**79**) and cyclopentadiene (**80**) gave a mixture of mono- **81** and bis-adducts **82**.

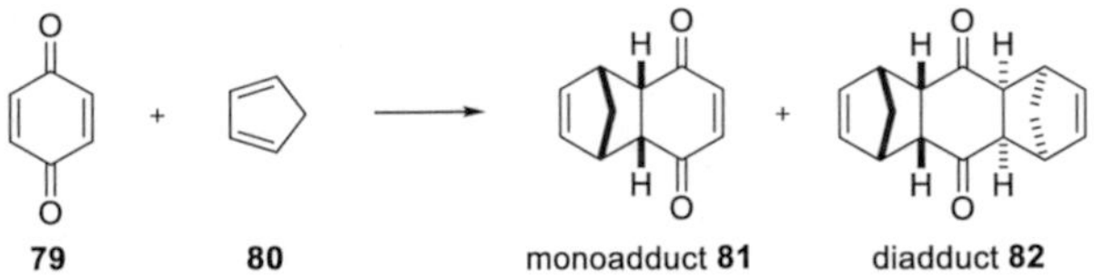

Scheme 17. Discovery of the Diels-Alder reaction of 1,4-benzoquinone (**79**) with cyclopentadiene (**80**) by Diels and Alder.[121]

The mechanism of the reaction proceeds *via* a [4+2]-cycloaddition between a conjugated diene and an alkene to form a cyclic component. Therefore, an overlap of the highest occupied molecular orbital (HOMO) of the diene and the lowest unoccupied molecular orbital (LUMO) of the dienophile is required (Figure 18). Electron-withdrawing groups (EWG) at the dienophile lower the energy of its LUMO, thus minimize the energy gap and facilitate the reaction. EDGs on the diene facilitate the reaction by increasing the energy of its HOMO.

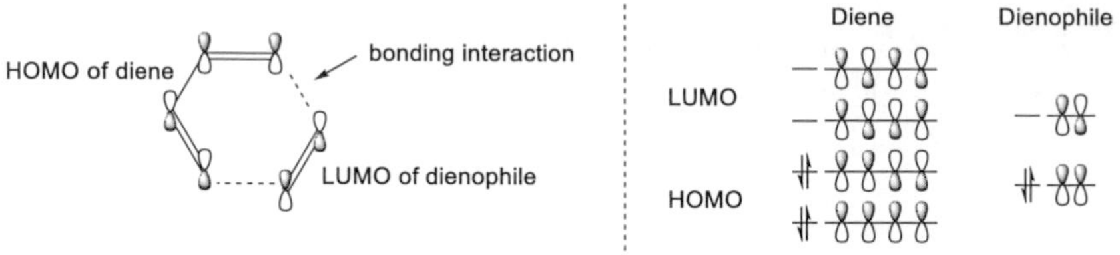

HOMO of diene — bonding interaction

LUMO of dienophile

Diene   Dienophile

LUMO

HOMO

Figure 18. Frontier molecular orbitals showing orbital overlap in Diels-Alder [4+2] cycloadditions.

5.4.1.1 Stereospecificity and Stereoselectivity

During Diels-Alder reactions the stereochemical information of dienophile and diene is retained, meaning the reaction is a stereospecific *syn* addition. This was demonstrated for various substrates, among them the simplest possible example: *cis*-1,2-dideuterio ethylene (**83**) with 1,1,4,4-tetradeuterio-1,3-butadiene (**84**) (Scheme 18).[124]

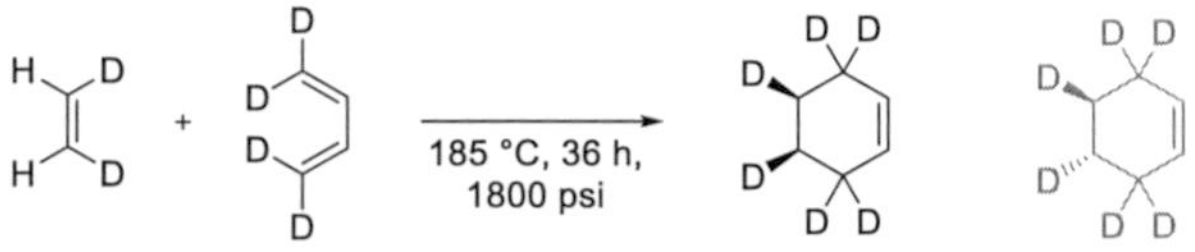

83   **84**   185 °C, 36 h, 1800 psi   *cis* **85**   *trans* **85** (not observed)

Scheme 18. Stereospecific Diels-Alder reaction of deuterated ethene (**83**) with deuterated butadiene (**84**).[124]

When substituted dienophiles react with substituted dienes, diastereomers can be formed, better known as *exo* and *endo* products. In the *endo* transition state, the substituent on the dienophile is oriented towards the π-orbitals of the diene, while in the *exo* transition state the opposite orientation is present (Figure 19). The *endo* and *exo* transition state can give two different stereoisomeric products for a lot of substituted butadienes. Due to secondary orbital interactions in the transition state, usually the major product is the *endo* product even though it is sterically more congested than the *exo* product.[125]

endo   *exo*

Figure 19. *Endo* and *exo* transition states of the Diels-Alder reaction.

5.4.1.2 Regioselectivity

In general, the regioselectivity of Diels-Alder reactions follows the *ortho-para* rule (Figure 20).[125] The position of the substituents in the cyclohexene product is described with the terms *ortho*, *meta* and *para* in analogy to disubstituted arenes. In the case of dienes bearing an EDG at C2, such as isoprene (**74b**), maximization of orbital interaction is true for the "*para*" product which represents the major product since the HOMO has its largest coefficient at C1 while the dienophile with the EWG at C1 has its largest LUMO coefficient at C2 (Figure 20, A). For 1-substituted dienes the largest HOMO coefficient is at C4 (Figure 20, B). If there is a substituent present at the 1- as well as the 2-position of the diene, the substituent at C1 usually is the directing group.

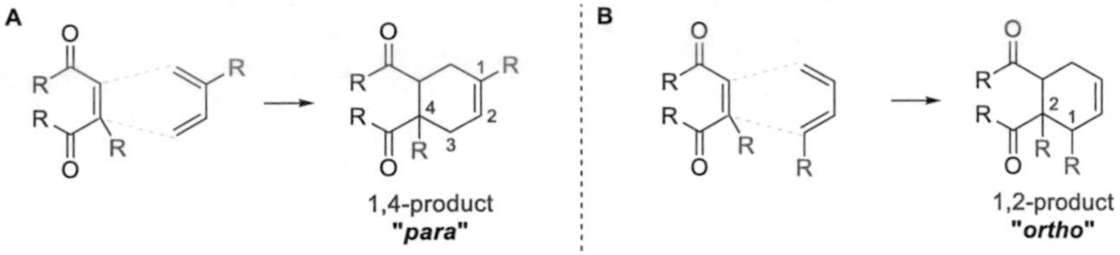

Figure 20. The *ortho-para* rule explaining the regioselectivity of the Diels-Alder reaction.

5.4.2 Anthraquinone Model System

It was envisioned to investigate the cycloaddition reaction with easily accessible precursors. For the development of such a model system, a simple, non-functionalized diene as well as poorly functionalized dienophiles **73a** and **73b** were used for the cycloadditions. The simplest conjugated diene is 1,3-butadiene (**74a**), however for laboratory use this compound is inconvenient as it is gaseous and carcinogenic. A solid, stable and non-toxic source for 1,3-butadiene (**74a**) is 3-sulfolene (**75**) which upon heating releases SO_2 **86** as well as the desired gas **74a** (Scheme 19).

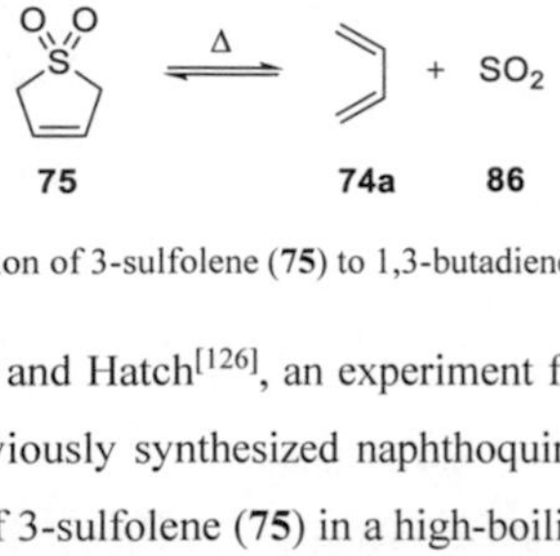

Scheme 19. Reversible decomposition of 3-sulfolene (**75**) to 1,3-butadiene (**74a**) and sulfur dioxide (**86**).

Inspired by the work of Sample and Hatch[126], an experiment for the application of 3-sulfolene (**75**) in cycloadditions with previously synthesized naphthoquinone derivatives was developed. For this purpose, a suspension of 3-sulfolene (**75**) in a high-boiling solvent was heated to 125 °C for half an hour to enable the decomposition of **75**. The thereby released gaseous 1,3-butadiene (**74a**) was led into a cooled reaction vessel containing a solution of naphthoquinone **73a** or **73b** in dichloromethane. After completion of the gas evolution, as indicated by terminated bubbling, the cooled vessel containing diene **74a** and the solution of dienophile **73a** or **73b**, was slowly warmed to room temperature and stirred at this temperature for about two hours. After removing residual gas by flushing the reaction system with argon, a small workup was done and the crude was purified *via* column chromatography on silica gel to obtain compound **87aa** or **87ba**. Both the iodinated **87aa** and the brominated anthraquionone derivative **87ba** were successfully synthesized in good yields, following this procedure (Scheme 20).

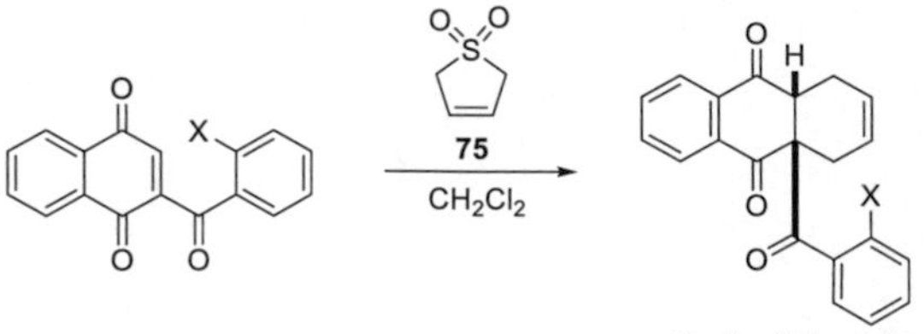

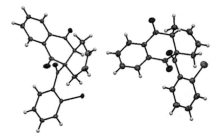

Scheme 20. Diels-Alder reaction of naphthoquinone derivatives **87a** and **87b** using 3-sulfolene (**75**) as a source for 1,3-butadiene (**74a**).

For both compounds single crystals suitable for X-ray diffraction could be obtained, which confirmed their structures in the solid state containing a sterically congested all-carbon quaternary stereocenter (Figure 21). Rings A and B build a plane due to the conjugated system, while ring C is twisted to it.

Figure 21. Molecular structures of anthraquinone derivatives **87aa** (left) and **87ba** (right) determined by single-crystal X-ray diffraction. Displacement parameters are drawn at 50% probability level. Crystallographic data can be found in chapter 7.3.

5.4.3   Diels-Alder Reaction with Cyclic Dienes

Following the historic cycloaddition procedure published by Diels and Alder[121], the reaction of naphthoquinone derivative **73a** with cyclopentadiene (**80**) was investigated (Scheme 21). In a first attempt the general procedure for cycloadditions, as described in chapter 7.2.1, was applied. However, it turned out that the reaction conditions were too harsh, as no product was isolated and the diene in the mixture was degraded or polymerized. Thus, for further studies the reactions were run at a lower temperature of 0 °C and only freshly distilled cyclopentadiene (**80**) was used. By applying these conditions, the desired product **89** was isolated in a poor yield of 26%, which can be explained by the observed low stability of the product. Already after half an hour, NMR spectra showed decomposition of **89** which is why no IR and MS data are given in the experimental part of the thesis (chapter 7.2.5).

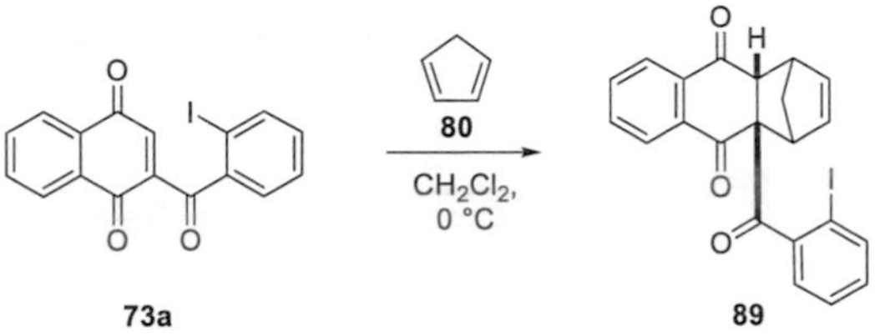

Scheme 21. Cycloaddition of naphthoquinone derivative **73a** and cyclopentadiene (**80**).

For comparison, the usage of furan (**90**) as a cyclic diene was considered (Scheme 22). Because of its high stability due to aromaticity, [4+2]-cycloaddition reactions with furan (**90**) usually require harsh conditions such as high pressure and elevated temperatures.[127] Nevertheless, the reaction of naphthoquinone **73a** and furan (**90**) at 40 °C was investigated. Instead of the formation of the cycloaddition product **91** however, an electrophilic aromatic substitution reaction towards **92**, as depicted in Scheme 22, was observed.

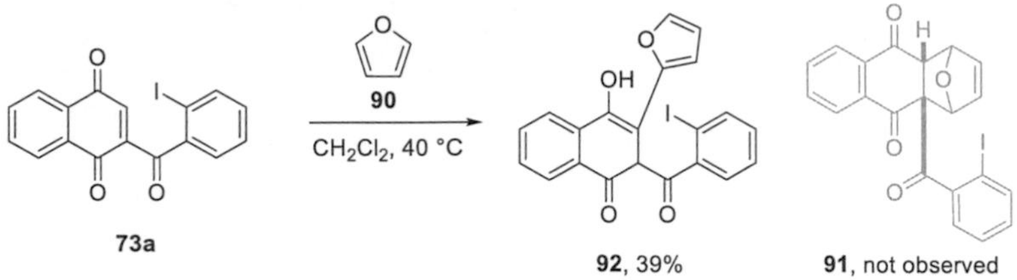

Scheme 22. Attempted Diels-Alder reaction of iodinated naphthoquinone **73a** and furan (**90**) and the instead observed electrophilic aromatic substitution reaction to **92**.

5.4.4 Scope of the Cycloaddition

After the successful establishment of a procedure towards anthraquinone model systems **87aa** and **87ba**, it was decided to increase the complexity of the molecules. For this purpose, easily accessible 2-substituted naphthoquinone derivatives **73a–b** (syntheses described in chapter 5.2) and dienes bearing functional groups (syntheses described in chapter 5.3) were applied in cycloaddition reactions (Scheme 23). In general, the formation of four different products is possible. The two anthraquinone regioisomers are assigned as *ortho* and *para* products which in each case can further be divided into two diastereomers, being the *endo* and *exo* product of the Diels-Alder reaction.

For the synthesis of the anthraquinone subunit of beticolin 0 (**1**), the desired cycloaddition products are represented by the *para* products as they facilitate the installation of a protected hydroxy functionality at the required position in ring C.

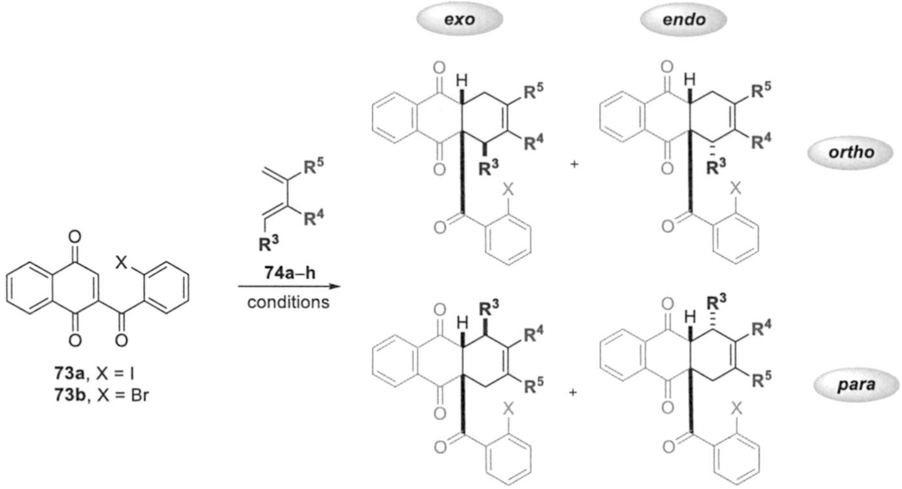

Scheme 23. General scheme for the Diels-Alder reaction between 2-substituted naphthoquinone derivatives **73a–b** and functionalized dienes **74a–h** with all possible products and the respective classification in *exo/endo* diastereomers as well as *ortho/para* regioisomers.

Due to the expected different reactivity of the applied substrates, initially the reaction conditions had to be explored. To our delight, it turned out that thanks to the highly activated reactants, the cycloaddition reactions even proceeded at −78 °C and in relatively short time. When the reaction mixtures were heated, the reaction time could be shortened, but as most of the employed dienes are volatile compounds temperatures above 40 °C were not examined to prevent a loss in yield. Therefore, in a typical Diels-Alder reaction a solution of naphthoquinone derivative **73a–b** (1.0 equiv.) and diene **74a–h** (3.0 – 5.0 equiv.) in dichloromethane was stirred at 40 °C for 2 – 6 hours in a high-pressure vial under Schlenk conditions and after a small aqeous workup the crude was purified *via* column chromatography on silica gel. In the case of symmetrical dienes, only one regioisomer can be formed in the cycloaddition, whereas by the usage of non-symmetrical dienes two regioisomers can occur during the reaction.

The results obtained for the cycloaddition reactions are summarized in Table 2, with the general structures for *exo* (**87**) and *endo* (**88**) products shown in the respective scheme. The names given for the products contain the compound's number and two letters that each stem from a letter of the name of the dienophile and the diene. For example, dienophile **73a** and diene **74a** result in product **87aa**.

Table 2. Scope for the Diels-Alder reactions of naphthoquinone derivatives **73a–b** towards functionalized anthraquinone derivatives **87aa–bf** and **88ad–bf**.

R^5, R^4, R^3 **74a–h** conditions → **87aa–bf** + **88ad–bf**

73a, X = I
73b, X = Br

Entry	X	R^3	R^4	R^5	Product (Yield [%]), {ratio of regioisomers}
1	I	H	H	H	**87aa**[a] (70)
2	Br	H	H	H	**87ba**[a] (82)
3	I	H	CH$_3$/H	H/CH$_3$	**87ab**[b], [c] (63), {7.1:1}
4	Br	H	CH$_3$/H	H/CH$_3$	**87bb**[b], [c] (53), {7.7:1}
5	I	H	CH$_3$	CH$_3$	**87ac**[a] (70)
6	Br	H	CH$_3$	CH$_3$	**87bc**[a] (88)
7	I	OTMS	H	H	**87ad**[a] (20), **88ad**[a] (68)
8	Br	OTMS	H	H	**87bd**[a] (20), **88bd**[a] (61)
9	I	OTMS	CH$_3$	H	**87ae**[a] (26), **88ae**[a] (48)
10	I	OTBDMS	CH$_3$	H	**87af**[a] (29), **88af**[a] (38)
11	Br	OTBDMS	CH$_3$	H	**87bf**[a] (27), **87/88bf** (53)

Reaction conditions: argon atmosphere, dienophile (1.0 equiv.), diene (3.0 – 5.0 equiv.), CH$_2$Cl$_2$, 40 °C, 3 – 6 h. [a] Stereochemistry determined by X-ray diffraction. [b] The product was isolated as a non-separable mixture of diastereomers, {ratio of regioisomers as estimated by ^{1}H NMR}. [c] For R^4 and R^5 two options are given, since the exact structure could not be resolved by analysis of the NMR spectra.

For reasons of completeness, the results of the anthraquinone model system described in chapter 5.4.2 are included in Table 2 (entries 1, 2). As these reactions with 3-sulfolene (**75**) proceeded smoothly, the complexity of the dienes was increased stepwise for further investigations on the reaction.

By considering the structure of beticolin 0 (**1**) and its bridged bicyclic system, isoprene (**74b**) could be a suitable candidate for the incorporation of the methyl group into the system. Diene **74b** is commercially available and was used for reactions with the iodinated as well as the brominated dienophile (entries 3, 4). In both cases the reaction gave the products as a mixture of regioisomers in good yield, with one product formed in large excess. Attempts to separate the two isomers in the product mixtures by either column chromatography on silica gel, preparative thin layer chromatography (prepTLC) on silica gel or high-pressure liquid chromatography (HPLC) failed.

In addition, the exact structure of the products could not be assigned by analyzing the NMR spectra. The expected regioselectivity of the reaction with isoprene (**74b**) would give the "*para*" products **87ab** and **87bb** with R^5 being represented by a methyl group. Based on the results of the X-ray analysis of the other isoprene Diels-Alder products (see chapter 7.3), the isolated molecules most likely show the expected regioselective outcome. The desired product however would be the "*meta*" compound. Concerning attempts for regioselective control of the Diels-Alder reactions, see chapter 5.4.5.

Subsequently, the reaction of 2,3-dimethyl-1,3-butadiene (**74c**) with dienophiles **73a** and **73b** was investigated (entries 5, 6). The reaction ran smoothly and gave very good yields of 70% and 88%, respectively. For both products **87ac** and **87bc** the molecular structures in the solid state were verified by X-ray analysis (see Figure 22 and chapter 7.3).

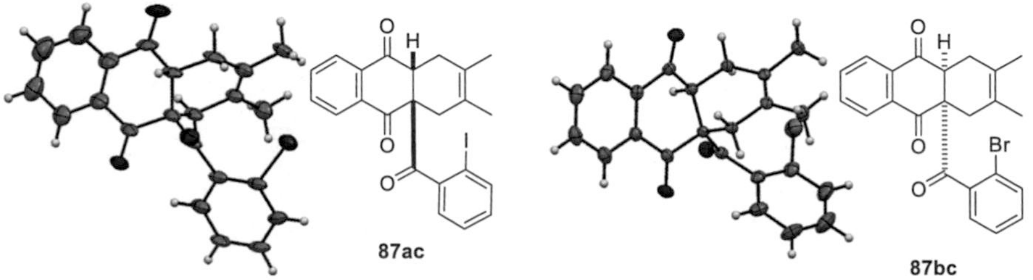

Figure 22. Molecular structures of **87ac** (left), and **87bc** (right) determined by single-crystal X-ray diffraction. Displacement parameters are drawn at 50% probability level. Crystallographic data can be found in chapter 7.3.

Regarding the incorporation of a protected alcohol functionality into the anthraquinone core (ring C, Figure 14), commercially available (trimethylsiloxy)-1,3-butadiene (**74d**) was used for cycloaddition reactions with the two simple naphthoquinone derivatives **73a** and **73b** (entries 7, 8). In both cases a mixture of diastereomers **87ad** (*exo*) and **88ad** (*endo*) (ratio 1 : 3.5) or **87bd** (*exo*) and **88bd** (*endo*) (ratio 1 : 3.0) was isolated. Following general rules, in this reaction the *endo* products are favored. The product mixtures derived from the iodinated as well as the brominated precursor were successfully separated to give pure *exo* and *endo* products *via* column chromatography on silica gel. Moreover, single crystals of all four products were obtained and X-ray analyses confirmed the expected regioselectivity of the reactions with the "*ortho*" products being favored (two structures exemplarily shown in Figure 23). For the incorporation of a silyl ether in ring C of the anthraquinone core of beticolin 0 (**1**), the corresponding regioisomer would be the desired product of the reaction, however the presence of the bulky OTMS group did not affect the regioselectivity. Studies on the regioselective control of the reaction can be found in chapter 5.4.5.

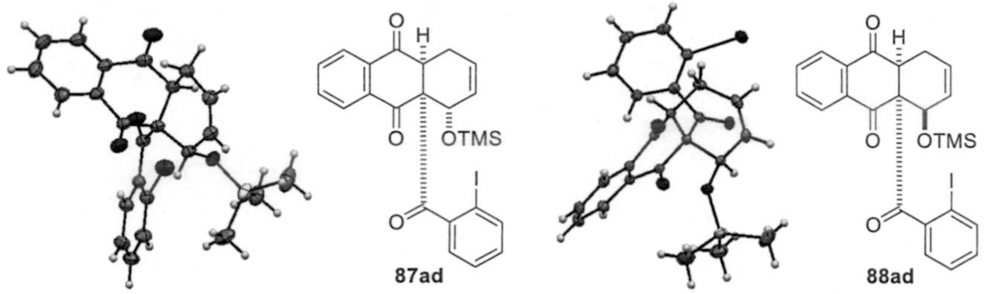

Figure 23. Molecular structures of **87ad** (left) and **88ad** (right) determined by single-crystal X-ray diffraction. Displacement parameters are drawn at 50% probability level. Crystallographic data can be found in chapter 7.3.

Diene **74e** contains a TMS protected hydroxy group as well as a methyl group and was easily accessible from *trans*-2-methyl-2-butenal (**76**) (see chapter 5.3). With this building block, the effect of an additional residue on the diene regarding the regioselectivity of the reaction was investigated. The reaction was performed with 2-(2-iodobenzoyl)naphthalene-1,4-dione (**73a**) as the dienophile and gave the two diastereomers **87ae** and **88ae** (ratio 1 : 1.9) as products which were separated *via* column chromatography on silica gel (entry 9). Since the functionality at the 1-position of the diene presumably was the directing group in the reaction, exclusively the "*ortho*" products were isolated, while substituents at the diene did not affect the regioselectivity of the reaction.

For the investigation of the effect of sterically more demanding residues on the regioselectivity, the OTMS group of diene **74e** was replaced by an OTBDMS group, thus using diene **74f**. The cycloaddition gave a mixture of diastereomers for both the iodinated **73a** and the brominated dienophile **73b** (entries 10, 11). The products were obtained with very similar yields and ratios of isomers in comparison to the cycloaddition with OTMS diene **74e**. Also, as the silyl ether again acts as the directing group, the "*para*" products were selectively formed. As a result, the electronic nature of the compounds seems to have a bigger impact on the regioselective outcome of the reaction than steric effects. X-ray diffraction provided the molecular structure of three of the successfully synthesized products (Figure 24 and chapter 7.3).

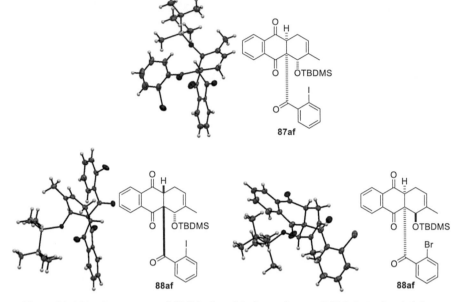

Figure 24. Molecular structures of **87af** (top), and both enantiomers of **88af** (bottom) and their corresponding structures determined by single-crystal X-ray diffraction. Displacement parameters are drawn at 50% probability level. Crystallographic data can be found in chapter 7.3.

Although the installation of a silyl ether at the desired position could not be realized, a modular Diels-Alder approach facilitated easy access to various anthraquinones bearing functional groups. As outlined in the retrosynthetic analysis in chapter 5.1, the next step was supposed to be an intramolecular cyclization in order to generate the pentacyclic subunit of beticolin 0 (**1**). The above described highly functionalized anthraquinone derivatives **87aa–bf** and **88ad–bf** represent interesting precursors for such intramolecular coupling reactions to create unique bicyclo[3.3.1]- or bicyclo[3.2.2]nonane ring systems which are interesting natural product skeletons. Investigations on the latter will be described in chapter 5.5.

5.4.5 Studies on the Control of the Selectivity in Diels-Alder Reactions

As the Diels-Alder reaction is a useful tool in synthetic chemistry, a lot of research was done on the development of catalysts to enhance the reaction and facilitate highly selective outcome. Diels-Alder reactions predominantly lead to the formation of the corresponding *endo* cycloadducts.[128] Several factors like secondary orbital effects, inductive effects and electrostatic attractions are assumed to stabilize the transition state in the reaction and thus enable a stereochemical outcome.[129-131] However, steric hindrance for example can offset the stabilizing interactions and *exo* selectivity can be achieved. Due to the observed lack of regioselectivity and/or *exo/endo* selectivity for some of the Diels-Alder reactions described in chapter 5.4.4, methods to control the selectivity were investigated.

5.4.5.1 Low Temperature Experiments

In first experiments, the reaction temperature was reduced to 0 °C or −78 °C to examine the effect on the outcome of the reaction. In the case of the cycloaddition between iodo dienophile **73a** and isoprene (**74b**) the reaction was run at −78 °C which increased the regioselectivity by shifting the product ratio from 7.1 : 1 to 12.5 : 1 with consistent yield. When the reaction of iodo dienophile **73a** with TMS diene **74d** was performed at −78 °C, the reaction was completed after 4.5 h with the yield for the mixture of diastereomers dropping to 73%. In comparison to the reaction at 40 °C, the reaction time was prolonged while the yield was lowered, and the ratio of isomers did not change. The Diels-Alder reaction between naphthoquinone **73a** and 2-chloro-3-methylbuta-1,3-diene (**93**) at 0 °C gave a regioisomeric mixture of products in a ratio of 1 : 1.9. In comparison to the reaction at 40 °C described in chapter 5.10.3, no improvement of the selectivity was achieved. To conclude, lowering the reaction temperature did not induce selectivity for most of the examined reactions, however in the case of the cycloaddition between iodo dienophile **73a** and isoprene (**74b**) the regioselectivity was improved.

5.4.5.2 Lewis Acid Catalysis

In organic reactions metal-based lewis acids can increase the reactivity of a substrate by accepting electron pairs. Lewis acids like $SnCl_4$, $ZnCl_2$, $TiCl_4$ or BF_3 can catalyze Diels-Alder cycloadditions *via* complexation to the dienophile, resulting in an enhancement of the reaction and/or improvement of its regio- and stereoselectivity.[132-135] It is assumed that the donor–acceptor interaction established between the dienophile and the lewis acid results in a significant stabilization of the LUMO of the dienophile, which translates into a smaller energy gap between the HOMO of the diene and the LUMO of the dienophile and, as a consequence, lowers the reaction's energy barrier.[136-137]

In previous studies the impact of several lewis acids on cycloadditions was investigated.[138] The application of $TiCl_4$ as well as the application of $ZnCl_2$ resulted in the decomposition of iodinated naphthoquinone **73a**. When $B(OAc)_3$ was used in the reaction of iodo dienophile **73a** with TBDMS diene **74f**, the yield of the reaction was increased to 78% while the ratio of isomers did not change. The addition of $BF_3 \cdot OEt_2$ to the cycloaddition between dienophile **73a** and TMS diene **74d** at −20 °C resulted in a decreased total yield of 52% without affecting the isomeric ratio of products.

As the applied lewis acids did not give rise to improved results, the method was not further investigated.

5.4.5.3 Organocatalysts

Organocatalysts like Brønsted acids, chiral amines, thioureas or diols were found to catalyze Diels-Alder reactions in a highly enantioselective fashion.[139-143] In studies on naphthoquinone monoketal Diels–Alder reactions published by Bräse et al., Schreiner's catalyst **94** was reported to facilitate the reaction between sterically hindered monoketal dienophile **54b** and TBDMS diene **55a** and exclusively give product **56a** in 63% yield (Scheme 24).[144] In contrast to the uncatalyzed reaction, selectivity towards the exo product was achieved, although longer reaction times were a drawback. The usage of Jacobsen's catalyst **95** equally promoted exo selectivity but resulted in a lower yield of 29%.

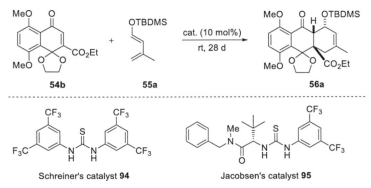

Scheme 24. Thiourea-catalyzed naphthoquinone monoketal Diels–Alder reaction performed by Bräse et al.[144]

In previous studies the above described organocatalysts **94** and **95** were deployed in cycloadditions of naphthoquinone **73a** with TBDMS diene **74f**.[138] Through the addition of Schreiner's catalyst **94** the reaction time was shortened to 2 h with the yield as well as the ratio of isomers not being affected. The application of Jacobsen's catalyst **95** resulted in a slightly increased selectivity towards endo product **88af** (ratio **87af** to **88af** 1 : 1.4) with consistent yield. It thus was assumed that both catalysts do not form a complex with dienophile **73a** in a way that enables significant stereoselectivity of the reaction.

In conclusion, the investigated methods for the control of the selectivity of the Diels-Alder reactions, including temperature decrease, lewis acid catalysis and organocatalysis did not result in substantial improvements. For the reaction between isoprene **74b** and iodo dienophile **73a** at low temperature however, the regioselectivity was enhanced. In upcoming experiments, the methods were not further pursued.

5.5   Intramolecular Cyclization of Anthraquinone Derivatives

After having established a synthetic route towards derivatives of the anthraquinone subunit of beticolin 0 (**1**), the aim was to facilitate intramolecular cyclization reactions for the synthesis of the central motif of the natural product (Scheme 25). The bicyclo[3.2.2]nonane ring system is a unique structural motif, explaining why there is no general synthetic route for its construction. The synthesis of this exceptional bicyclo ring system represents the most challenging part and therefore the key step of the total synthesis of beticolin 0 (**1**).

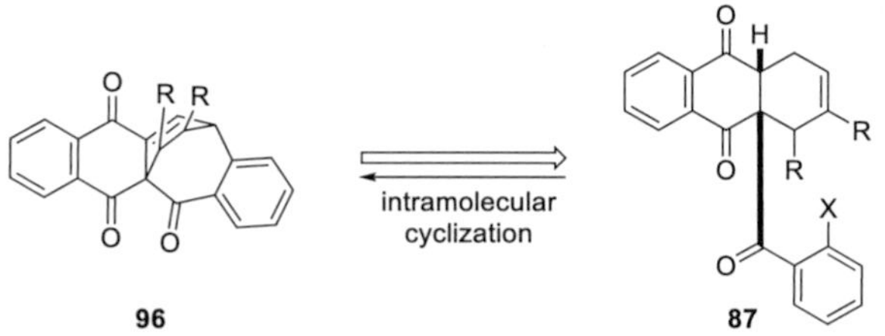

Scheme 25. Attempted cyclization of anthraquinone derivative **87** towards the desired [3.2.2]bicycle **96**.

In 1998 a study on the development of a radical-induced route for the synthesis of [3.2.2]bicycles was published by Duffault and Tellier.[145] Another literature-known approach towards this motif consists of an intramolecular Heck reaction of a bromo acyl pyrrole that generates product mixtures of [3.3.1]- and [3.2.2]bicycles in different ratios depending on the applied reaction conditions.[146]

5.5.1   Radical Cyclization

As the name already depicts, radical cyclizations proceed *via* radical intermediates and give rise to mono- and polycyclic systems.[147] Through radical generation and subsequent intramolecular addition to a multiple bond, almost any ring size can be built.

One of the very few publications dealing with synthetic strategies towards the synthesis of beticolins made use of radical cyclization reactions.[145] In this study Duffault and Tellier describe the application of radical-mediated intramolecular cyclizations in the synthesis of simple [3.2.2]bicycles. In the presence of tributyltin hydride (Bu$_3$SnH) and azobisisobutyronitrile (AIBN) a radical species is generated and subsequently cyclization takes place. The product **98**, based on a seven-membered ring, was formed *via* a 7-*endo-trig* process as the major product of the reaction, with the conditions shown in Scheme 26. Product **98** was synthesized *via* addition of the radical to the electron poor position of the α,β-unsaturated ketone **97** while the observed side reactions were a 6-*exo-trig* cyclization and an H abstraction.

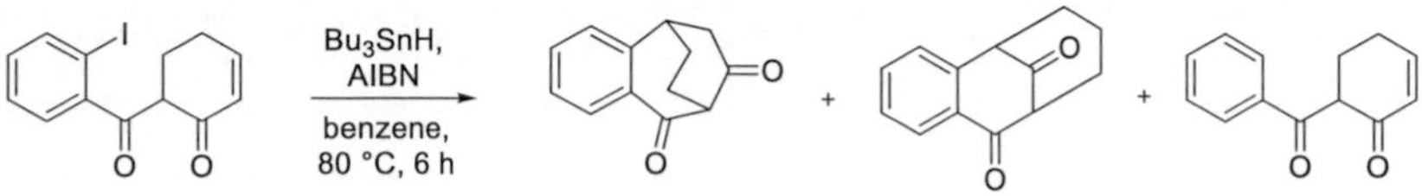

97 **98** 7-*endo* 78% **99** 6-*exo* 12% **100** H abstraction 10%

Scheme 26. Development of a synthetic method towards a [3.2.2]bicycle **98** *via* radical cyclization.[145]

In this thesis the behavior of anthraquinone derivatives **87aa** and **88af** (see chapter 5.4) in radical cyclization reactions was investigated. By applying the conditions described in the publication of Duffault and Tellier, the formation of the desired bicyclo[3.2.2]systems was attempted. The performed reactions were closely monitored by mass spectrometry, but the formation of neither of the two possible products (**96, 101**) was observed. Instead only starting material was present in the mixture (Scheme 27).

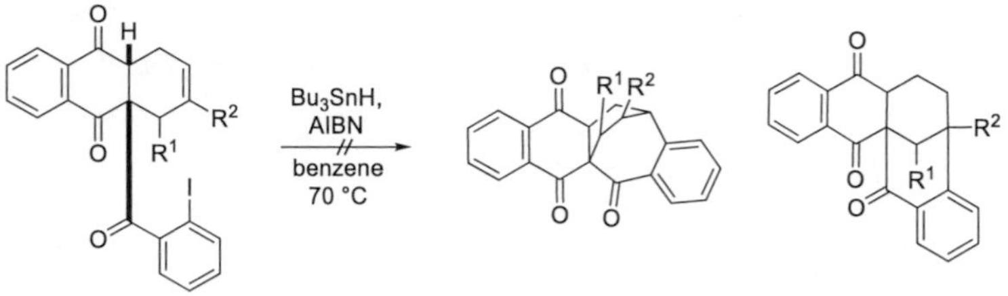

87aa $R^1 = R^2 = H$
88af R^1 = OTBDMS
R^2 = Me

96 **101**

Scheme 27. Approach for the radical cyclization of anthraquinone derivatives **87aa** and **88af** to generate the desired [3.2.2] product **96**.

5.5.2 First Studies on Intramolecular Heck Reactions

A competing method to the radical cyclization is the Heck reaction. As it can be used to form rings of a variety of sizes and topologies, its application in the intramolecular coupling of anthraquinone derivatives was investigated in this project.

5.5.2.1 The Intramolecular Heck Reaction

The Heck reaction, discovered in 1968, is a palladium-catalyzed coupling reaction forming a C–C bond between an aryl halide and an olefin.[148-150] High functional group tolerance and mild reaction conditions make it an important tool which enables its application at late stages in synthetic routes. In comparison to transition-metal-mediated cross-coupling reactions, the Heck reaction is unique as it constructs sp^2–sp^3 bond linkages while generating stereogenic centers.[151-152]

The intramolecular version of the reaction was realized for the first time in the formation of indole **103** by Mori *et al.* in 1977 (Scheme 28).[153] Since then, it was extensively utilized to construct various ring sizes and therefore became a key step in the total synthesis of many natural products, such as (–)-morphine and ibogamine.[154-156]

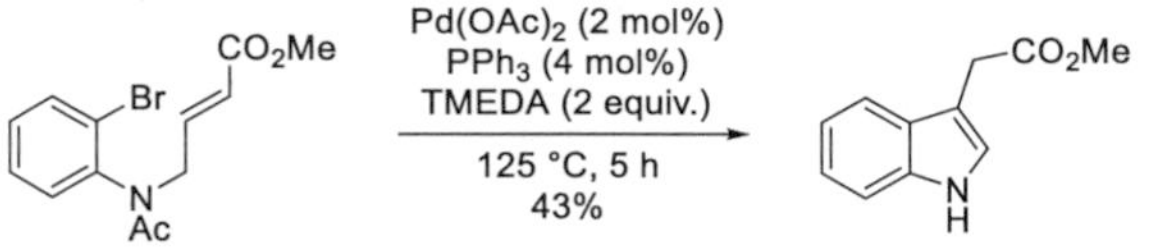

Scheme 28. First described intramolecular Heck reaction towards indole **103**.[153] TMEDA = tetramethylethylenediamine.

5.5.2.2 Establishment of Bicyclo[3.2.2]systems *via* Heck Reaction

Only a few examples on the establishment of bicyclo[3.2.2]systems *via* an intramolecular Heck reaction can be found in literature. For their studies on the synthesis of the natural product huperzine A, the working group of Mann developed methods towards bicyclo[3.3.1]systems **106a/b** (Table 3). By applying Heck conditions to various enones **104**, they obtained reduced bicyclo[3.2.2]system **105** as well as [3.3.1]bicycles **106a** and **106b**.

Table 3. Results of the intramolecular Heck reaction towards bicyclo[3.3.1]- and bicyclo[3.2.2]systems performed by Mann *et al.*[157]

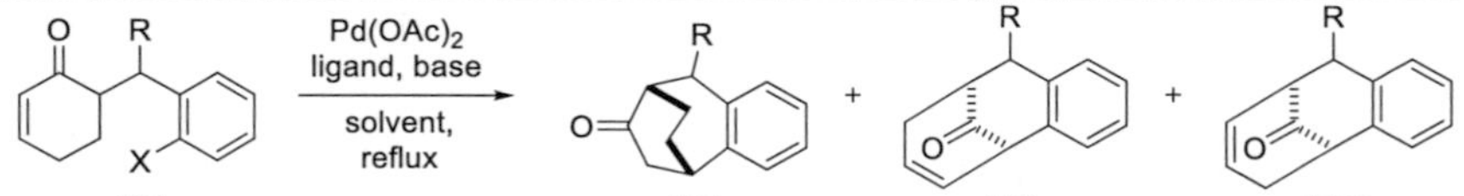

Entry	X	R	Ligand (equiv.)	Base (equiv.)	Solvent	Product ratio [%] 105 : 106a : 106b
1	Br	OTBDMS	PPh₃ (0.4)	Et₃N (12)	MeCN	23 : 21 : 0
2	I	OTBDMS	PPh₃ (0.4)	Et₃N (12)	MeCN	11 : 30 : 0
3	I	OTBDMS	POT (0.2)	PMP (2)	DMA	11 : 60 : 0
4	Br	H	PPh₃ (0.4)	PMP (2)	DMF	44 : 5 : 0
5	Br	H	PPh₃ (0.4)	Et₃N (12)	DMA	63 : 25 : 6

POT = tris(*o*-tolyl)phosphine, PMP = pentamethylpiperidine.

With systematical variation of the applied conditions, the reaction was optimized to give the desired product **106a** in large excess (Table 3, entries 1 – 3). When enone **104** was simplified by replacing the TBDMS ether with a proton, reduced bicyclo[3.2.2]system **105** was obtained as the major product, with the source for the hydride ion not being identified (entries 4, 5). Moreover, double bond migration was observed in [3.3.1]bicycle **106b** (entry 5). By comparing the data given for the screening of different conditions, the impact of the C7 substituent R, ligand, solvent as well as base on the outcome of the reaction becomes clear. Subtle modification of the conditions employed in the Heck reactions allows for the formation of various novel ring systems.

For investigations on the synthesis of dragmacidin F, Stoltz *et al.* performed intramolecular Heck reactions to construct [3.3.1]bicycles **108**.[146] During their studies the formation of bicyclo[3.2.2]system **109** as a side product of the reaction was observed (Scheme 29). By significantly reducing the palladium concentration, the amount of bicyclo[3.2.2]system **109** was increased. The impact of the base was demonstrated when the reaction exclusively gave **109** by applying pyridine instead of *N,N*-dicyclohexylmethylamine (Cy$_2$NMe). Another finding of this investigation is the change of the ratio of products during the reaction, indicating a change of selectivity with increasing concentration of trialkylammonium salt R$_3$NH$^+$Br$^-$.

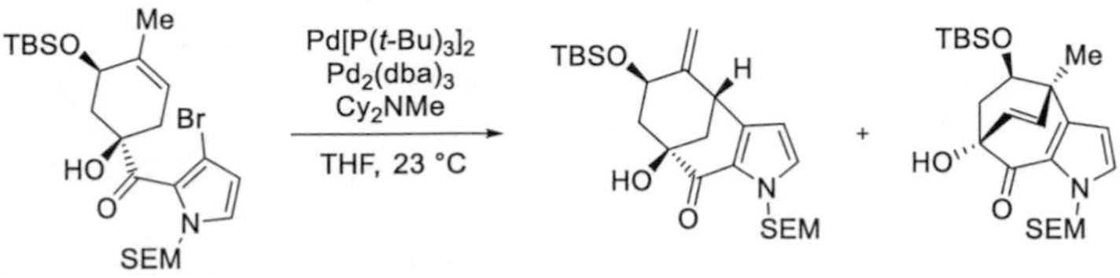

Scheme 29. Intramolecular Heck cyclization towards bicycles **108** and **109** for the synthesis of dragmacidin F by Stoltz *et al.*[146] Pd$_2$(dba)$_3$ = tris(dibenzylideneacetone)dipalladium(0).

On the basis of the above described studies on Heck cyclizations giving [3.2.2]bicycles, it was decided to investigate palladium-catalyzed intramolecular coupling reactions as an alternative strategy to the radical cyclization.

To initiate the studies, model anthraquinone derivatives **87aa** and **88ba** were subjected to standard Heck protocols where a mixture of Pd-catalyst, phosphine ligand and base was heated and stirred under Schlenk conditions. Due to the fact that palladium(II) acetate is cheap and can be reduced to palladium(0) *in situ* with phosphine, it was used for the first experiment (Table 4, entry 1). As a product anthraquinone **111** bearing a novel [3.3.1]ring system was isolated in good yield. A single crystal of the product **111a** was obtained, and the structure was confirmed by X-ray diffraction (Figure 25). As depicted, after the coupling the previously halogenated benzene ring is positioned approximately perpendicular (106°) to the anthraquinone scaffold. The formation of the desired [3.2.2]product **110** was not observed under these conditions. By increasing the amount of ligand to 0.80 equivalents and applying *i*-Pr$_2$NEt instead of PMP (pentamethylpiperidine), the reaction as well showed selectivity for the [3.3.1]product **111** with consistent yield (entry 2).

Table 4. Heck cyclization with anthraquinone model systems **87aa** and **87ba**.

		87aa / ba		110		111
Entry	X	Catalyst (equiv.)	Ligand (equiv.)	Base (equiv.)	Solvent	Product, yield
1	I	Pd(OAc)$_2$ (0.2)	PPh$_3$ (0.4)	PMP (2)	DMA	**111**, 72%
2	I	Pd(OAc)$_2$ (0.2)	PPh$_3$ (0.8)	i-Pr$_2$NEt (12)	DMA	**111**, 76%
3	Br	Pd(OAc)$_2$ (0.2)	PPh$_3$ (0.4)	PMP (2)	DMA	**111**, 59%
4	I	Pd(PPh$_3$)$_4$ (0.2)	---	PMP (2)	DMA	**111**, 95%
5	I	Pd(OAc)$_2$ (0.1)	POT (0.2)	Et$_3$N (12)	DMF	**111**, 73%
6	I	Pd$_2$(dba)$_3$ (0.2)	---	PMP (2)	DMA	**111**, 86%
7	I	Pd(OAc)$_2$ (0.2)	P(Cy)$_3$ (0.4)	PMP (2)	DMA	**111**, 54%
8	I	Herrmann-Beller (0.2)	---	PMP (2)	DMA	**111**, 80%
9	I	Pd(OAc)$_2$ (0.1)	dppe (0.2)	PMP (2)	DMA	mixture[a]
10	I	Pd(OAc)$_2$ (0.1)	dppf (0.2)	PMP (2)	DMA	mixture[a]
11	I	Pd(OAc)$_2$ (0.1)	BINAP (0.2)	PMP (2)	DMA	mixture[b]

[a] non-separable mixture of starting material **87aa/ba** and epimer thereof.
[b] non-separable mixture of starting material **87aa/ba**, epimer thereof and [3.3.1] product **111**.
Dppe = 1,2 bis(diphenylphosphino)ethane, dppf = 1,1' bis(diphenylphosphino)ferrocene, BINAP = 2,2'-bis(diphenylphosphino)-1,1'-binaphthyl.

When using brominated anthraquinone derivative **87ba** and the above described standard coupling conditions, the same [3.3.1]ring system **111** was obtained. However, the yield dropped to 59%, presumably since iodine represents the better leaving group (entry 3). Figure 25 depicts the three-dimensional arrangement of the coupled anthraquinone **111a** (racemic) with two planes arranged perpendicular to each other.

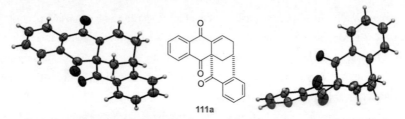

Figure 25. Molecular structure of **111a** (racemic) from two different perspectives, determined by single-crystal X-ray diffraction. Displacement parameters are drawn at 50% probability level. Crystallographic data can be found in chapter 7.3.

In order to establish the conditions required for the intramolecular coupling towards desired product **110**, model studies on anthraquinone derivative **87aa** were carried out. For this purpose, catalyst, ligand and base were systematically varied (entries 4 – 8). Interestingly, most of the conducted experiments exclusively gave the [3.3.1] product **111** with the best yield obtained by applying tetrakis(triphenylphosphine)palladium(0) (Pd(PPh3)4) and PMP (Table 4, entry 4). Most reactions gave good yields, only the approach with tricyclohexylphosphine (P(Cy)3) as a ligand yielded 54% (entry 7).

The Herrmann Beller catalyst is popular in Heck reactions, as it shows an increased activity and stability with high turnover numbers in comparison to conventional catalytic systems (Figure 26).[158-159] As described in Table 4, anthraquinone derivative **87aa** was applied in a coupling using this catalyst (entry 8) which exclusively gave the bicyclo[3.3.1]product **111** in very good yield.

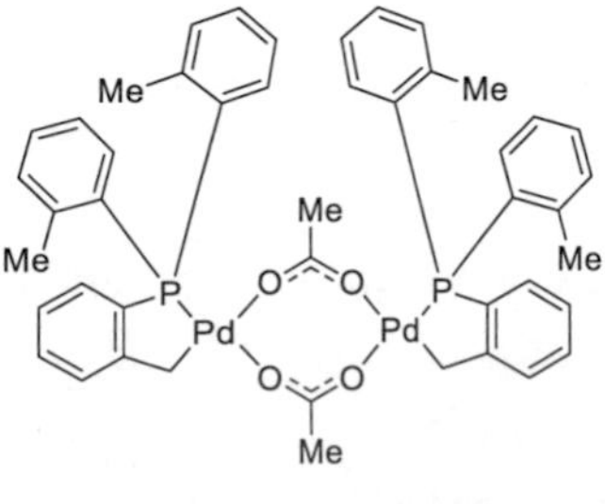

Figure 26. Structure of the Herrmann Beller catalyst.

Bidentate ligands are known to be able to enhance stereoselectivity in asymmetric reactions.[160-162] By applying ligands like dppe (1,2-bis(diphenylphosphino)ethane), dppf (1,1'-bis(diphenylphosphino)ferrocene) or BINAP (2,2'-bis(diphenylphosphino)-1,1'-binaphthyl) in this study, non-separable mixtures of starting material, its epimer and the coupled product were obtained, as estimated by mass spectrometry (entries 9 – 11). When a purification of the product mixture by crystallization was attempted, the epimer of the starting material **87aa-epi** crystallized from the crude. Moreover, the conversion to epimer **87aa-epi** was favored and a ratio of 1 : 3.0 (**87aa** : **87aa-epi**), as estimated by ¹H NMR, was achieved by increasing the amount of palladium acetate and BINAP. This finding can be explained with the accessibility of *trans*-fused quinone cycloadducts by treatment with strong acids or bases.[163-164] A single crystal of the epimer **87aa-epi** was obtained and the structure was confirmed by X-ray diffraction. Figure 27 shows a comparison of the molecular structures of anthraquinone derivative **87aa** and its epimer **87aa-epi** which illustrates that in **87aa-epi** ring system ABC is planar in contrast to the tricyclic system in **87aa**.

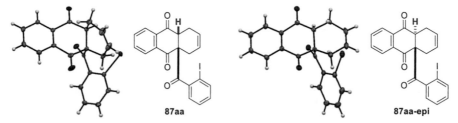

Figure 27. Comparison of the molecular structures of compound **87aa** (left) and its epimer **87aa-epi** (right) determined by single-crystal X-ray diffraction. In **87aa-epi** the ABC ring system is planar in contrast to the same tricycle in anthraquinone derivative **87aa**. Displacement parameters are drawn at 50% probability level. Crystallographic data can be found in chapter 7.3.

To investigate the influence of the modified three-dimensional arrangement in anthraquinone **87aa-epi**, the molecule was used as a reactant in a standard Heck coupling. As a result the same bicyclo[3.3.1]system **111** as in the previously described studies was isolated. Thus, no influence of the three-dimensional structure of the ABC ring system on the outcome of the Heck reaction was observed.

Heck reactions under "Jeffery conditions" involve simple palladium salts, an inorganic base and tetraalkylammonium salts but lack the addition of a ligand.[165-166] Depending on the substrates, the addition of ammonium salts to phosphine-free reaction mixtures can result in higher catalytic activities.[167] By applying these conditions to anthraquinone derivative **87aa**, two different fractions were isolated from the reaction after column chromatography on silica gel (Scheme 30). One of these contained a mixture of starting material **87aa** and its epimer **87aa-epi** in a ratio of 1 : 2.9 (total isolated amount: 48%). The other fraction contained the [3.3.1]system **111** in 41% yield, with the desired compound **110** not being obtained.

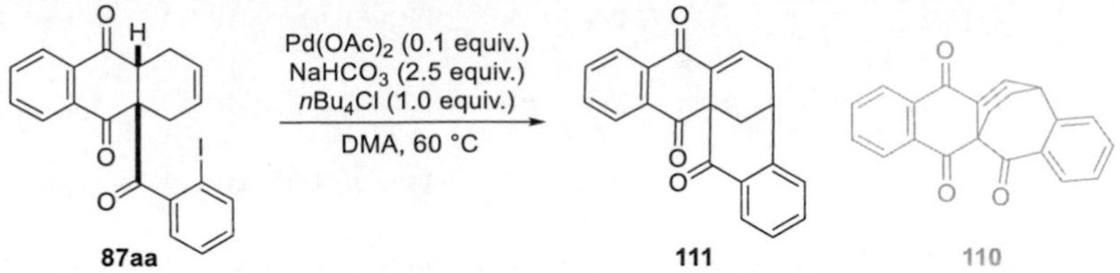

Scheme 30. Jeffery conditions applied to anthraquinone model system **87aa**.

In summary, the above described Heck reactions with anthraquinone model systems **87aa** and **87ba** did not result in an effective construction of the bicyclo[3.2.2]nonane ring system **110**. Systematic variation of reaction conditions like catalyst, ligand, base or solvent biased the yields of the reactions, but not the selectivity. Even epimer **87aa-epi** resulting from the Heck reactions with bidentate ligands despite its modified, planar ABC scaffold gave bicyclo[3.3.1]product **111**.

Based on the above described results, in following experiments the behavior of anthraquinone derivatives containing functional groups was tested, with systematic variation of the conditions of the coupling. The related results are shown in Table 5.

Table 5. Scope for the Heck reactions on anthraquinone derivatives **87/88aa – 87/88bf** towards highly functionalized [3.2.2] **112** or [3.3.1]systems **113**.

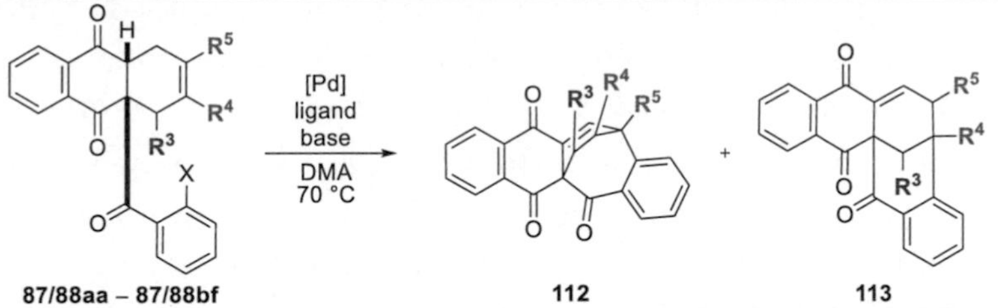

Entry	X	R^3	R^4	R^5	Catalyst (equiv.)	Ligand (equiv.)	Base (equiv.)	Product, yield
1	I	H	CH_3/ H	H/ CH_3	Pd(OAc)$_2$ (0.2)	PPh$_3$ (0.4)	PMP (2)	mixture[a]
2	I	H	CH_3/ H	H/ CH_3	Pd(OAc)$_2$ (0.2)	PPh$_3$ (0.8)	i-Pr$_2$NEt (12)	mixture[a]
3	I	H	CH_3/ H	H/ CH_3	Pd(OAc)$_2$ (0.1)	BINAP (0.2)	PMP (2)	mixture[b]
4	Br	H	CH_3/ H	H/ CH_3	Pd(OAc)$_2$ (0.2)	PPh$_3$ (0.4)	PMP (2)	mixture[a]
5	I	H	CH_3	CH_3	Pd(OAc)$_2$ (0.2)	PPh$_3$ (0.4)	PMP (2)	mixture[a]
6	Br	H	CH_3	CH_3	Pd(OAc)$_2$ (0.2)	PPh$_3$ (0.4)	PMP (2)	mixture[a]

7	I	OTMS	H	H	Pd(OAc)$_2$ (0.2)	PPh$_3$ (0.4)	PMP (2)	**113ad**, 58%
8	I	OTMS	H	H	Pd(OAc)$_2$ (0.1)	BINAP (0.2)	PMP (2)	mixture[b]
9	Br	OTMS	H	H	Pd(OAc)$_2$ (0.2)	PPh$_3$ (0.4)	PMP (2)	**113bd**, 50%
10	I	OTMS	CH$_3$	H	Pd(OAc)$_2$ (0.2)	PPh$_3$ (0.4)	PMP (2)	**113ae**, 70%
11	I	OTBDMS	CH$_3$	H	Pd(OAc)$_2$ (0.2)	PPh$_3$ (0.4)	PMP (2)	**113af**, 99%
12	Br	OTBDMS	CH$_3$	H	Pd(OAc)$_2$ (0.2)	PPh$_3$ (0.4)	PMP (2)	**113bf**, 41%
13	Br	OTBDMS	CH$_3$	H	Pd(OAc)$_2$ (0.2)	PPh$_3$ (0.4)	PMP (2)	**114bf**, 37%

[a] non-separable mixture of starting material and product(s).
[b] non-separable mixture of starting material, epimer thereof and [3.3.1] product **113**.

Initially, Heck reactions with the iodinated **87ab** as well as the brominated anthraquinone derivative **87bb** derived from isoprene were studied (Table 5, entries 1 – 4). All tested conditions gave complex mixtures of different products that could not be separated by column chromatography on silica gel, preparative thin layer chromatography or HPLC. Mass analysis though verified the existence of at least one of the desired products in each of the mixtures and therefore a purification of the crude product by crystallization was attempted. Indeed, this approach was successful and the presence of two different isomers **113ab** (racemic) and **115ab** containing the [3.3.1]motif was verified by X-ray analysis (Figure 28). Moreover, NMR measurements of the crystals gave good spectra. As depicted, double bond migration was observed. In addition to the two isomers, the epimer of the starting material (**87ab-epi**) crystallized from the crude product of the reaction with bidentate ligand BINAP (Table 5, entry 3). The conditions of the performed Heck reactions lead to a small amount of decomposed starting material; as indicated by the isolation of traces of 2-methylanthracene-9,10-dione (**116**).

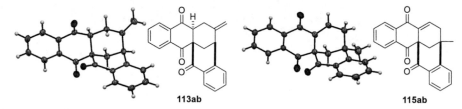

113ab   **115ab**

Figure 28. Molecular structures of **113ab** (left) and **115ab** (right) determined by single-crystal X-ray diffraction. **113ab** crystallized as an enantiomeric mixture. Displacement parameters are drawn at 50% probability level. Crystallographic data can be found in chapter 7.3.

For the separation of unsaturated compounds, silica gel impregnated with silver ions can give satisfying results where other purification methods fail.[168-170] Due to the extremely similar polarity of the starting material and the product mixture in the above described reaction with **87ab**, a separation of the crude product *via* a silver nitrate (AgNO₃) coated silica plate was examined. For this purpose, a preparative TLC plate was incubated with a 10% solution of silver nitrate in acetonitrile, then dried and activated with a heat gun.[171] For the separation process, the TLC plate was loaded with crude product and developed in a glass chamber with a suitable solvent mixture. After the isolation of different fractions, it turned out that still complex mixtures are present which can not be identified due to overlapping signals in the corresponding NMR spectra. The described method was not further pursued as the isolated fractions showed no significant improvement in separation efficiency.

Then, anthraquinone derivatives **87ac** and **87bc** derived from dimethyl butadiene **74c** were investigated regarding their behavior in intramolecular C–C coupling reactions (entries 5, 6). The Heck reactions were performed under standard conditions and again gave non-separable product mixtures for both the iodinated **87ac** and the brominated **87bc** starting material. The [3.3.1]product **113ac** crystallized from the mixtures and was analyzed by X-ray crystallography (Figure 29). Here the same three-dimensional arrangement as in the other [3.3.1]products was present, and again double bond migration was observed. The desired product **112** could not be verified. A decomposition of the starting material to quinone derivative **117** was observed in trace amounts.

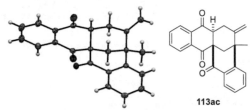

113ac

Figure 29. Molecular structure of **113ac** determined by single-crystal X-ray diffraction. Displacement parameters are drawn at 50% probability level. Crystallographic data can be found in chapter 7.3.

To investigate the effect of space demanding groups on the process of the palladium-catalyzed coupling, the anthraquinone derivative **88ad** containing an OTMS group was applied in the reaction (entry 7). Under standard Heck conditions, the reaction gave the [3.3.1]product **113ad** in 58% yield. The molecular structure of the compound was verified by X-ray analysis (Figure 30). The application of BINAP as a ligand for palladium, yielded a non-separable product mixture (entry 8). When brominated TMS anthraquinone derivative **88bd** was coupled under standard conditions, the [3.3.1]product **113ad** was isolated in slightly decreased yield (entry 9).

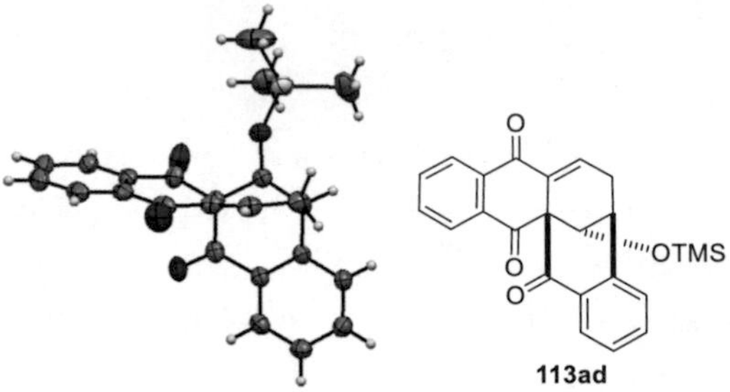

113ad

Figure 30. Molecular structure of **113ad** determined by single-crystal X-ray diffraction. Displacement parameters are drawn at 50% probability level. Crystallographic data can be found in chapter 7.3.

Anthraquinone derivative **88ae** contains a methyl residue in addition to the OTMS group on ring C. Intramolecular coupling with this compound exclusively gave bicyclo[3.3.1]nonane ring system **113ae** in 70% yield. A single crystal of **113ae** was obtained and the structure was confirmed by X-ray diffraction (Figure 31).

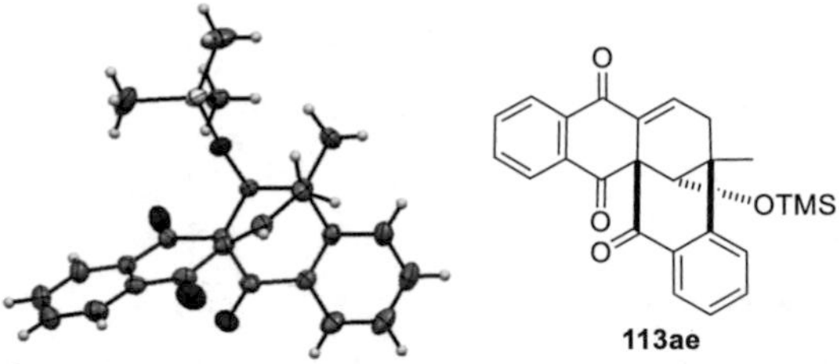

113ae

Figure 31. Molecular structure of **113ae** determined by single-crystal X-ray diffraction. Displacement parameters are drawn at 50% probability level. Crystallographic data can be found in chapter 7.3.

Further increase in complexity was provided by compounds **87/88af** and **87/88bf** bearing a bulky OTBDMS group. A Pd(0)-catalyzed intramolecular Heck reaction of **88af** selectively yielded the [3.3.1]product **113af** in an excellent yield (entry 11). The coupling reactions with brominated diastereomers **87bf** and **88bf** gave the two diastereomers **113bf** and **114bf** as products (entry 12, 13). For both products single crystals were successfully obtained and the molecular structures were verified by X-ray analysis (Figure 32). The two different diastereomers gave two different Heck products, which is evidence for the impact of the starting material on the stereochemistry of the Heck reaction. The desired [3.2.2]systems however have not been isolated from any of the product mixtures.

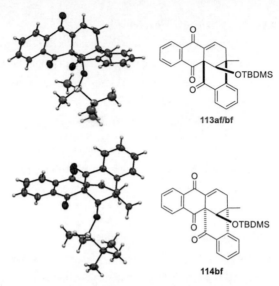

Figure 32. Molecular structures of **113af/bf** (top) and **114bf** (bottom) determined by single-crystal X-ray diffraction. Displacement parameters are drawn at 50% probability level. Crystallographic data can be found in chapter 7.3.

The significant influence of modifying the reaction conditions as well as the subtrates, on the selectivity of the Heck reaction is well-known. For the above described couplings, a variation of the substrates, as well as a screening of the reaction parameters such as catalyst, ligand, base, and the ratio of palladium to phosphine however did not deliver the desired [3.2.2]scaffold **112** in synthetically useful yield. The separation of the obtained isomeric mixtures *via* common methods like column chromatography on silica gel, prepTLC or HPLC was extremely challenging due to the very similar polarity of the starting materials and the products. Moreover, the crude NMR spectra were too complex to be able to differentiate between different structures. In such cases where the formation of one single product was clearly favored, a pure product was provided. Anthraquinone **88af** bearing a TBDMS residue, for example gave rise to a [3.3.1]bicycle in high yield and with good selectivity. The syntheses of various compounds bearing a novel scaffold with a bicyclo[3.3.1]nonane ring system **113** were successfully accomplished, the obtained products were fully characterized and a lot of them could be crystallized in order to verify their structures by X-ray diffraction.

As a result, the successful formation of the desired [3.2.2]product can not be excluded, but due to difficulties in its verification and purification, it was not isolated from the reactions.

5.6 Further Studies on Diels-Alder Reactions

Due to the fact that neither by radical cyclizations nor by intramolecular Heck reactions the desired [3.2.2]products **112** were successfully isolated so far, further adjustments of the reaction parameters had to be considered. It is known that also steric effects can have a high impact on the outcome of Heck reactions which is why the incorporation of sterically demanding substituents was pursued to further modify the anthraquinone derivatives.

In addition, the basis for the integration of the tetrahydroxanthone part into the anthraquinone derivative was laid by employing naphthoquinone derivatives bearing functionalized halogenated benzene rings in the Diels-Alder reactions. The reactivity and behavior of these dienophiles was compared with their related, non-functionalized compound **73b**.

5.6.1 Towards Highly Functionalized Anthraquinone Derivatives

In first attempts, dienes containing bulky substituents were applied in cycloaddition reactions. This time the functionalities were also positioned at R^5. The obtained results on the scope of the reaction are summarized in Table 6.

Table 6. Scope for the Diels-Alder reactions with naphthoquinone derivatives **73a–g** towards highly functionalized anthraquinone derivatives **87aa–gh** and **88cd–gh**.

Entry	dienophile		diene		Product (Yield [%]), [*dr* value], {ratio of regioisomers}
1	I	**73a**	OTIPS	**74g**	**87ag**[a] (75)
2	I	**73a**	OTBDPS	**74h**	**87ah** (72)
3	Br	**73b**	OTBDPS	**74h**	**87bh**[a] (79)

4	Br, OMe structure	73c	diene structure	74b	87cb[b], [c] (51), {10:1}
5	Br, OMe structure	73c	diene structure	74c	87cc (65)
6	Br, OMe structure	73c	diene OTMS structure	74d	87/88cd (63), [5.6:1]
7	Br, OMe, OMe structure	73d	diene structure	74b	87db[b], [c] (70), {6.7:1}
8	Br, OMe, OMe structure	73d	diene structure	74c	87dc (39)
9	Br, OH, OH structure	73e	diene structure	74c	87ec (54)
10	Br, NH-C(O) structure	73f	diene structure	74b	87fb[b], [c] (38), {7.7:1}
11	Br, NH-C(O) structure	73f	diene structure	74c	87fc[a] (68)
12	Br, NH-C(O)CF$_3$ structure	73g	diene structure	74b	87gb[b], [c] (79), {7.7:1}
13	Br, NH-C(O)CF$_3$ structure	73g	diene structure	74c	87gc[a] (84)
14	Br, NH-C(O)CF$_3$ structure	73g	diene OTMS structure	74d	87/88gd[a] (76), [3.7:1]

Reaction conditions: argon atmosphere, dienophile (1.0 equiv.), diene (3.0 – 5.0 equiv.), CH$_2$Cl$_2$, 40 °C, 3 – 5 h. [a] Relative stereochemistry determined by X-ray diffraction. [b] The product was isolated as a non-separable mixture of diastereomers, [ratio of isomers as estimated by ^{1}H NMR]. [c] For R^4 and R^5 two options are given since the exact structure could not be resolved by NMR spectra analysis.

An increase in complexity and steric demand provided the previously synthesized TIPS diene **74g**. The cycloaddition reaction with iodo dienophile **73a** proceeded regioselectively and exclusively afforded *para* naphthoquinone derivative **87ag** in a very good yield (Table 6, entry 1). The structure of the product was verified by X-ray diffraction (Figure 33 and chapter 7.3).

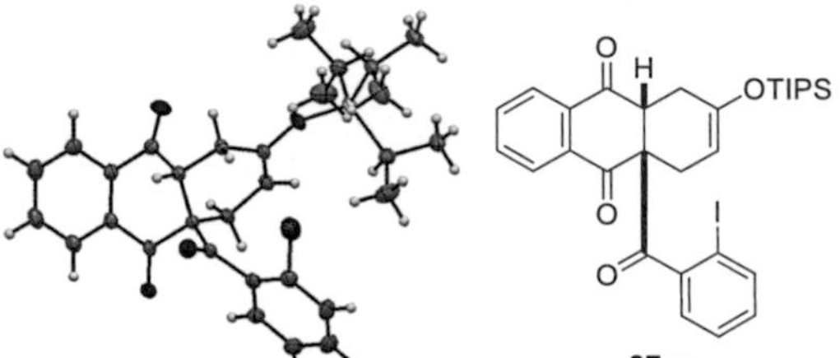

87ag

Figure 33. Molecular structure of **87ag** determined by single-crystal X-ray diffraction. Displacement parameters are drawn at 50% probability level. Crystallographic data can be found in chapter 7.3.

Another example for a Diels-Alder reaction that afforded a single regioisomer is the cycloaddition with TBDPS diene **74h** and **73a/73b** (entries 2 and 3). The reaction gave up to 79% yield with the brominated structure **87bh** being verified by X-ray crystallography (Figure 34).

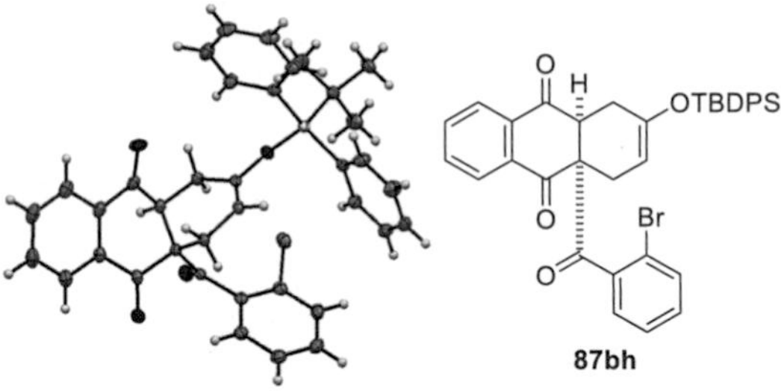

87bh

Figure 34. Molecular structure of **87bh** determined by single-crystal X-ray diffraction. Displacement parameters are drawn at 50% probability level. Crystallographic data can be found in chapter 7.3.

In consideration of the structure of beticolin 0 (**1**), different functionalities were supposed to be integrated into the halogenated aromatic ring of the naphthoquinone derivatives. This way the basis for the construction of the tetrahydroxanthone subunit of the natural product ought to be laid. The related investigations were commenced with the incorporation of a simple methoxy functionality by applying 2-bromo-5-methoxybenzoic acid (**71c**) in the acylation step of the naphthoquinone derivative synthesis. The Diels-Alder reaction with isoprene (**74b**) and methoxy dienophile **73c** gave a very similar yield and ratio of products in comparison to the reactions with dienophiles **73a** and **73b** (entry 4). Dienophile **73c** was also reacted with 2,3-dimethyl-1,3-butadiene (**74c**) which provided naphthoquinone derivative **87cc** in 65% yield (entry 5). The cycloaddition reaction between methoxy dienophile **73c** and TMS diene **74e** resulted in a minor decrease in yield with the ratio of products being shifted from 1 : 3.1 to 1 : 5.6, in comparison to the reaction between

TMS diene **74e** and non-functionalized bromo dienophile **73b** (entry 6). This indicates an effect of the methoxy group on the stereoselectivity of the reaction with one diastereomer forming in large excess. The exact structure of the products of the reaction could not be verified by X-ray crystallography. However, it is assumed that the same selectivity for the *ortho* products with the major product being the *endo* anthraquinone is observed as in the reaction between **73b** and **74e**, since comparison of the NMR spectra gives the exact same characteristic splitted peaks for the CH$_2$ group.

As outlined before, dimethoxy dienophile **73d** was employed in Diels-Alder cycloadditions with isoprene (**74b**) as well as dimethyl butadiene **74c**. By applying the standard conditions, the reaction with isoprene (**74b**) resulted in an increased yield of 70% of a non-separable mixture of isomers with consistent ratio of regioisomers (6.7 : 1), in comparison to the cycloaddition between **73b** and **74b** (entry 7). In the reaction of dienophile **73d** with diene **74c** only 39% of product **87dc** were obtained (entry 8).

To increase the polarity of the compounds, the methyl ethers of dimethoxy dienophile **73d** were cleaved to provide the free hydroxy groups. The resulting dienophile **73e** reacted in a Diels-Alder addition with dimethyl butadiene **74c** to give anthraquinone **87ec** in an acceptable yield (entry 9).

Additionally, an *N*-acetyl residue was successfully integrated into the naphthoquinone derivative **73f**. The latter was applied in a reaction with isoprene (**74b**) to give anthraquinone **87fb** in 38% yield with a mixture of regioisomers in a ratio of 7.7 : 1 (entry 10). The Diels-Alder cycloaddition of dimethyl diene **74c** with *N*-acetylated naphthoquinone **73f** proceeded smoothly to afford **87fc** in good yield (entry 11).

An *N*-trifluoroacetyl residue was installed into the dienophile precursor, when 5-amino-2-bromobenzoic acid (**71f**) was applied in the acylation reaction with 1,4-dimethoxynaphthalene (**70**) in TFAA. The obtained dienophile **73g** was employed in cycloadditions with various dienes. Firstly, it was reacted with isoprene (**74b**) which resulted in the trifluoromethylated anthraquinone derivative **87gb** in a significantly improved yield in comparison to the reaction between **73b** and **74b**, however the same non-separable mixture of products (ratio 7.7 : 1) was observed (entry 12). Additionally, dienophile **73g** underwent a reaction with dimethyl butadiene **74c** to give **87gc** in a very good yield of 84% (Table 6, entry 13), which was successfully crystallized and its molecular structure identified (Figure 35). The Diels-Alder reaction with TMS diene **74d** resulted in a diastereomeric mixture of products, which showed the expected regiochemistry (entry 14). *Ortho* anthraquinone derivatives **88gd** and **87gd** were isolated in 76% total yield in an *endo/exo* ratio of

3.7 : 1. The structure of the *endo* product **88gd** was verified by X-ray crystallography (Figure 35 and chapter 7.3).

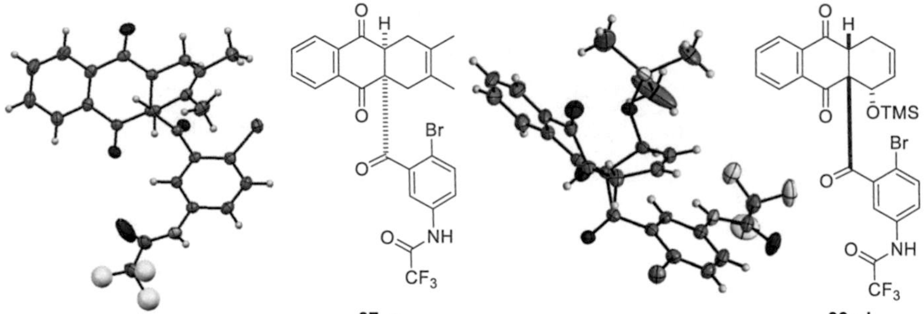

Figure 35. Molecular structures of **87gc** (left) and **88gd** (right) determined by single-crystal X-ray diffraction. Displacement parameters are drawn at 50% probability level. Crystallographic data can be found in chapter 7.3.

To conclude, a broad range of substrates was tolerated in the [4+2]-cycloadditions which allowed for the synthesis of highly substituted anthraquinone derivatives. Among these, a significant amount was successfully crystallized, and their structure analyzed in the solid state with single-crystal X-ray diffraction. Most of the compounds underwent the cycloaddition very smoothly, in short time and under mild conditions, since 1,4-naphthoquinone derivatives as well as dienes bearing EDGs represent highly activated starting materials. The obtained yields were mostly good to excellent, giving rise to anthraquinone derivatives comprising up to three stereogenic centers. The regiochemistry of the Diels-Alder reactions turned out to be as expected, with the 2-substituted dienes yielding *para* anthraquinones and dienes bearing substituents at C1 providing *ortho* products. Moreover, *ortho* anthraquinones were isolated as a separable mixture of diastereomers, with an excess of *endo* products since there is less steric clash in the *endo* transition state of the cycloaddition. This is in accordance with what was expected according to the *endo* rule and also demonstrates that the regiochemistry was not affected by space demanding residues.

In general, the substituents at the dienophiles did not render significant impact on the outcome of the Diels-Alder reactions. In some reactions decreased yields were observed, presumably due to steric hindrance, whereas for the cycloadditions with isoprene mostly improved yields were obtained in comparison to the reactions with non-functionalized dienophiles.

The incorporation of functionalities such as methoxy groups into the anthraquinone derivatives paves the way for the installation of a tetrahydroxanthone subunit *via* a domino oxa-Michael–aldol condensation reaction with a suitable cyclohexanone.

however a suitbale separation method is required in order to be able to analyze the obtained compounds.

5.8   Modifications of Anthraquinones

5.8.1   Silyl Ether Deprotection

By subjecting naphthoquinones to Diels-Alder reactions, highly functionalized anthraquinone derivatives bearing diverse silyl ethers were synthesized. In order to modify these intramolecular coupling precursors, and to investigate the effect of an increased polarity of the compounds on the following reactions, suitable methods for the cleavage of the silyl ethers were examined.

The standard protocol applied for the cleavage of trimethylsilyl ethers is the reaction with acids or fluorides such as tetrabutylammonium fluoride (TBAF). However, for the anthraquinones **88ad** and **88bd** depicted below, the reaction with TBAF resulted in decomposition of the starting material. The application of acidic conditions to anthraquinones **88ad** and **88bd** in a mixture of CH_2Cl_2 and MeOH resulted in ether cleavage together with double bond migration to form enone **120a/b**. Moreover, acid catalyzed addition of methanol to the cyclic dione yielding acetals **120a** and **120b**, possibly *via* formation of a hemiacetal, was observed (Scheme 32). The desired product **121** was not obtained.

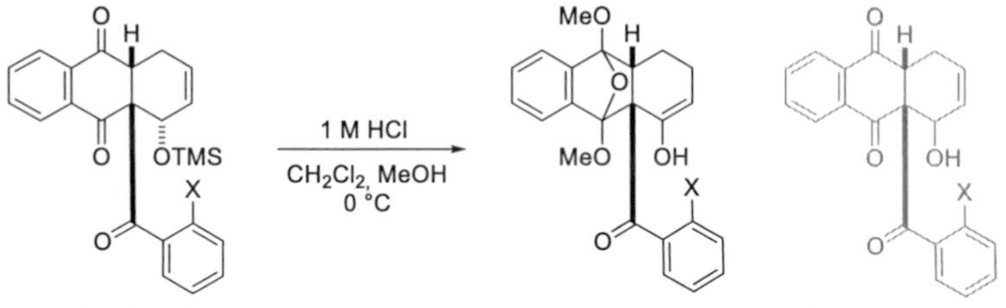

Scheme 32. Attempted TMS ether cleavage of **88ad** and **88bd** to give **121** and the instead observed formation of acetals **120a** and **120b**.

Acetals **120a** and **120b** were verified by NMR as well as HRMS analysis. A characteristic of the structure is the signal of the enol proton in the ^1H NMR spectrum, which is located far downfield at around 13.9 ppm, presumably due to its close proximity to the methoxy group. Surprisingly aqueous workup did not result in reversion of the acetal formation, however it is expected to occur by treatment with an aqueous acid and an excess of water.

By following a literature known procedure for the cleavage of TBDMS ethers, it was attempted to cleave the TBDPS ether in anthraquinone **87ah** using a catalytic amount of acetyl chloride in dry methanol.[172] Presumably the reaction took place as expected, however subsequent acid catalyzed addition of methanol occurred, resulting in acetal **122** (Scheme 33). Here as well the acetal formation is expected to be reversible.

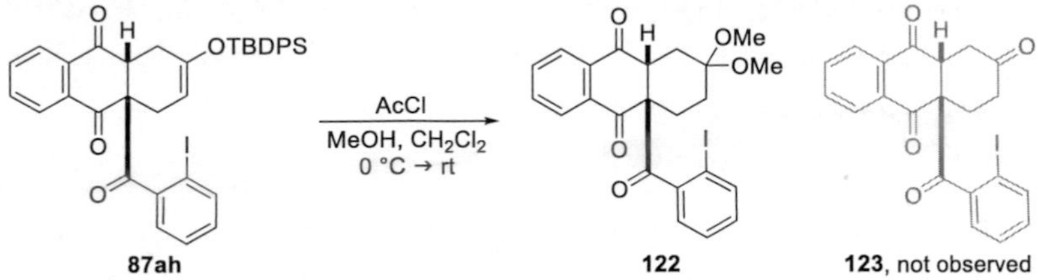

Scheme 33. Attempted TBDPS ether cleavage of **87ah** to give **123** and the instead observed formation of acetal **122**.

5.9 Synthesis of a Xanthone Dienophile

As outlined before, beticolin 0 (**1**) consists of an anthraquinone subunit linked to a tetrahydroxanthone. To integrate a tetrahydroxanthone moiety into the synthetic strategy described in chapter 5.1, the synthesis of a naphthoquinone derivative bearing a xanthone subunit, was investigated. For reasons of simplicity, the tetrahydroxanthone that *de facto* represents a subunit of beticolin 0 (**1**), was replaced by a xanthone derivative.

For the development of the synthetic strategy, a xanthone with specific functionalities was required. By incorporating a halogen residue as well as a carboxylic acid into the xanthone skeleton **126**, it could be applied as a reactant in the acylation step of the naphthoquinone synthesis route (see chapter 5.2). After subsequent oxidative demethylation, the thereby generated naphthoquinone derivatives bearing a xanthone subunit **125** can be used as a dienophile in Diels-Alder cycloadditions to give anthraquinone derivative **124** (Scheme 34). In the same way as described in chapter 5.5, this anthraquinone derivative **124** is supposed to be applied in intramolecular coupling reactions in order to generate the desired [3.2.2]system. The impact of the xanthone residue on the outcome of the coupling should be investigated.

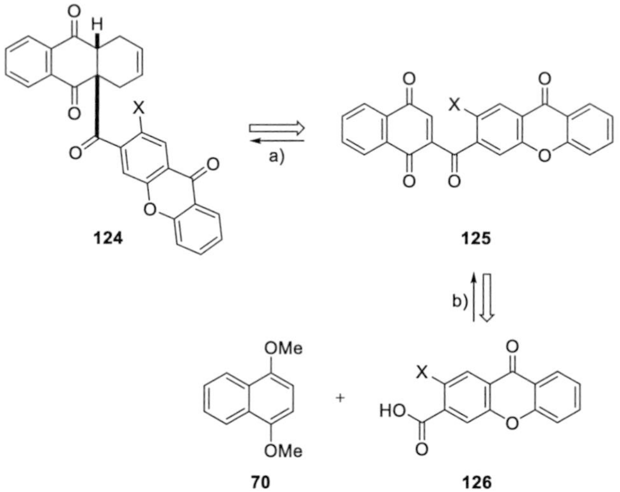

Scheme 34. Retrosynthetic analysis of the synthesis of anthraquinone **124** bearing a xanthone subunit. a) Diels-Alder reaction; b) acylation, oxidative demethylation.

First, the reaction conditions were explored by establishing a model system. In an Ullmann type coupling, commercially available bromoterephthalic acid **127** was reacted with phenol **128** in the presence of a catalytic amount of copper, copper iodide and a base to give *o*-phenylsalicylic acid derivative **129** (see Scheme 35).[173] Ferguson *et al.* discovered the need for the usage of the non-nucleophilic base 1,8-diazabicyclo[5.4.0]undec-7-ene (DBU) for the successful formation of the diarylether. The cyclization to functionalized xanthone **130** proceeded in an 80% aqueous solution of sulfuric acid *via* formation of a carbonium ion that attacks the non-carboxylated ring. By pouring the crude product on ice and subsequent purification *via* column chromatography on silica gel, product **130** was isolated in 23% yield. Due to the free carboxylic acid residue in the molecule, the purification turned out to be difficult which presumably is the reason for the poor yield of the cyclization.

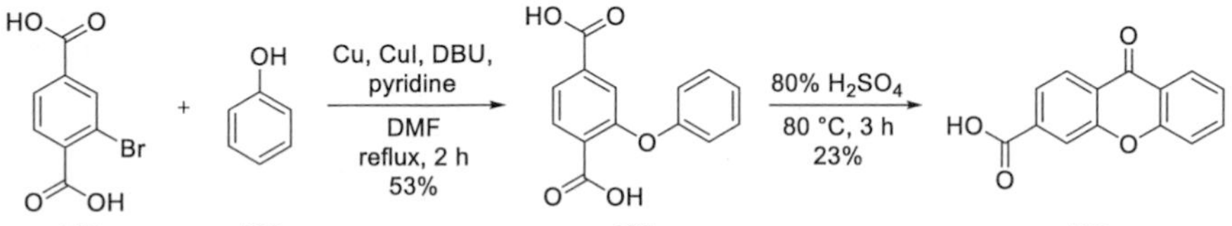

Scheme 35. Synthesis route towards xanthone **130** bearing a carboxylic acid residue.[173]

In the following step the functionalized xanthone **130** was reacted with 1,4-dimethoxynaphthalene (**70**) in TFAA (Scheme 36). The acylation was monitored *via* APCI (atmospheric pressure chemical ionization) mass spectrometry and quenched after 24 h by the addition of water. Despite great efforts, the purification of the desired product **131** could not be realized which is why the obtained NMR spectra displayed a mixture of compounds. Nevertheless, the formation of xanthone derivative **131** was verified by HRMS measurement (see chapter 7.2.8).

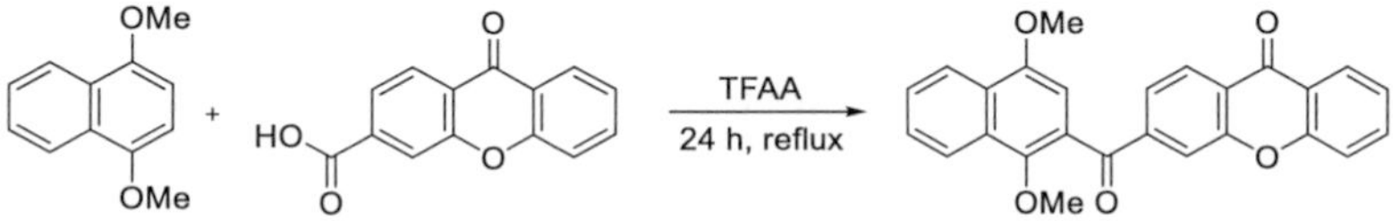

Scheme 36. Acylation of the functionalized xanthone **130** with 1,4-dimethoxynaphthalene (**70**).

Since the feasibility of the acylation of xanthone derivative **130** was proven, and there were no concerns regarding the subsequent oxidative demethylation, it was decided to continue the investigation by integrating the desired halogen into **130**. For this purpose, commercially available 2,5-dibromoterephthalic acid (**132**) was used as a starting material for the copper catalyzed Ullmann type reaction with phenol (**128**) (Scheme 37). As a result, mostly undesired disubstituted product **134** was isolated which is presumably favored due to the electron-donating group installed after the first coupling. Moreover, the separation of the product mixture of mono- **133a** and disubstituted **134** compound was rather challenging. To simplify the purification of the product mixture *via* column chromatography on silica gel, the free carboxylic acid functionalities of dibromoterephthalic acid **132** were capped *via* esterification (Scheme 37). However, the Ullmann type reaction with dimethyl 2,5-dibromoterephthalate (**135**) and phenol **128** gave the same mixture of disubstituted **137** and monosubstituted **136** products as before. In addition, the purification of the crude product isolated from the Ullmann reaction remained challenging, thus again no pure Ullmann product **136** was isolated.

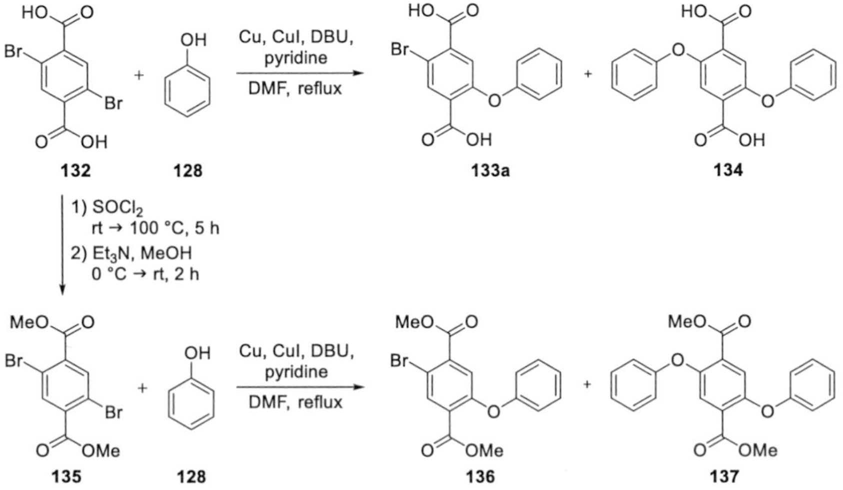

Scheme 37. Ullmann type reaction of 2,5-dibromoterephthalic acid **132** with the corresponding products and esterification of **132** to **135** with following Ullmann type reaction.

These results clearly show the necessity to implement selectivity into the Ullmann type reaction. In literature, a procedure decribing the synthesis of 2-iodo-5-bromoterephthalic acid (**142**) which in theory should facilitate a chemoselective process of the Ullmann type reaction, was found. According to the procedure by Cohen *et al.*, initially dimethyl-2-aminoterephthalate **138** was brominated with *N*-bromosuccinimide (NBS) (Scheme 38).[174] As products two different regioisomers **139** and **140** were obtained which were successfully separated with column chromatography on silica gel. Following a two-step procedure described by Yoshida *et al.*[175], dimethyl 2-amino-5-bromoterephthalate (**140**) was iodinated with potassium iodide (KI) *via* a diazonium salt intermediate, followed by saponification, to give the desired 2-iodo-5-bromoterephthalic acid (**142**) in excellent yield.

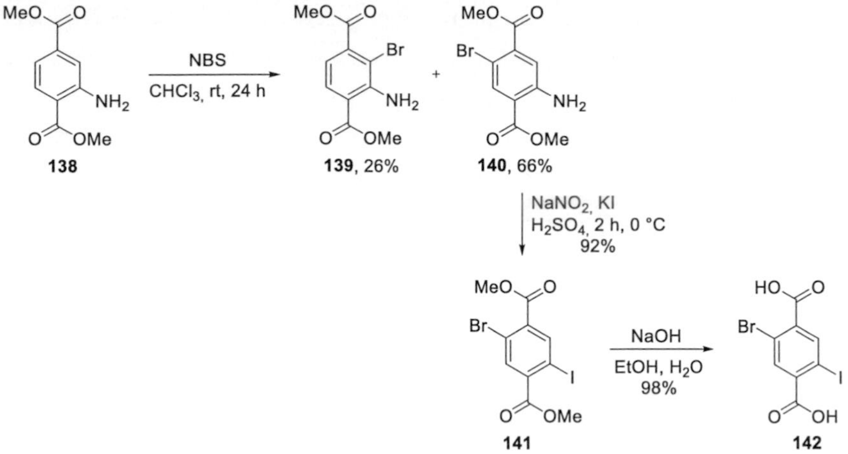

Scheme 38. Three step synthesis of 2-iodo-5-bromoterephthalic acid (**142**) from dimethyl-2-aminoterephthalate (**138**).[174-175]

The synthesized 2-iodo-5-bromoterephthalic acid (**142**) was then applied in a copper-catalyzed Ullmann type reaction with phenol (**128**). Under standard conditions, the Ullmann type reaction gave a mixture of all three possible products, even though a coupling with the iodinated position of bromoterephthalic acid derivative **142** should be favored (Table 7, entry 1).

Table 7. Variation of the reaction conditions of the Ullmann type reaction with 2-iodo-5-bromoterephthalic acid (**142**) and phenol (**128**).

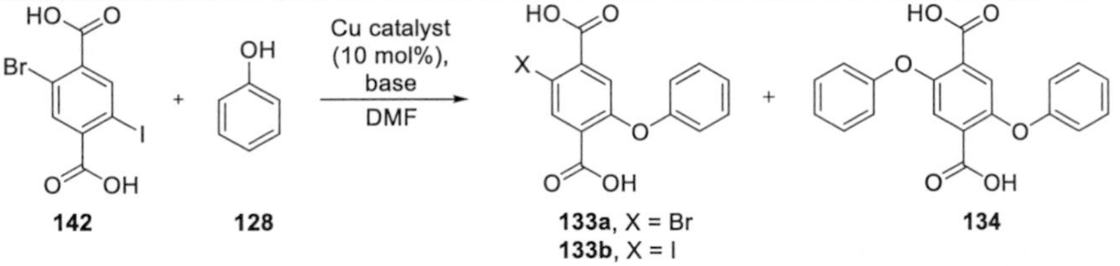

Entry	Phenol equiv.	Catalyst	Base	Temp.	Time	Products
1	1.6	Cu(0), Cu(I)I	DBU, pyridine	153 °C	16 h	**133 – 134**
2	1.3	Cu(0), Cu(I)I	DBU, pyridine	rt	16 h	**133 – 134**
3	1.0	Cu(0)	DBU	rt	16 h	**133 – 134**
4	1.3	Cu(0)	DBU	rt	3 h	**133a, 134** (total yield: 64%)

Then, *via* a screening with APCI mass spectrometry, in further experiments suitable reaction conditions enabling a chemoselective reaction were examined. By lowering the reaction temperature to room temperature and reducing the equivalents of phenol (**128**) to 1.3 equiv., no improvement of selectivity was detected (entry 2). Subsequently, the amount of phenol was even more reduced, and the use of copper iodide was neglected since it might promote a copper catalyzed halogen exchange (entry 3). Nevertheless, the reaction gave a mixture of all three possible products **133–134**. Shortening the reaction time from sixteen to three hours, finally suppressed the formation of the unwanted disubstituted product **134** (entry 4) and the desired products **133a** and **133b** were isolated in 64% total yield.

In conclusion, even though a chemoselective reaction was not possible, the suppression of the unwanted side product **134** was achieved. Thus, the reaction was successfully directed towards halogenated xanthone precursors **133a** and **133b**. During the workup of the reaction, it was observed that the different products can be separated from each other in the following way: First, the iodinated product **133b** precipitated when the crude was poured on ice water. After separation of that precipitate, the remaining aqueous solution was extracted with ethyl acetate which enabled the isolation of the brominated Ullmann product **133a**. The two Ullmann products **133a** and **133b** can be distinguished by means of ^{13}C NMR spectroscopy, as well as mass spectrometry.

Next, the formation of xanthone derivatives **126** was envisioned. Therefore, the Ullmann products **133a** and **133b** were cyclized in an 80% aqueous solution of sulfuric acid at 80 °C (Scheme 39).

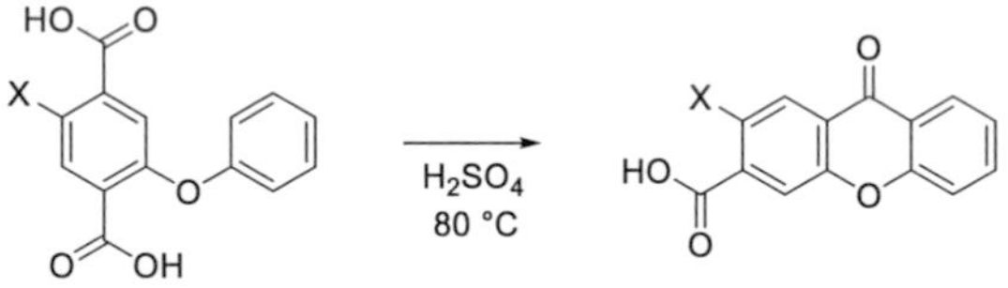

133a, X = Br
133b, X = I

126a, X = Br
126b, X = I

Scheme 39. Cyclization of Ullmann products **133a/b** towards xanthone derivatives **126a/b**.

Despite the above described separation of both Ullmann products **133a** and **133b**, a brominated **126a** as well as an iodinated xanthone derivative **126b** were isolated from the cyclization reaction. To some extent, the starting material for the cyclization must have been a mixture. Hence, since the product mixture of the cyclization could not be separated *via* column chromatography on silica gel, an HPLC method suitable for the separation was developed. This was realized with 35% acetonitrile in water (isocratic) with the major drawback being the isolation of only a small amount of desired products **126a** and **126b**. Mass measurements and ^{1}H NMR experiments proved the existence of product **126a**, however the amount of product was not sufficient for further analytics. Hence, experiments on the synthesis of the desired dienophile **125** could not be pursued.

Since halogenated xanthone derivatives **126a** and **126b** both are suitable precursors for the synthesis of dienophiles bearing a xanthone residue, in future experiments a mixture of them could be used for the acylation with 1,4-dimethoxynaphthalene (**70**).

5.10 Alternative Coupling Strategies: Suzuki-Miyaura Coupling

Since all attempts regarding intramolecular radical cyclization reactions as well as the screening for suitable Heck reaction conditions, did not give the desired result, the development of a new strategy was envisioned. In order to facilitate a selective intramolecular coupling and control the outcome of the reaction, the so far pursued synthetic route was redesigned to pave the way for Suzuki-Miyaura reactions.

The Suzuki reaction is a palladium-catalyzed cross-coupling reaction between organohalides and boronic acids, forming a new carbon-carbon bond.[176-178] Since the reaction has advantages like mild reaction conditions and functional group tolerance, it is widely applied. In order to make it applicable for the intramolecular cross-coupling of anthraquinone derivatives, the general synthetic route described in chapter 5.1 was adjusted, and procedures to generate appropriate substrates for Suzuki cross-coupling reactions were investigated.

As depicted in route A in Scheme 40, it was planned to apply halogentated anthraquinone **143** in Suzuki-Miyaura reactions with bis(pinacolato)diboron (B$_2$pin$_2$) **144** to enable intramolecular cross-coupling for the generation of the bicyclo[3.2.2]nonane ring system in **145**. Alternatively, an anthraquinone **146** derived from a borylated naphthoquinone derivative and iodinated diene **154** were planned to apply in an intramolecular coupling towards **145**, as shown in route B.

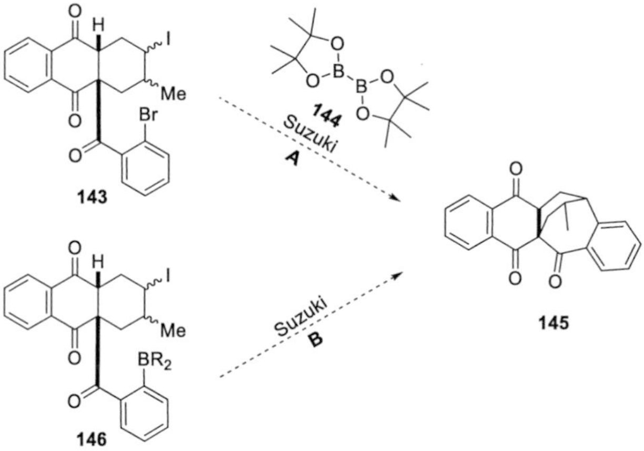

Scheme 40. Planned strategies towards bicyclo[3.2.2]nonane ring system **145** through Suzuki-Miyaura cross-coupling.

5.10.1 Halogenated Dienes

Initially, the synthesis of halogenated dienes was adapted. With their application in Diels-Alder reactions with naphthoquinone derivatives (see chapter 5.2), a clearly defined position in ring C of the formed anthraquinones will be halogenated and undergo the subsequent cross-coupling reaction.

According to a literature known procedure, 2-chloro-3-methylbuta-1,3-diene **149** was synthesized in two steps from 2-methyl-3-yn-2-ol **147** (Scheme 41).[179] Since dienes are rather unstable compounds and distillation was necessary to purify the product, only small amounts of desired diene **149** were isolated.

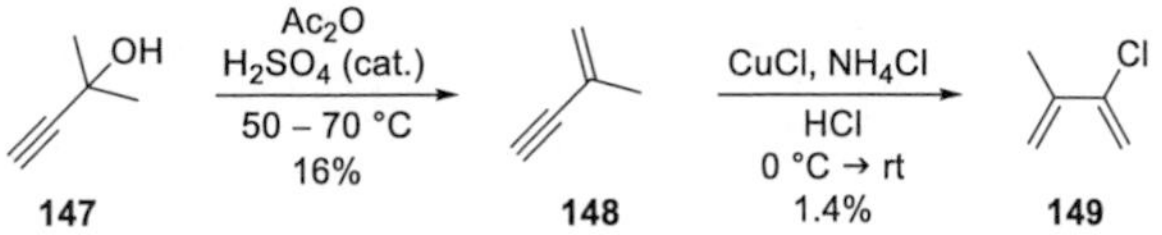

Scheme 41. Synthesis of chlorinated diene **149** from 2-methyl-3-yn-2-ol **147**.[179]

Iodinated diene **154** was synthesized from but-2-yn-1-ol **150** in a three-step procedure (Scheme 42).[180-181] Tosylation of alcohol **150**, and subsequent Grignard reaction with **152** in the presence of CuCN and LiCl gave 1-trimethylsilyl-2,3-butadiene **153**. Reaction with iodine followed by TBAF provided iodinated diene **154** which was isolated in 23% yield after distillation. Here as well the poor yield is owed to the low stability of dienes and the required purification method.

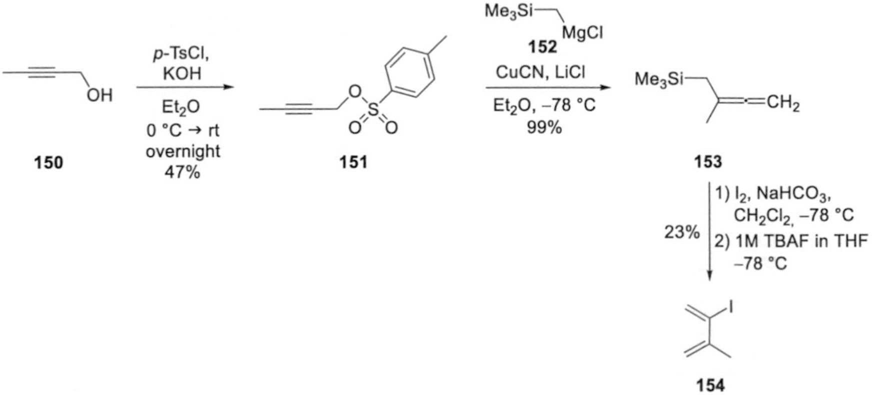

Scheme 42. Synthesis of iodinated diene **154** from but-2-yn-1-ol **150**.[180-181] p-TsCl = 4-toluenesulfonyl chloride.

With these two halogenated dienes (**149** and **154**) bearing a methyl group in hand, the development of building blocks containing boronic acids was pursued.

5.10.2 Borylated Dienophiles

As a counterpart for the halogenated carbon in the anthraquinone, a borylated position is required to facilitate Suzuki-Miyaura cross-coupling reactions. In a first attempt the carboxylation of 2-tolylboronic acid **155** with potassium permanganate in aqueous solution was pursued following a literature known procedure (Scheme 43).[182]

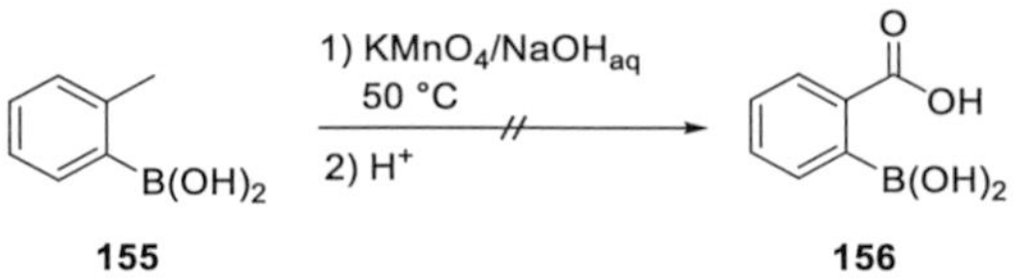

Scheme 43. Attempted synthesis of 2-carboxyphenylboronic acid **156**.[182]

Despite several attempts, the desired product could not be isolated. The reaction is reported to show a strong temperature dependence since a temperature increase of around 10 K already yields a mixture of 2-carboxyphenylboronic acid **156** and diphenic acid while a decrease of around 10 K results in no conversion to product **156**.

The synthesis of naphthoquinone derivative **157** bearing a boronic acid was attempted by following two different approaches. One method makes use of B$_2$pin$_2$ **144** and palladium catalysis while the other is a lithiation followed by the addition of trimethylborate and acidic workup (Scheme 44).[183-184] Both reactions did not afford the desired product **157**. Presumably compound **73b** intramolecularly reacts with its double bond after lithiation or a 1,2-addition takes place.

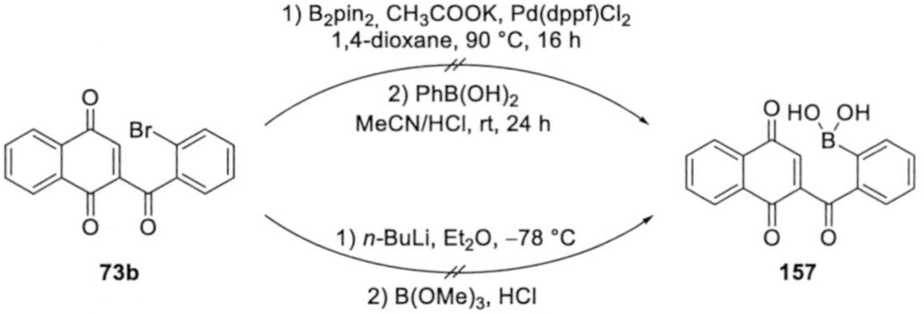

Scheme 44. Attempted synthesis of **157** *via* two different routes.[183]

Thus, the incorporation of pinacolato boron was considered since it is a lot more stable and products can be purified by column chromatography on silica gel. Following the synthetic route towards naphthoquinone derivatives described in chapter 5.2, commercially available 2-carboxyphenylboronic acid pinacol ester **158** was subjected to an acylation reaction with 1,4-dimethoxynaphthalene (**70**) in TFAA (Scheme 45). Since compound **159** was not isolated from the reaction, the stability of benzoic acid derivative **158** in TFAA was investigated by stirring it at 40 °C for 24 h.

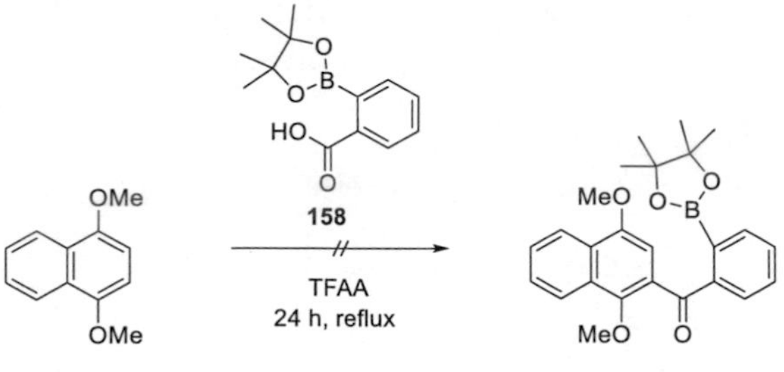

Scheme 45. Attempted acylation of 1,4-dimethoxynaphthalene (**70**) with 2-carboxyphenylboronic acid pinacol ester **158**.

Since only decomposed material was isolated, bromo dienophile **73b** was used as a precursor and applied in a Miyaura borylation with B₂pin₂ **144** (Scheme 46).[178] Despite being a widely used reaction in the Bräse working group, the borylation did not give the desired product **160**. The reaction might be too sensitive towards humidity although performed under Schlenk conditions. Another possible explanation is the existence of a spatially neighbouring carbonyl group which might be disturbing and end in the formation of an acetal.

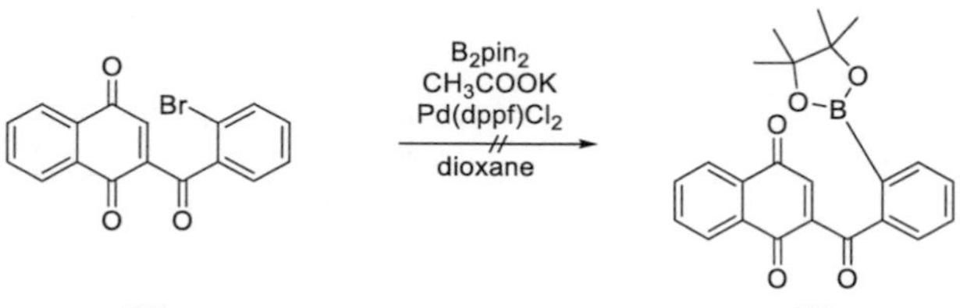

Scheme 46. Attempted Miyaura borylation of dienophile **73b**.

Since no procedure towards borylated naphthoquinone derivatives turned out to be successful, no further studies were performed for this project.

5.10.3 Diels-Alder Reactions with Halogenated Dienes

As depicted below, naphthoquinone derivative **73a** was applied in cycloaddition reactions with 2-chloro-3-methylbuta-1,3-diene **149** (Scheme 47). After workup and column chromatography on silica gel a non-separable product mixture of isomers **161a** and **161b** was isolated in a ratio of about 1 : 2 as estimated by ^{1}H NMR analysis.

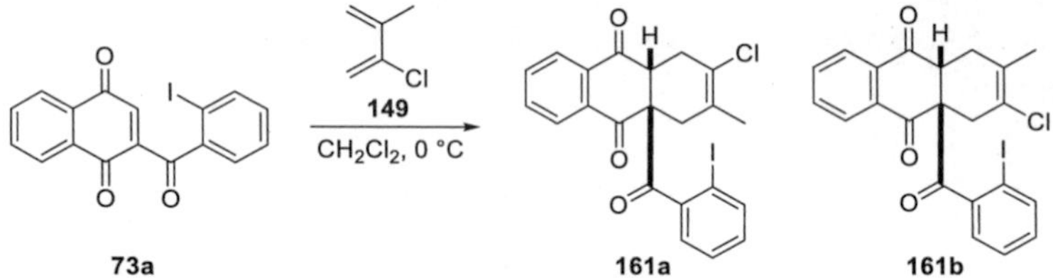

Scheme 47. Diels-Alder cycloaddition with 2-chloro-3-methylbuta-1,3-diene **149**.

One isomer crystallized from the mixture and was analyzed by X-ray crystallography giving the molecular structure **161a** shown below (Figure 39).

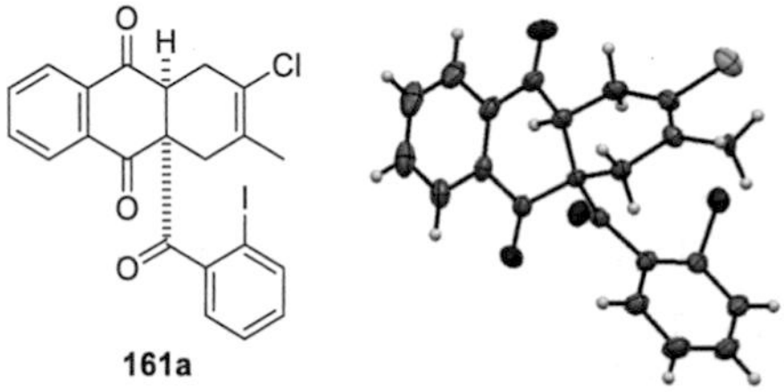

161a

Figure 39. Molecular structure of **161a** determined by single-crystal X-ray diffraction. Displacement parameters are drawn at 50% probability level. Crystallographic data can be found in chapter 7.3.

Similarly, iodinated diene **154** was applied in Diels-Alder reactions with naphthoquinone derivative **73b** (Scheme 48).

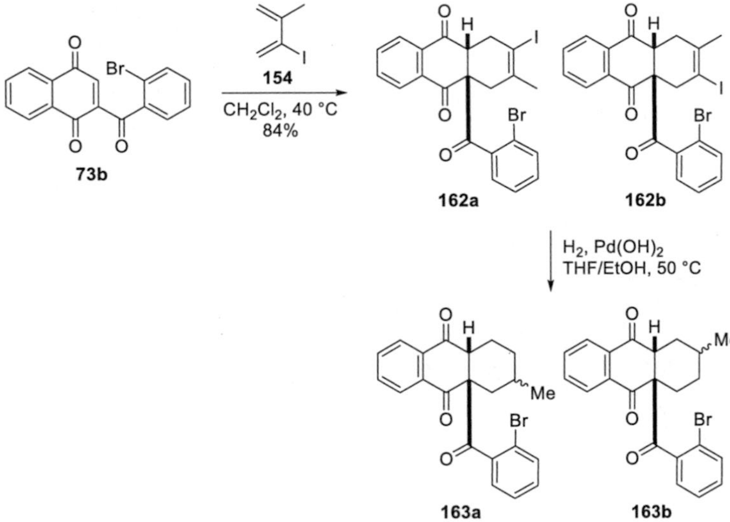

Scheme 48. Cycloaddition of naphthoquinone derivative **73b** and iodinated diene **154**. Subsequent hydrogenation yielded a mixture of **163a** and **163b**.

After workup and column chromatography on silica gel a non-separable product mixture of isomers **162a** and **162b** was isolated in a ratio of 1 : 1 as estimated by ^{1}H NMR analysis.

According to Bredt's rule, at the bridgehead of a bridged ring system a double bond can not be placed.[185-186] Hence, hydrogenation of the cyclohexene ring in anthraquinone **162a/b** was required. By applying palladium (hydroxide) on carbon as well as hydrogen pressure to the isomeric mixture of anthraquinone derivatives **162a** and **162b**, deiodination towards products **163a** and **163b** took place, as expected.

For future experiments it is envisioned to first convert the iodine into a pinacol boronic ester and then perform the hydrogenation, for example catalyzed by a cyclic (alkyl)(amino)carbene (CAAC)-ligated rhodium complex.[187]

6 Summary

Anthraquinone-xanthone heterodimers are an emerging class of natural products with attractive biological activities. The molecules share a characteristic structure built on a unique bicyclo[3.2.2]nonane skeleton which is rarely found in natural products.

Among these, beticolins represent interesting compounds due to the synthetic challenge resulting from the molecular complexity, as well as their reported properties such as a broad cytotoxic profile and antibiotic activity. It was envisioned to develop a synthetic route towards the unique structure of beticolin 0 (**1**), since it can easily be converted to its related compounds. The design of the general synthetic path that was established, is depicted in Scheme 49.

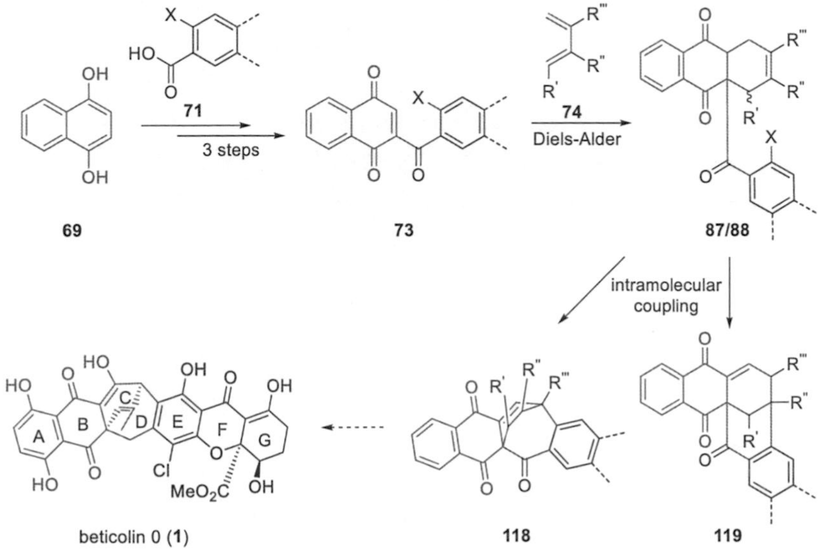

Scheme 49. Design of the general path that was established towards the construction of the unique scaffold of beticolin 0 (**1**).

Starting from commercially available 1,4-dihydroxynaphthalene **69**, a simple three step synthetic route towards diverse naphthoquinone derivatives **73** was established. By following this straightforward procedure, seven different 2-substituted naphthoquinones **73a–g** containing various functionalities were successfully synthesized in mostly good yields (Scheme 50). These compounds represent ideal dienophiles bearing electron-withdrawing groups in conjugation with the alkene.

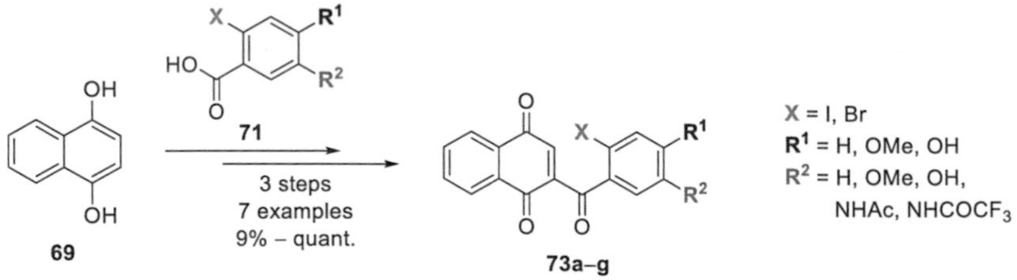

Scheme 50. Summary of the synthesized functionalized dienophiles **73**.

For the formation of the ABC tricycle of the anthraquinone subunit, Diels-Alder cycloadditions with different dienes were performed. Initially, the formation of an anthraquinone model system was realized by making use of 3-sulfolene **75** as a convenient precursor for the generation of 1,3-butadiene **74a**. The thereby obtained anthraquinones **87/88** were designed to serve as reactants in intramolecular coupling reactions for the construction of [3.3.1] as well [3.2.2] scaffolds. By applying an iodinated **87aa** as well as a brominated anthraquinone model system **87ba** with systematic variation of the reaction conditions, the reactivity and selectivity of the molecules in Heck couplings were studied. It turned out that the formation of bicyclo[3.3.1]systems **111a** is favored when anthraquinones **87aa** and **87ba** are coupled intramolecularly (Scheme 51).

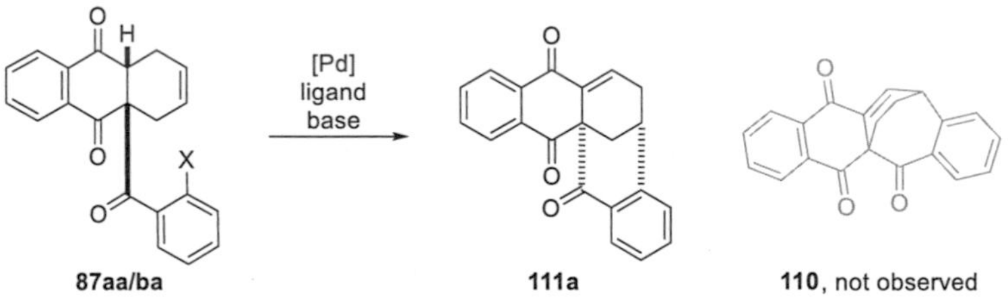

Scheme 51. Intramolecular Heck coupling of model anthraquinone systems **87aa** and **87ba** resulting in [3.3.1]bicycle **111a**.

Thereupon, the complexity of the anthraquinones was significantly increased by applying substituted dienes **74** as well as functionalized naphthoquinones **73** in the cycloadditions (Scheme 52). The regioselectivity of the Diels-Alder reactions followed the *ortho/para* rule, and with the cycloadditions being governed by the stereoelectronic nature of the reactants, *endo* products were formed in excess. Since only highly activated substrates were applied, the cycloadditions proceeded smoothly and under mild conditions, resulting in good to excellent yields. A wide variety of stable, highly functionalized anthraquinones containing a sterically congested quaternary center was synthesized which represent derivatives of the subunit of beticolin 0 (**1**). Among these, many derivatives were successfully crystallized, and the molecular structures could be confirmed by X-ray diffraction.

With the incorporation of functionalities such as methoxy groups into the halogenated benzene ring attached to the anthraquinone core, the basis for a domino oxa-Michael-aldol condensation reaction with cyclohexenones, to buil the tetrahydroxanthone subunit, was laid.

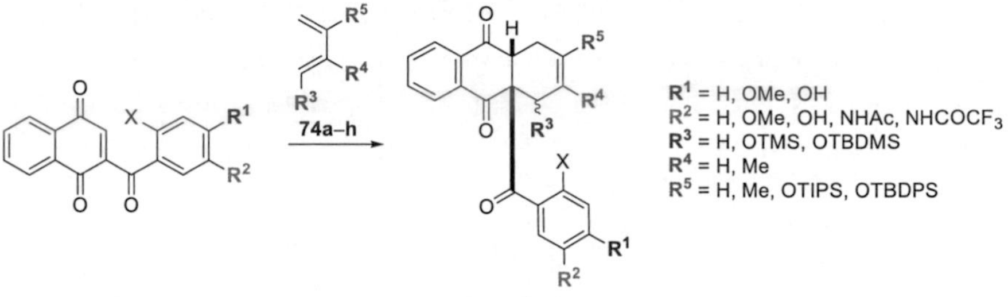

R^1 = H, OMe, OH
R^2 = H, OMe, OH, NHAc, NHCOCF$_3$
R^3 = H, OTMS, OTBDMS
R^4 = H, Me
R^5 = H, Me, OTIPS, OTBDPS

73a–g   **87aa–88gh**

Scheme 52. Summary of the synthesized highly functionalized anthraquinone derivatives **87aa–88gh**.

For the purpose of constructing the bicyclo[3.2.2]nonane ring system of beticolin 0 (**1**), various intramolecular couplings were performed with the anthraquinone derivatives. The application of reaction conditions for radical cyclizations did not result in conversion to the desired products. Although many approaches towards the desired scaffold were conducted by using different anthraquinones under diverse Heck coupling conditions, its formation turned out to be challenging. In addition, a lot of the reactions provided complex product mixtures with very similar polarity, which rendered the purification of single compounds a tough task. As a result, the successful formation of the desired [3.2.2]product can not be excluded, but due to difficulties in its verification and purification, it was never isolated from the reactions. However, many novel bicyclo[3.3.1]systems **119** representing a no less interesting class of scaffolds was obtained, with their biological properties remaining to be investigated (for exemplary structures see Figure 40).

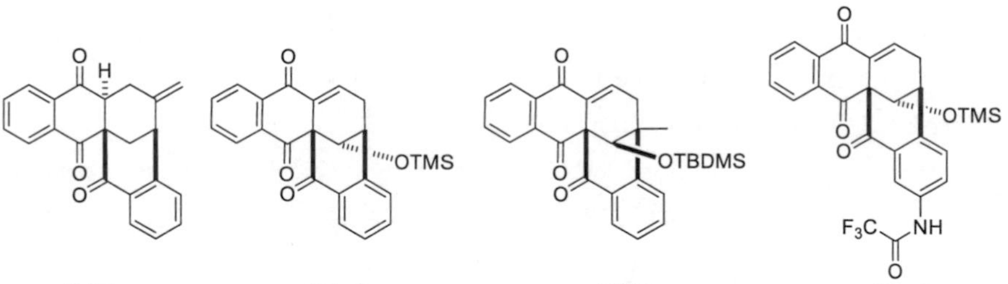

113ab   **113ad**   **113bf**   **119gd**

Figure 40. Examples of successfully synthesized bicyclo[3.3.1]nonane ring systems.

For the incorporation of a xanthone moiety into the anthraquinone derivatives, a synthetic procedure making use of a chemoselective Ullmann type reaction with phenol (**128**) and a following cyclization was designed in order to synthesize functionalized xanthones **126** (Scheme 53). The latter can be attached to dimethoxynaphthalene **70** *via* acylation, and subsequently be converted to halogenated naphthoquinone dienophiles **125**, which is supposed to be investigated in future experiments. With cycloadditon reactions of the latter, the formation of an anthraquinone xanthone-heterodimer **124** is envisioned. Important steps towards the synthesis of such compounds were accomplished by establishing a procedure towards functionalized xanthones **126a** and **126b**.

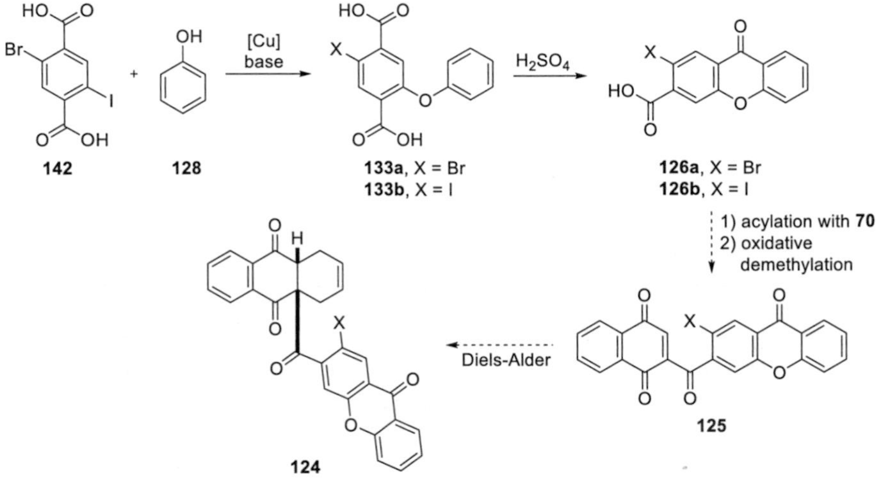

Scheme 53. Envisioned synthetic route towards xanthone bearing dienophile **125** with subsequent cycloaddition to obtain anthraquinone-xanthone heterodimer **124**.

In order to facilitate a selective intramolecular cross-coupling, such as a Suzuki-Miyaura reaction of the anthraquinone derivatives, halogenated dienes **149** and **154** were synthesized, and subsequent Diels-Alder reactions were performed. The obtained anthraquinone products **162a/b** represent precursors, which after borylation to afford pinacol boronic esters, and reduction are envisaged to facilitate access to bicyclo[3.2.2]systems.

Hydroboration at the anthraquinone double bond could equally provide a precursor for Suzuki couplings.

With the insertion of a temporary regiochemistry controller like $B(OH)_2$ at C3 of the naphthoquinone dienophiles, access to the unexpected *meta* Diels-Alder products could be facilitated.[188]

Since numerous biological and physicochemical studies on beticolins and related structures revealed interesting biological activities, the behavior of the synthesized anthraquinone derivatives as well as of the [3.3.1]bicyclo Heck products remains to be investigated in appropriate studies.

7 Experimental Part

7.1 General Remarks

7.1.1 Preparative Work

The starting materials, solvents and reagents were purchased from abcr, Acros, Alfa Aesar, Carbolution, ChemPUR, Fluka, Iris, Merck, Riedel-de Haën, TCI, Thermo Fisher Scientific, Sigma Aldrich and used without further purification. Dienes were distilled before use.

All reactions containing air- and moisture-sensitive compounds were performed under argon atmosphere using oven-dried glassware applying Schlenk-techniques. Liquids were added *via* steel cannulas and solids were added directly in powdered shape. Reactions were accomplished at room temperature, if nothing else is mentioned. For low reaction temperatures flat dewars with ice/water or isopropanol/dry ice mixtures were used. The solvents were removed with a rotary evaporator at 40 °C under reduced pressure. For solvent mixtures each solvent was measured volumetrically. If nothing else is mentioned, saturated, aqueous solutions of inorganic salts were used. Celite® for filtrations was purchased from Alfa Aesar (Celite® 545, treated with Na_2CO_3).

Crude products were purified *via* flash chromatography using Merck silica gel 60 (0.040 × 0.063 mm, 230 – 400 mesh ASTM) and quartz sand (glowed and purified with hydrochloric acid). Therefore, the eluents were distilled or used directly in *p.a.* quality bought from Merck.

7.1.2 Solvents and Reagents

Solvents of technical quality have been purified by distillation or with the solvent purification system MB SPS5 from MBRAUN prior to use. Solvents of the grade *p.a.* have been purchased (Acros, Fisher Scientific, Sigma Aldrich, Roth, Riedel-de Haën) and were used without further purification. Absolute solvents have been dried, using the methods listed in Table 8 and were stored under argon afterwards or have been purchased from a commercial supplier (abs. acetonitrile (Acros, <0.005% water), anhydrous *N,N*-dimethylformamide (Sigma Aldrich, <0.005% water), anhydrous dimethylsulfoxide (Sigma Aldrich, <0.005% water), abs. methanol (Fischer, <0.005% water), anhydrous *N,N*-dimethylacetamide (Sigma Aldrich, <0.005% water)).

Table 8. Methods for the absolutizing of solvents. All distillations were carried out under argon atmosphere.

Solvent	Method
CH_2Cl_2	heating to reflux over CaH_2, distilled over a packed column or MB SPS5
THF	heating to reflux over Na metal (benzophenone as an indicator), distilled over a packed column or MB SPS5
Et_2O	heating to reflux over Na metal (benzophenone as an indicator), distilled over a packed column or MB SPS5

7.1.3   Analytics and Equipment

7.1.3.1   Nuclear Magnetic Resonance (NMR)

NMR spectra have been recorded using the following machines:

^{1}H NMR: Bruker *Avance AV 300*   (300 MHz),   Bruker *Avance 400*   (400 MHz),   Bruker *Avance DRX 500* (500 MHz). The chemical shift δ is expressed in parts per million (ppm) where the residual signal of the solvent has been used as reference: chloroform-d_1 ($\delta = 7.26$ ppm), dimethyl sulfoxide-d_6 ($\delta = 2.50$ ppm) and methanol-d_4 ($\delta = 4.87$ ppm).[189] The spectra were analysed according to first order.

^{13}C NMR: Bruker *Avance 300*   (76 MHz), Bruker *Avance 400*   (101 MHz), Bruker *Avance DRX 500* (126 MHz). The chemical shift δ is expressed in parts per million (ppm) where the residual signal of the solvent has been used as reference: chloroform-d_1 ($\delta = 77.0$ ppm), dimethyl sulfoxide-d_6 ($\delta = 39.4$ ppm) and methanol-d_4 ($\delta = 49.0$ ppm).[189] The spectra were ^{1}H-decoupled and characterization of the ^{13}C NMR spectra ensured through the DEPT-technique (DEPT = Distortionless Enhancement by Polarization Transfer) and are stated as follows: DEPT: "+" = primary or secondary carbon atoms (positive DEPT-signal), "–" = secondary carbon atoms (negative DEPT-signal), C_q = quaternary carbon atoms (no DEPT-signal).

All spectra were obtained at room temperature. NMR-solvents were obtained from Eurisotop and Sigma Aldrich: chloroform-d_1, dimethyl sulfoxide-d_6, methanol-d_4. For central symmetrical signals the midpoint and for multiplets the range of the signal region are given. The multiplicities of the signals are abbreviated as follows: s = singlet, d = doublet, t = triplet, q = quartet, hept = heptet, brs = broad singlet, m = multiplet, b = broad and combinations thereof, apparent multiplicity of NMR signals = app. All coupling constants J are stated as modulus in Hertz [Hz].

In some cases the signals were assigned using ^{1}H-^{13}C-HSQC (Heteronuclear Single Quantum Coherence) and ^{1}H-^{13}C-HMBC (Heteronuclear Multiple Quantum Correlation) techniques.

7.1.3.2 Infrared Spectroscopy (IR)

IR spectra were recorded on a Bruker Alpha P and a Bruker IFS 88. Measurements of the samples were conducted via attenuated total reflection (ATR). Position of the absorption bands is given as wavenumber $\tilde{v}$ with the unit $[cm^{-1}]$.

7.1.3.3 Mass Spectrometry (EI-MS, FAB-MS, ESI-MS, HRMS)

EI-MS and **FAB-MS**: The measurements were recorded with a Finnigan MAT 95 (70 eV). Ionization was achieved through either EI (electron ionization), FAB (fast atom bombardment).

ESI-MS: The measurements were recorded with a ThermoFisher QExactive Plus (4 kV) with a ThermoFisher LT Orbitrap XL. Ionization was achieved through ESI (electrospray ionization).

HR-MS (high resolution-mass spectrometry): The measurements were either recorded with a Finnigan MAT 95 (EI/FAB) or with a ThermoFisher QExactive Plus (ESI). The following abbreviations were used: calc. = expected value (calculated); found = value found in analysis.

Notation of molecular fragments is given as mass to charge ratio (m/z); the intensities of the signals are noted in percent relative to the base signal (100%). As abbreviation for the ionized molecule $[M]^+$ is used. Characteristic fragmentation peaks are given as $[M–fragment]^+$ and $[fragment]^+$.

7.1.3.4 Single-Crystal X-ray Diffraction (XRD)/ Powder X-ray Diffraction (PXRD)

Two different diffractometers were used in this work:

Bruker D8 Venture diffractometer with Photon100 detector at 123(2) K using Cu-Kα radiation ($\lambda = 1.54178$ Å). Dual space methods (SHELXT)[190] were used for structure solution and refinement was carried out using SHELXL-2014 (full-matrix least-squares on F^2).[191] Hydrogen atoms were localized by difference electron density determination and refined using a riding model (H(O) free). Semi-empirical absorption corrections were applied.

STOE STADIVARI diffractometer at 200 K with monochromated Ga-Kα radiation ($\lambda = 1.34143$ Å). Using Olex2,[192] the structure was solved with the ShelXS[193] structure solution program using Direct Methods and refined with the ShelXL[191] refinement package using Least Squares minimization. Refinement was performed with anisotropic temperature factors for all non-hydrogen atoms; hydrogen atoms were calculated on idealized positions.

7.1.3.5 Analytical HPLC

The determination of the purity of the compounds was carried out on an Agilent 1100 series HPLC system with a G1322A degasser, a G1311A pump, G1313A autosampler, a G1316A column oven and a G1315B diode array detector. The flow rate was 1 mL/min. The stationary phase used was a VDSpher C18-M-SE (VDS Optilab) C18 column (5 µm, 4.0 mm × 250 mm). The runs were carried out with a linear gradient of A: 5% acetonitrile, 0.1% TFA in water to B: 95% acetonitrile, 0.1% TFA in water within 30 min. The purity was determined by integration of the signals at 218 nm or 256 nm.

7.1.3.6 Preparative HPLC

The purification of some compounds was carried out on a Shimadzu SPD-M10AVP series HPLC system with LC-8A pumps, a SCL-10AVP diode array detector and a type 202 fraction collector from Gilson. The flow rate was 15 mL/min. The stationary phase used was a VDSphereR C18-E (VDS Optilab) C18 column (5 µm, 20 mm × 250 mm). The runs were adjusted to the corresponding compounds. All separations were performed with the following eluents: A = ddH$_2$O, 0.1% TFA; B = acetonitrile, 0.1% TFA. The separation of the crude products was detected at 230 nm, 256 nm, 280 nm, 300 nm and 400 nm.

7.1.3.7 Thin Layer Chromatography (TLC)

All reactions were monitored by thin layer chromatography (TLC) using silica gel coated aluminum plates (Merck, silica gel 60, F$_{254}$). The detection was performed with UV light (254 nm) and/or by dipping the TLC into a solution of Seebach reagent (2.5% phosphor molybdic acid, 1.0% Cerium(IV) sulfate tetrahydrate and 6.0% sulfuric acid in H$_2$O, dipping solution) or a solution of potassium permanganate (1.5 g KMnO$_4$, 10 g K$_2$CO$_3$ and 1.25 mL 10% NaOH in 200 mL H$_2$O, dipping solution) and heating it with a heat gun.

7.1.3.8 Analytical scales

For mass determination balances from Mettler Toledo, AE163 and Radwag AS220.X2 were used.

7.2 Syntheses and Characterizations

7.2.1 General Procedures (GPs)

GP1: Acylation of 1,4-dimethoxynaphthalene

A mixture of trifluoroacetic anhydride (7.00 – 10.0 equiv.), 1,4-dimethoxynaphthalene (**70**) (1.00 equiv.) and a benzoic acid derivative **71** (1.00 – 1.20 equiv.) was heated to reflux under argon atmosphere. After 24 h the mixture was cooled to room temperature, quenched by the addition of water and the aqueous phase was extracted with EtOAc. The combined organic phases were washed with saturated aq. $NaHCO_3$ solution, dried over Na_2SO_4 and the solvents were removed under reduced pressure. The crude product was purified *via* column chromatography on silica gel.

GP2: Oxidative demethylation

Under argon atmosphere a 1 M solution of CAN (3.70 equiv.) in water was rapidly added to a 0.1 M solution of the 1,4-dimethoxynaphthalene derivative **72** (1.00 equiv.) in $MeCN/CH_2Cl_2$ (4:1) at −40 °C. The resulting reaction mixture was warmed to −20 °C over the course of 1 h and then poured into water. The aqueous phase was extracted with EtOAc and the combined organic phases were dried over Na_2SO_4. The solvents were removed under reduced pressure and the remaining crude product was dissolved in CH_2Cl_2. Cyclohexane was added, and the product was crystallized by the evaporation of CH_2Cl_2.

GP3: Diels-Alder cycloaddition

73a–g 87aa–gh 88aa–gh

In a crimp vial under argon atmosphere the dienophile **73a–g** (1.00 equiv.) was dissolved in dry CH$_2$Cl$_2$ and the diene **74a–h** (3.00 – 5.00 equiv.) was added. The reaction was stirred at 40 °C until the consumption of the dienophile was completed, as indicated by TLC. The solvent was removed under reduced pressure and the crude product was purified *via* column chromatography on silica gel.

GP4: Heck reaction

87/88aa–gh 118 119

Under argon atmosphere a mixture of the anthraquinone derivative **87/88aa–gh** (1.00 equiv.), Pd(OAc)$_2$ (20 mol%) and PPh$_3$ (40 mol%) was placed into a high-pressure glass tube. The mixture was dissolved in DMA, then PMP (2.00 equiv.) was added and the mixture was stirred at 70 °C. After completion of the reaction as indicated by TLC it was quenched by the addition of water. The aqueous phase was extracted with EtOAc and the combined organic phases were dried over Na$_2$SO$_4$. The solvents were removed under reduced pressure and the remaining crude product was purified *via* column chromatography on silica gel.

7.2.2   Precursor Synthesis

1,4-Dimethoxynaphthalene (70)

In a round-bottom flask, 1,4-dihydroxynaphthalene (**69**) (2.00 g, 12.5 mmol, 1.00 equiv.) was dissolved in anhydrous DMF (30 mL) under argon atmosphere. After the addition of K_2CO_3 (5.18 g, 37.5 mmol, 3.00 equiv.) and methyl iodide (1.96 mL, 4.44 g, 31.3 mmol, 2.50 equiv.) the mixture was stirred at 25 °C for 20 h. After completion of the reaction (TLC control), the mixture was filtered through Celite®. The filtrate was poured into water (60 mL) and the aqueous phase was extracted with EtOAc (2 × 60 mL). The combined organic extracts were dried over $MgSO_4$ and the solvent was removed under reduced pressure. The crude product was purified *via* column chromatography on silica gel (*c*Hex/EtOAc = 4:1). Product **70** was obtained as a beige solid (2.11 g, 11.2 mmol, 90%).

– R_f (*c*Hex/EtOAc = 4:1) = 0.76. – ^{1}H NMR (500 MHz, CDCl$_3$): δ = 8.23 (dd, ^3J = 6.4 Hz, ^4J = 3.3 Hz, 2H, CH_{Ar}), 7.52 (dd, ^3J = 6.4 Hz, ^4J = 3.3 Hz, 2H, CH_{Ar}), 6.69 (s, 2H, CH_{Ar}), 3.95 (s, 6H, OCH_3) ppm. – ^{13}C NMR (126 MHz, CDCl$_3$): δ = 149.6 (C$_q$, 2 × C$_{qAr}$), 126.4 (C$_q$, 2 × C$_{qAr}$), 125.9 (+, 2 × CH_{Ar}), 121.9 (+, 2 × CH_{Ar}), 103.3 (+, 2 × CH_{Ar}), 55.8 (+, 2 × OCH_3) ppm. – IR (ATR): $\tilde{v}$ = 3070 (w), 3011 (w), 2954 (w), 2838 (w), 1629 (w), 1593 (m), 1463 (m), 1445 (m), 1424 (w), 1383 (m), 1270 (m), 1237 (m), 1158 (m), 1085 (s), 1022 (m), 998 (m), 955 (m), 805 (m), 762 (s), 718 (m), 616 (m), 591 (w), 461 (w), 421 (w) cm^{-1}. – MS (EI, 70 eV), *m/z* (%): 188 (97) [M]$^+$, 173 (100) [M–CH$_3$]$^+$. – HRMS (EI, C$_{12}$H$_{12}$O$_2$): calc. 188.0837; found 188.0838. – Analytical data is in accordance with previously published literature.[115]

5-Acetamido-2-bromobenzoic acid (71g)

To a heterogeneous suspension of 5-amino-2-bromobenzoic acid (**71f**) (216 mg, 1.00 mmol, 1.00 equiv.) in water (5 mL) was added 6 M HCl (240–400 µL) until it became a homogeneous solution (~ pH 1.5). The resulting solution was cooled in an ice bath. Then acetic anhydride (0.09 mL, 102 mg, 1.00 mmol, 1.00 equiv.) was added and a pH value of around 5.5 was adjusted by the portionwise addition of powdered NaHCO$_3$. The solid was filtered, washed with water (2 × 1 mL) and dried under high vacuum. Product **71g** was obtained as a light grey solid (632 mg, 2.45 mmol, 61%).

– R_f (CH$_2$Cl$_2$/MeOH = 2:1) = 0.31. – ^{1}H NMR (400 MHz, DMSO-d_6): δ = 10.20 (s, 1H, CO$_2$H), 8.02 (s, 1H, CH_{Ar}), 7.62 (app. d, J = 2.3 Hz, 2H, CH_{Ar}), 2.05 (s, 3H, COCH_3) ppm. Signal missing (brs, NH). – ^{13}C NMR (101 MHz, DMSO-d_6): δ = 168.7 (C$_q$, 1 × C=O), 167.1 (C$_q$, 1 × C=O),

138.8 (C_q, 1 × C_{qAr}), 134.1 (+, 1 × CH_{Ar}), 133.6 (C_q, 1 × C_{qAr}), 122.6 (+, 1 × CH_{Ar}), 120.7 (+, 1 × CH_{Ar}), 112.7 (C_q, 1 × C_{qAr}), 24.0 (+, 1 × CH_3). – IR (ATR): $\tilde{v}$ = 3541 (vw), 3398 (vw), 3098 (vw), 1698 (w), 1661 (w), 1602 (w), 1543 (w), 1481 (w), 1416 (vw), 1375 (w), 1322 (w), 1295 (w), 1263 (w), 1225 (w), 1028 (vw), 982 (vw), 901 (w), 827 (w), 789 (w), 710 (vw), 657 (vw), 609 (vw), 542 (w), 398 (vw) cm^{-1}. – MS (APCI), m/z (%): 258/260 (100/98) [M+H]$^+$. – HRMS (APCI, $C_9H_9BrNO_3$): calc. 257.9766; found 257.9754. – Analytical data is in accordance with previously published literature.[194]

(1,4-Dimethoxynaphthalen-2-yl)(2-iodophenyl)methanone (72a)

According to **GP1** a mixture of trifluoroacetic anhydride (11.2 mL, 16.7 g, 79.5 mmol, 7.10 equiv.), 1,4-dimethoxynaphthalene (**70**) (2.11 g, 11.2 mmol, 1.00 equiv.) and 2-iodobenzoic acid (**71a**) (2.78 g, 11.2 mmol, 1.00 equiv.) was used. The crude product was purified *via* column chromatography on silica gel (*c*Hex/EtOAc = 4:1). Product **72a** was obtained as a yellow solid (3.14 g, 7.51 mmol, 67%).

– R_f (*c*Hex/EtOAc = 4:1) = 0.30. – ^{1}H NMR (500 MHz, CDCl$_3$): δ = 8.29 (dd, ^3J = 8.3 Hz, ^4J = 1.3 Hz, 1H, CH_{Ar}), 8.09 (dd, ^3J = 8.3 Hz, ^4J = 1.3 Hz, 1H, CH_{Ar}), 7.97 (d, ^3J = 8.0 Hz, 1H, CH_{Ar}), 7.63–7.60 (m, 1H, CH_{Ar}), 7.59–7.56 (m, 1H, CH_{Ar}), 7.44–7.38 (m, 2H, CH_{Ar}), 7.19–7.16 (m, 1H, CH_{Ar}), 7.11 (s, 1H, CH_{Ar}), 4.04 (s, 3H, OCH_3), 3.63 (s, 3H, OCH_3) ppm. – ^{13}C NMR (126 MHz, CDCl$_3$): δ = 196.9 (C_q, 1 × C=O), 152.3 (C_q, 1 × C_{qAr}), 152.1 (C_q, 2 × C_{qAr}), 145.5 (C_q, 1 × C_{qAr}), 140.2 (+, 1 × CH_{Ar}), 131.6 (+, 1 × CH_{Ar}), 129.6 (+, 1 × CH_{Ar}), 128.8 (C_q, 1 × C_{qAr}), 128.3 (+, 1 × CH_{Ar}), 127.8 (+, 1 × CH_{Ar}), 127.3 (+, 1 × CH_{Ar}), 125.7 (C_q, 1 × C_{qAr}), 123.6 (+, 1 × CH_{Ar}), 122.7 (+, 1 × CH_{Ar}), 103.2 (+, 1 × CH_{Ar}), 92.4 (C_q, 1 × C_{qAr}), 64.3 (+, 1 × OCH_3), 56.0 (+, 1 × OCH_3) ppm. – IR (ATR): $\tilde{v}$ = 3047 (vw), 2993 (w), 2930 (w), 2836 (w), 1650 (m), 1577 (m), 1460 (m), 1366 (m), 1271 (m), 1206 (m), 1091 (m), 1047 (m), 1017 (m), 999 (m), 955 (m), 858 (m), 803 (m), 783 (w), 761 (m), 746 (m), 688 (m), 665 (m), 648 (m), 634 (m), 449 (w) cm^{-1}. – MS (FAB, 3-NBA), m/z (%): 418 (100) [M]$^+$, 419 (95) [M+H]$^+$. – HRMS (EI, $C_{19}H_{15}{}^{127}IO_3$): calc. 418.0066; found 418.0066. – X-Ray: Crystallographic information on the product can be found in chapter 7.3.2. – Analytical data is in accordance with previously published literature.[116]

(1,4-Dimethoxynaphthalen-2-yl)(2-bromophenyl)methanone (72b)

According to **GP1** a mixture of trifluoroacetic anhydride (1.49 mL, 1.47 g, 7.00 mmol, 7.00 equiv.), 1,4-dimethoxynaphthalene (**70**) (188 mg, 1.00 mmol, 1.00 equiv.) and 2-bromobenzoic acid (**71b**) (201 mg, 1.00 mmol, 1.00 equiv.) was used. The crude product was purified *via* column chromatography on silica gel (*c*Hex/EtOAc = 15:1). Product **72b** was obtained as a yellow solid (310 mg, 835 µmol, 84%).

– R_f (*c*Hex/EtOAc = 15:1) = 0.36. – ^{1}H NMR (400 MHz, CDCl$_3$): δ = 8.33–8.25 (m, 1H, CH_{Ar}), 8.12–8.05 (m, 1H, CH_{Ar}), 7.69–7.64 (m, 1H, CH_{Ar}), 7.63–7.53 (m, 2H, CH_{Ar}), 7.47 (dd, ^3J = 7.5 Hz, ^4J = 2.0 Hz, 1H, CH_{Ar}), 7.43–7.29 (m, 2H, CH_{Ar}), 7.12 (s, 1H, CH_{Ar}), 4.04 (s, 3H, OCH_3), 3.62 (s, 3H, OCH_3) ppm. – ^{13}C NMR (101 MHz, CDCl$_3$): δ = 195.3 (C$_q$, 1 × C=O), 152.1 (C$_q$, 1 × C_{qAr}), 151.8 (C$_q$, 1 × C_{qAr}), 142.3 (C$_q$, 1 × C_{qAr}), 133.1 (+, 1 × CH_{Ar}), 131.1 (+, 1 × CH_{Ar}), 129.4 (+, 1 × CH_{Ar}), 129.3 (C$_q$, 1 × C_{qAr}), 128.5 (C$_q$, 1 × C_{qAr}), 128.0 (+, 1 × CH_{Ar}), 127.0 (+, 1 × CH_{Ar}), 126.8 (+, 1 × CH_{Ar}), 126.0 (C$_q$, 1 × C_{qAr}), 123.3 (+, 1 × CH_{Ar}), 122.4 (+, 1 × CH_{Ar}), 119.5 (C$_q$, 1 × C_{qAr}), 102.6 (+, 1 × CH_{Ar}), 63.8 (+, 1 × OCH_3), 55.7 (+, 1 × OCH_3) ppm. – IR (ATR): $\tilde{v}$ = 2931 (w), 1655 (m), 1583 (m), 1457 (m), 1366 (s), 1271 (m), 1206 (m), 1111 (m), 1092 (m), 1050 (m), 1027 (m), 1000 (m), 955 (m), 860 (m), 846 (m), 804 (m), 761 (m), 750 (s), 692 (m), 669 (m), 652 (m), 636 (m), 480 (w), 458 (w), 429 (w) cm^{-1}. – MS (FAB, 3-NBA), *m/z* (%): 371/373 (82/80) [M+H]$^+$, 370/372 (87/100) [M]$^+$. – HRMS (FAB, C$_{19}$H$_{15}$^{79}BrO$_3$): calc. 370.0205; found 370.0204. – X-Ray: Crystallographic information on the product can be found in chapter 7.3.1.

(2-Bromo-5-methoxyphenyl)(1,4-dimethoxynaphthalen-2-yl)methanone (72c)

According to **GP1** a mixture of trifluoroacetic anhydride (2.82 mL, 4.20 g, 20.0 mmol, 10.0 equiv.), 1,4-dimethoxynaphthalene (**70**) (376 mg, 2.00 mmol, 1.00 equiv.) and 2-bromo-5-methoxybenzoic acid (**71c**) (555 mg, 2.40 mmol, 1.20 equiv.) was used. The crude product was purified *via* column chromatography on silica gel (*c*Hex/EtOAc = 12:1). Product **72c** was obtained as a yellow solid (588 mg, 1.47 mmol, 73%).

– R_f (*c*Hex/EtOAc = 12:1) = 0.24. – ^{1}H NMR (400 MHz, CDCl$_3$): δ = 8.32–8.26 (m, 1H, CH_{Ar}), 8.13–8.07 (m, 1H, CH_{Ar}), 7.64–7.55 (m, 2H, CH_{Ar}), 7.52 (d, ^3J = 8.8 Hz, 1H, CH_{Ar}), 7.09 (s, 1H, CH_{Ar}), 7.00 (d, ^4J = 3.0 Hz, 1H, CH_{Ar}), 6.90 (dd, ^3J = 8.7 Hz, ^4J = 3.0 Hz, 1H, CH_{Ar}), 4.03 (s, 3H, OCH_3), 3.79 (s, 3H, OCH_3), 3.67 (s, 3H, OCH_3) ppm. – ^{13}C NMR (101 MHz, CDCl$_3$): δ = 195.5

(C$_q$, 1 × C=O), 158.8 (C$_q$, 1 × C$_{qAr}$), 152.4 (C$_q$, 1 × C$_{qAr}$), 152.1 (C$_q$, 1 × C$_{qAr}$), 143.4 (C$_q$, 1 × C$_{qAr}$), 134.1 (+, 1 × CH$_{Ar}$), 129.6 (C$_q$, 1 × C$_{qAr}$), 128.8 (C$_q$, 1 × C$_{qAr}$), 128.3 (+, 1 × CH$_{Ar}$), 127.3 (+, 1 × CH$_{Ar}$), 126.2 (C$_q$, 1 × C$_{qAr}$), 123.7 (+, 1 × CH$_{Ar}$), 122.8 (+, 1 × CH$_{Ar}$), 117.5 (+, 1 × CH$_{Ar}$), 114.9 (+, 1 × CH$_{Ar}$), 110.2 (C$_q$, 1 × C$_{qAr}$), 103.0 (+, 1 × CH$_{Ar}$), 64.1 (+, 1 × OCH$_3$), 56.0 (+, 1 × OCH$_3$), 55.8 (+, 1 × OCH$_3$) ppm. – IR (ATR): $\tilde{v}$ = 2935 (w), 2838 (w), 1650 (m), 1620 (m), 1592 (m), 1568 (m), 1458 (m), 1404 (m), 1366 (s), 1310 (m), 1281 (m), 1214 (s), 1162 (m), 1113 (m), 1094 (s), 1057 (m), 1020 (m), 962 (m), 853 (m), 813 (m), 767 (s), 709 (m), 669 (m), 603 (m), 486 (w), 430 (w) cm^{-1}. – MS (APCI), m/z (%): 401/403 (100/100) [M+H]$^+$. – HRMS (APCI, C$_{20}$H$_{18}$^{79}BrO$_4$): calc. 401.0388; found 401.0374. – X-Ray: Crystallographic information on the product can be found in chapter 7.3.2.

(2-Bromo-4,5-dimethoxyphenyl)(1,4-dimethoxynaphthalen-2-yl)methanone (72d)

According to **GP1** a mixture of trifluoroacetic anhydride (2.82 mL, 4.20 g, 20.0 mmol, 10.0 equiv.), 1,4-dimethoxynaphthalene (**70**) (376 mg, 2.00 mmol, 1.00 equiv.) and 2-bromo-4,5-dimethoxybenzoic acid (**71d**) (2.78 g, 11.2 mmol, 1.00 equiv.) was used. The crude product was purified via column chromatography on silica gel (cHex/EtOAc = 4:1). Product **72d** was obtained as a yellow solid (799 mg, 1.85 mmol, 77%).

– R_f (cHex/EtOAc = 4:1) = 0.27. – ^{1}H NMR (400 MHz, CDCl$_3$): δ = 8.32–8.27 (m, 1H, CH$_{Ar}$), 8.15–8.10 (m, 1H, CH$_{Ar}$), 7.64–7.54 (m, 2H, CH$_{Ar}$), 7.09 (s, 1H, CH$_{Ar}$), 7.07 (s, 1H, CH$_{Ar}$), 7.00 (s, 1H, CH$_{Ar}$), 4.02 (s, 3H, OCH$_3$), 3.95 (s, 3H, OCH$_3$), 3.83 (s, 3H, OCH$_3$), 3.70 (s, 3H, OCH$_3$) ppm. – ^{13}C NMR (101 MHz, CDCl$_3$): δ = 195.3 (C$_q$, 1 × C=O), 152.0 (C$_q$, 1 × C$_{qAr}$), 151.4 (C$_q$, 1 × C$_{qAr}$), 151.3 (C$_q$, 1 × C$_{qAr}$), 148.1 (C$_q$, 1 × C$_{qAr}$), 133.7 (C$_q$, 1 × C$_{qAr}$), 129.1 (C$_q$, 1 × C$_{qAr}$), 128.8 (C$_q$, 1 × C$_{qAr}$), 128.0 (+, 1 × CH$_{Ar}$), 127.3 (+, 1 × CH$_{Ar}$), 126.9 (C$_q$, 1 × C$_{qAr}$), 123.5 (+, 1 × CH$_{Ar}$), 122.7 (+, 1 × CH$_{Ar}$), 116.1 (+, 1 × CH$_{Ar}$), 113.3 (+, 1 × CH$_{Ar}$), 112.2 (C$_q$, 1 × C$_{qAr}$), 103.3 (+, 1 × CH$_{Ar}$), 64.1 (+, 1 × OCH$_3$), 56.4 (+, 1 × OCH$_3$), 56.3 (+, 1 × OCH$_3$), 56.0 (+, 1 × OCH$_3$) ppm. – IR (ATR): $\tilde{v}$ = 2932 (vw), 1634 (w), 1591 (w), 1504 (w), 1439 (w), 1402 (w), 1378 (w), 1335 (w), 1254 (w), 1209 (w), 1182 (w), 1157 (w), 1121 (w), 1095 (w), 1053 (w), 1029 (w), 994 (w), 965 (w), 864 (w), 839 (w), 772 (w), 738 (w), 713 (w), 649 (vw), 626 (w), 551 (vw), 430 (vw) cm^{-1}. – MS (FAB, 3-NBA), m/z (%): 433/431 (66/68) [M+H]$^+$, 432/430 (100/90) [M]$^+$, 351 (12) [M–Br]$^+$. – HRMS (FAB, C$_{21}$H$_{19}$^{79}BrO$_5$): calc. 430.0416; found 430.0417. – X-Ray: Crystallographic information on the product can be found in chapter 7.3.2.

N-(4-Bromo-3-(1,4-dimethoxy-2-naphthoyl)phenyl)acetamide (72f)

According to **GP1** a mixture of trifluoroacetic anhydride (2.11 mL, 3.15 g, 15.0 mmol, 10.0 equiv.), 1,4-dimethoxynaphthalene **(70)** (376 mg, 2.00 mmol, 1.00 equiv.) and 5-acetamido-2-bromobenzoic acid **(71e)** (464 mg, 1.80 mmol, 1.20 equiv.) was used. The crude product was purified *via* column chromatography (*c*Hex/EtOAc = 1:1). Product **72e** was obtained as a yellow solid (560 mg, 1.31 mmol, 65%).

– R_f (*c*Hex/EtOAc = 1:1) = 0.32. – ^{1}H NMR (500 MHz, CDCl$_3$): δ = 8.28 (dd, ^3J = 8.2 Hz, ^4J = 1.3 Hz, 1H, CH_{Ar}), 8.08 (dd, ^3J = 8.2 Hz, ^4J = 1.3 Hz, 1H, CH_{Ar}), 7.70 (dd, ^3J = 8.7 Hz, ^4J = 2.6 Hz, 1H, CH_{Ar}), 7.63–7.55 (m, 3H, CH_{Ar}), 7.43 (d, ^4J = 2.6 Hz, 1H, CH_{Ar}), 7.36 (brs, 1H, NH), 7.07 (s, 1H, CH_{Ar}), 4.02 (s, 3H, OCH_3), 3.65 (s, 3H, OCH_3), 2.13 (s, 3H, CH_3) ppm. – ^{13}C NMR (126 MHz, CDCl$_3$): δ = 195.2 (C$_q$, 1 × C=O), 168.5 (C$_q$, 1 × C=O), 152.5 (C$_q$, 1 × C$_{qAr}$), 152.2 (C$_q$, 1 × C$_{qAr}$), 142.8 (C$_q$, 1 × C$_{qAr}$), 137.3 (C$_q$, 1 × C$_{qAr}$), 134.0 (+, 1 × CH$_{Ar}$), 129.7 (C$_q$, 1 × C$_{qAr}$), 128.8 (C$_q$, 1 × C$_{qAr}$), 128.4 (+, 1 × CH$_{Ar}$), 127.4 (+, 1 × CH$_{Ar}$), 126.0 (C$_q$, 1 × C$_{qAr}$), 123.6 (+, 1 × CH$_{Ar}$), 122.8 (+, 1 × CH$_{Ar}$), 122.7 (+, 1 × CH$_{Ar}$), 120.6 (+, 1 × CH$_{Ar}$), 113.9 (C$_q$, 1 × C$_{qAr}$), 102.8 (+, 1 × CH$_{Ar}$), 64.3 (+, 1 × OCH$_3$), 56.0 (+, 1 × OCH$_3$), 24.7 (+, 1 × CH$_3$) ppm. – IR (ATR): $\tilde{v}$ = 3306 (w), 2935 (w), 2250 (vw), 1664 (m), 1593 (m), 1531 (m), 1459 (m), 1394 (m), 1366 (s), 1309 (m), 1242 (m), 1207 (m), 1163 (w), 1117 (m), 1095 (s), 1047 (m), 1028 (m), 993 (m), 963 (w), 907 (m), 821 (w), 800 (w), 769 (m), 727 (s), 647 (m), 587 (w), 483 (w), 415 (w) cm^{-1}. – MS (FAB, 3-NBA), *m/z* (%): 430/428 (56/55) [M+H]$^+$, 429/427 (69/56) [M]$^+$. – HRMS (FAB, C$_{21}$H$_{18}$O$_4$N^{79}Br): calc. 427.0419; found 427.0421. – X-Ray: Crystallographic information on the product can be found in chapter 7.3.2.

N-(4-Bromo-3-(1,4-dimethoxy-2-naphthoyl)phenyl)-2,2,2-trifluoroacetamide (72g)

According to **GP1** a mixture of trifluoroacetic anhydride (1.41 mL, 2.10 g, 10.0 mmol, 10.0 equiv.), 1,4-dimethoxynaphthalene **(70)** (188 mg, 1.00 mmol, 1.00 equiv.) and 5-amino-2-bromobenzoic acid **(71f)** (216 mg, 1.00 mmol, 1.00 equiv.) was used. The crude product was purified *via* column chromatography on silica gel (*c*Hex/EtOAc = 9:1). Product **72f** was obtained as a yellow solid (341 mg, 707 µmol, 71%).

– R_f (*c*Hex/EtOAc = 9:1) = 0.20. – ^{1}H NMR (400 MHz, CDCl$_3$): δ = 8.45 (brs, 1H, NH), 8.28 (d, ^3J = 8.3 Hz, 1H, CH_{Ar}), 8.06 (d, ^3J = 8.3 Hz, 1H, CH_{Ar}), 7.72 (dd, ^3J = 8.8 Hz, ^4J = 2.6 Hz, 1H,

CH_{Ar}), 7.64–7.53 (m, 4H, CH_{Ar}), 7.07 (s, 1H, CH_{Ar}), 4.01 (s, 3H, OCH_3), 3.62 (s, 3H, OCH_3) ppm. – ^{13}C NMR (101 MHz, CDCl$_3$): $\delta = 194.9$ (C$_q$, 1 × C=O), 155.1 (q, $^2J = 38.0$ Hz, 1 × COCF$_3$), 152.9 (C$_q$, 1 × C_{qAr}), 152.3 (C$_q$, 2 × C_{qAr}), 143.5 (C$_q$, 1 × C_{qAr}), 134.6 (C$_q$, 1 × C_{qAr}), 134.3 (+, 1 × CH_{Ar}), 129.9 (C$_q$, 1 × C_{qAr}), 128.7 (+, 1 × CH_{Ar}), 127.5 (+, 1 × CH_{Ar}), 125.6 (C$_q$, 1 × C_{qAr}), 123.6 (+, 1 × CH_{Ar}), 123.3 (+, 1 × CH_{Ar}), 122.9 (+, 1 × CH_{Ar}), 121.3 (+, 1 × CH_{Ar}), 116.3 (C$_q$, 1 × C_{qAr}), 115.7 (q, $^1J = 287.0$ Hz, 1 × CF_3), 102.4 (+, 1 × CH_{Ar}), 64.3 (+, 1 × OCH_3), 56.0 (+, 1 × OCH_3) ppm. – IR (ATR): $\tilde{v} = 3067$ (vw), 2944 (vw), 2843 (vw), 1723 (w), 1618 (w), 1578 (w), 1541 (w), 1458 (w), 1405 (w), 1371 (m), 1297 (w), 1225 (w), 1168 (w), 1119 (w), 1096 (w), 1045 (w), 1029 (w), 994 (w), 965 (w), 901 (vw), 877 (vw), 856 (w), 834 (w), 796 (w), 765 (w), 685 (w), 651 (w), 598 (w), 479 (w), 430 (vw), 389 (w) cm^{-1}. – MS (APCI), m/z (%): 482/484 (97/100) [M+H]$^+$. – HRMS (APCI, C$_{21}$H$_{16}$^{79}BrF$_3$NO$_4$): calc. 482.0215; found 482.0198. – X-Ray: Crystallographic information on the product can be found in chapter 7.3.2.

7.2.3   Synthesis of Naphthoquinone Derivatives

2-(2-Iodobenzoyl)naphthalene-1,4-dione (73a)

Following **GP2** the crude product was obtained from CAN (14.5 g, 26.5 mmol, 3.70 equiv.) and (1,4-dimethoxynaphthalen-2-yl)(2-iodophenyl)methanone (**72a**) (3.00 g, 7.17 mmol, 1.00 equiv.). Product **73a** was isolated as an orange solid (2.78 g, 7.17 mmol, quant.).

– R_f (cHex/EtOAc = 4:1) = 0.25. – ^{1}H NMR (300 MHz, CDCl$_3$): δ = 8.14–8.10 (m, 2H, CH_{Ar}), 7.96 (dd, ^3J = 7.8 Hz, ^4J = 1.1 Hz, 1H, CH_{Ar}), 7.83–7.80 (m, 2H, CH_{Ar}), 7.57 (dd, ^3J = 7.8 Hz, ^4J = 1.8 Hz, 1H, CH_{Ar}), 7.48 (td, ^3J = 7.8 Hz, ^4J = 1.1 Hz, 1H, CH_{Ar}), 7.29–7.20 (m, 1H, CH_{Ar}), 7.18 (s, 1H, C=CH) ppm. – ^{13}C NMR (101 MHz, CDCl$_3$): δ = 193.6 (C$_q$, 1 × C=O), 185.0 (C$_q$, 1 × C=O), 182.7 (C$_q$, 1 × C=O), 145.0 (C$_q$, 1 × C$_{qAr}$), 142.0 (C$_q$, 1 × C$_{qAr}$), 140.9 (+, 1 × C=CH), 137.8 (+, 1 × CH_{Ar}), 134.8 (+, 1 × CH_{Ar}), 134.5 (+, 1 × CH_{Ar}), 133.3 (+, 1 × CH_{Ar}), 132.1 (C$_q$, 1 × C$_{qAr}$), 131.9 (C$_q$, 1 × C$_{qAr}$), 131.1 (+, 1 × CH_{Ar}), 128.5 (+, 1 × CH_{Ar}), 127.1 (+, 1 × CH_{Ar}), 126.6 (+, 1 × CH_{Ar}), 93.1 (C$_q$, 1 × C$_{qAr}$) ppm. – IR (ATR): $\tilde{v}$ = 3041 (vw), 1653 (m), 1579 (w), 1424 (w), 1282 (w), 1252 (w), 1014 (w), 973 (w), 943 (w), 773 (w), 748 (w), 691 (w), 673 (w), 633 (w), 591 (w), 453 (vw), 413 (vw) cm^{-1}. – MS (FAB, 3-NBA), m/z (%): 390 (62) [M+H]$^+$, 389 (69) [M]$^+$. – HRMS (EI, C$_{17}$H$_{10}$^{127}IO$_3$): calc. 388.9675; found 388.9677. – X-Ray: Crystallographic information on the product can be found in chapter 7.3.1. – Analytical data is in accordance with previously published literature.[116]

2-(2-Bromobenzoyl)naphthalene-1,4-dione (73b)

Following **GP2** the crude product was obtained from CAN (10.1 g, 18.5 mmol, 3.70 equiv.) and (1,4-dimethoxynaphthalen-2-yl)(2-bromophenyl)methanone (**72b**) (2.09 g, 5.63 mmol, 1.00 equiv.). Product **73b** was isolated as an orange solid (1.92 g, 5.63 mmol, quant.).

– R_f (cHex/EtOAc = 9:1) = 0.30. – ^{1}H NMR (400 MHz, CDCl$_3$): δ = 8.15–8.08 (m, 2H, CH_{Ar}), 7.83–7.79 (m, 2H, CH_{Ar}), 7.71–7.69 (m, 1H, CH_{Ar}), 7.61 (d, ^3J = 7.8 Hz, 1H, CH_{Ar}), 7.48 (t, J = 7.4 Hz, 1H, CH_{Ar}), 7.45–7.41 (m, 1H, CH_{Ar}), 7.18 (s, 1H, C=CH) ppm. – ^{13}C NMR (101 MHz, CDCl$_3$): δ = 192.6 (C$_q$, 1 × C=O), 185.1 (C$_q$, 1 × C=O), 182.7 (C$_q$, 1 × C=O), 145.9 (C$_q$, 1 × C$_{qAr}$), 139.0 (C$_q$, 1 × C$_{qAr}$), 137.3 (+, 1 × C=CH), 134.7 (+, 1 × CH_{Ar}), 134.5 (+, 1 × CH_{Ar}), 133.9 (+, 1 × CH_{Ar}), 133.6 (+, 1 × CH_{Ar}), 132.2 (C$_q$, 1 × C$_{qAr}$), 131.8 (C$_q$, 1 × C$_{qAr}$), 131.3 (+, 1 × CH_{Ar}), 128.0 (+, 1 × CH_{Ar}), 127.1 (+, 1 × CH_{Ar}), 126.6 (+, 1 × CH_{Ar}), 121.0 (C$_q$, 1 × C$_{qAr}$) ppm.

– IR (ATR): $\tilde{v}$ = 3041 (vw), 1653 (m), 1584 (w), 1465 (w), 1430 (w), 1352 (w), 1328 (w), 1283 (m), 1252 (m), 1101 (w), 1051 (w), 1025 (w), 975 (w), 947 (w), 843 (w), 799 (w), 773 (w), 750 (m), 713 (w), 696 (m), 634 (w), 592 (m), 463 (w), 403 (vw) cm^{-1}. – MS (FAB, 3-NBA), m/z (%): 342/344 (14/14) [M+H]$^+$, 341/343 (16/21) [M]$^+$. – HRMS (FAB, C$_{17}$H$_{10}$^{79}BrO$_3$): calc. 340.9813; found 340.9814. – X-Ray: Crystallographic information on the product can be found in chapter 7.3.

2-(2-Bromo-5-methoxybenzoyl)naphthalene-1,4-dione (73c)

Following **GP2** the crude product was obtained from CAN (10.1 g, 18.4 mmol, 3.70 equiv.) and (2-bromo-5-methoxyphenyl)(1,4-dimethoxynaphthalen-2-yl)methanone (**72c**) (2.00 g, 4.98 mmol, 1.00 equiv.). Product **73c** was isolated as an orange solid (1.37 g, 3.69 mmol, 74%).

– R_f (cHex/EtOAc = 4:1) = 0.44. – ^{1}H NMR (400 MHz, CDCl$_3$): δ = 8.16–8.08 (m, 2H, CH$_{Ar}$), 7.84–7.78 (m, 2H, CH$_{Ar}$), 7.47 (d, ^3J = 8.8 Hz, 1H, CH$_{Ar}$), 7.23 (d, ^4J = 3.1 Hz, 1H, CH$_{Ar}$), 7.17 (s, 1H, C=CH), 6.98 (dd, ^3J = 8.8 Hz, ^4J = 3.1 Hz, 1H, CH$_{Ar}$), 3.86 (s, 3H, OCH$_3$) ppm. – ^{13}C NMR (101 MHz, CDCl$_3$): δ = 192.5 (C$_q$, 1 × C=O), 185.1 (C$_q$, 1 × C=O), 182.6 (C$_q$, 1 × C=O), 159.3 (C$_q$, 1 × C$_{qAr}$), 146.1 (C$_q$, 1 × C$_{qAr}$), 139.6 (C$_q$, 1 × C$_{qAr}$), 137.2 (+, 1 × CH$_{Ar}$), 134.7 (+, 1 × CH$_{Ar}$), 134.7 (+, 1 × CH$_{Ar}$), 134.5 (+, 1 × CH$_{Ar}$), 132.2 (C$_q$, 1 × C$_{qAr}$), 131.9 (C$_q$, 1 × C$_{qAr}$), 127.1 (+, 1 × CH$_{Ar}$), 126.6 (+, 1 × CH$_{Ar}$), 120.3 (+, 1 × CH$_{Ar}$), 115.9 (+, 1 × C=CH), 111.6 (C$_q$, 1 × C$_{qAr}$), 55.9 (+, 1 × OCH$_3$) ppm. – IR (ATR): $\tilde{v}$ = 2934 (w), 1655 (m), 1589 (m), 1460 (m), 1399 (w), 1349 (w), 1281 (m), 1236 (m), 1097 (w), 1018 (w), 993 (w), 920 (w), 885 (w), 818 (m), 765 (m), 720 (w), 667 (w), 607 (w), 584 (m), 456 (w), 413 (vw) cm^{-1}. – MS (FAB, 3-NBA), m/z (%): 371/373 (13/17) [M+H]$^+$, 291 (17) [M–Br]$^+$. – HRMS (FAB, C$_{18}$H$_{12}$^{79}BrO$_4$): calc. 370.9919; found 370.9918.

2-(2-Bromo-4,5-dimethoxybenzoyl)naphthalene-1,4-dione (73d)

Following **GP2** the crude product was obtained from CAN (620 mg, 1.12 mmol, 3.70 equiv.) and (2-bromo-4,5-dimethoxyphenyl)(1,4-dimethoxynaphthalen-2-yl)methanone (**72d**) (130 mg, 300 μmol, 1.00 equiv.). Product **73d** was isolated as an orange solid (88.0 mg, 220 μmol, 73%).

– R_f (cHex/EtOAc = 4:1) = 0.22. – ^{1}H NMR (400 MHz, CDCl$_3$): δ = 8.15–8.10 (m, 2H, CH_{Ar}), 7.83–7.79 (m, 2H, CH_{Ar}), 7.39 (s, 1H, CH_{Ar}), 7.10 (s, 1H, CH_{Ar}), 7.01 (s, 1H, C=CH), 3.95 (s, 3H, OCH_3), 3.94 (s, 3H, OCH_3) ppm. – ^{13}C NMR (101 MHz, CDCl$_3$): δ = 191.2 (C$_q$, 1 × C=O), 185.1 (C$_q$, 1 × C=O), 183.0 (C$_q$, 1 × C=O), 153.4 (C$_q$, 1 × C_{qAr}), 148.9 (C$_q$, 1 × C_{qAr}), 147.9 (C$_q$, 1 × C_{qAr}), 136.1 (+, 1 × CH_{Ar}), 134.6 (+, 1 × CH_{Ar}), 134.5 (+, 1 × CH_{Ar}), 132.2 (C$_q$, 1 × C_{qAr}), 132.0 (C$_q$, 1 × C_{qAr}), 130.5 (C$_q$, 1 × C_{qAr}), 127.0 (+, 1 × CH_{Ar}), 126.6 (+, 1 × CH_{Ar}), 116.2 (+, 1 × CH_{Ar}), 114.8 (C$_q$, 1 × C_{qAr}), 113.7 (+, 1 × CH_{Ar}), 56.6 (+, 1 × OCH_3), 56.4 (+, 1 × OCH_3) ppm. – IR (ATR): $\tilde{v}$ = 3009 (vw), 3045 (vw), 2925 (vw), 2840 (vw), 1659 (m), 1581 (m), 1506 (m), 1435 (w), 1377 (w), 1344 (w), 1294 (w), 1255 (m), 1211 (m), 1174 (m), 1131 (w), 1012 (m), 909 (w), 882 (w), 848 (w), 783 (w), 759 (m), 719 (w), 655 (w), 631 (w), 589 (w), 563 (w), 454 (vw), 423 (vw) cm^{-1}. – MS (FAB, 3-NBA), m/z (%): 401/403 (4/7) [M+H]$^+$. – HRMS (FAB, C$_{19}$H$_{14}$^{79}BrO$_5$): calc. 401.0025; found 401.0025. – X-Ray: Crystallographic information on the product can be found in chapter 7.3.2.

2-(2-Bromo-4,5-dihydroxybenzoyl)naphthalene-1,4-dione (73e)

BBr$_3$ (7.50 mL, 7.48 mmol, 1.87 g, 10.0 equiv., 1 M in CH$_2$Cl$_2$) was dropwise added to a solution of 2-(2-bromo-4,5-dimethoxybenzoyl)naphthalene-1,4-dione (**73d**) (300 mg, 750 µmol, 1.00 equiv.) in 21 mL CH$_2$Cl$_2$ at –78 °C over the course of 0.5 h. The resulting solution was allowed to warm to room temperature and was stirred for 4 h. The reaction mixture was poured into ice cold water (30 mL) and extracted with CH$_2$Cl$_2$ (2 × 30 mL). The combined organic phases were washed with brine (35 mL), dried over Na$_2$SO$_4$ and the solvent was removed under reduced pressure. After column chromatography on silica gel (cHex/EtOAc = 1:1) product **73e** was obtained as a dark red oil (33.0 mg, 88.4 µmol, 12%).

– R_f (cHex/EtOAc = 1:1) = 0.26. – ^{1}H NMR (500 MHz, MeOD): δ = 8.15–8.08 (m, 2H, CH_{Ar}), 7.88 (dd, ^3J = 5.5 Hz, ^4J = 2.8 Hz, 2H, CH_{Ar}), 7.31 (s, 1H, C=CH), 7.03 (s, 1H, CH_{Ar}), 7.01 (s, 1H, CH_{Ar}) ppm. Signals missing (2H, OH). – ^{13}C NMR (126 MHz, MeOD): δ = 192.0 (C$_q$, 1 × C=O), 186.1 (C$_q$, 1 × C=O), 184.3 (C$_q$, 1 × C=O), 152.7 (C$_q$, 1 × C_{qAr}), 149.4 (C$_q$, 1 × C_{qAr}), 146.4 (C$_q$, 1 × C_{qAr}), 136.4 (+, 1 × CH_{Ar}), 135.6 (+, 1 × CH_{Ar}), 135.5 (+, 1 × CH_{Ar}), 133.4 (C$_q$, 1 × C_{qAr}), 133.2 (C$_q$, 1 × C_{qAr}), 129.8 (C$_q$, 1 × C_{qAr}), 127.5 (+, 1 × CH_{Ar}), 127.2 (+, 1 × CH_{Ar}), 121.6 (+, 1 × CH_{Ar}), 119.7 (+, 1 × CH_{Ar}), 113.7 (C$_q$, 1 × C_{qAr}) ppm. – IR (ATR): $\tilde{v}$ = 3306 (w), 2922 (w), 2851 (w), 1660 (w), 1585 (m), 1503 (w), 1414 (w), 1347 (w), 1276 (m), 1179 (m), 1039 (w), 875 (w), 812 (vw), 767 (w), 719 (w), 624 (vw), 577 (w), 446 (vw) cm^{-1}. – HRMS (FAB,

$C_{17}H_7{}^2H_2O_5{}^{79}Br$): calc. 373.9753; found 373.9755. The HRMS was measured from the MeOD NMR sample, therefore the two OH protons exchanged with deuterium.

N-(4-Bromo-3-(1,4-dioxo-1,4-dihydronaphthalene-2-carbonyl)phenyl)acetamide (73f)

Following **GP2** the crude product was obtained from CAN (2.51 g, 4.58 mmol, 3.70 equiv.) and N-(4-bromo-3-(1,4-dimethoxy-2-naphthoyl)phenyl)acetamide (**72e**) (530 mg, 1.24 mmol, 1.00 equiv.). Product **73f** was isolated as an orange solid (389 mg, 977 µmol, 79%).

– R_f (cHex/EtOAc = 1:1) = 0.47. – ^{1}H NMR (500 MHz, CDCl$_3$): δ = 8.14–8.07 (m, 2H, CH$_{Ar}$), 7.84–7.80 (m, 3H, CH$_{Ar}$), 7.69 (d, ^4J = 2.7 Hz, 1H, CH$_{Ar}$), 7.53 (d, ^3J = 8.8 Hz, 1H, CH$_{Ar}$), 7.51 (brs, 1H, NH), 7.18 (s, 1H, C=CH), 2.19 (s, 3H, COCH$_3$) ppm. – ^{13}C NMR (126 MHz, CDCl$_3$): δ = 192.3 (C$_q$, 1 × C=O), 185.0 (C$_q$, 1 × C=O), 182.7 (C$_q$, 1 × C=O), 168.7 (C$_q$, 1 × C=O), 145.7 (C$_q$, 1 × C$_{qAr}$), 139.2 (C$_q$, 1 × C$_{qAr}$), 138.1 (C$_q$, 1 × C$_{qAr}$), 137.5 (+, 1 × CH$_{Ar}$), 134.8 (+, 1 × CH$_{Ar}$), 134.6 (+, 1 × CH$_{Ar}$), 134.5 (+, 1 × CH$_{Ar}$), 132.2 (C$_q$, 1 × C$_{qAr}$), 131.8 (C$_q$, 1 × C$_{qAr}$), 127.1 (+, 1 × CH$_{Ar}$), 126.6 (+, 1 × CH$_{Ar}$), 124.7 (+, 1 × CH$_{Ar}$), 121.7 (+, 1 × CH$_{Ar}$), 114.9 (C$_q$, 1 × C$_q$), 24.8 (+, 1 × COCH$_3$) ppm. – IR (ATR): $\tilde{v}$ = 3367 (vw), 2926 (vw), 1654 (w), 1581 (w), 1530 (w), 1466 (w), 1392 (w), 1295 (w), 1250 (w), 1233 (w), 1099 (vw), 1046 (vw), 1013 (w), 892 (vw), 826 (vw), 776 (w), 758 (w), 723 (vw), 661 (vw), 574 (vw), 487 (vw), 416 (vw) cm^{-1}. – MS (FAB, 3-NBA), m/z (%): 398 (5) [M+H]$^+$, 307 (35), 154 (100). – HRMS (FAB, $C_{19}H_{13}O_4N^{79}Br$): calc. 398.0028; found 398.0027.

N-(4-Bromo-3-(1,4-dioxo-1,4-dihydronaphthalene-2-carbonyl)phenyl)-2,2,2-trifluoroacetamide (73g)

Following **GP2** the crude product was obtained from CAN (1.43 g, 2.61 mmol, 3.70 equiv.) and N-(4-bromo-3-(1,4-dimethoxy-2-naphthoyl)phenyl)-2,2,2-trifluoroacetamide (**72f**) (340 mg, 710 µmol, 1.00 equiv.). Product **73g** was isolated as an orange solid (296 mg, 655 µmol, 92%).

– R_f (cHex/EtOAc = 4:1) = 0.33. – ^{1}H NMR (400 MHz, CDCl$_3$): δ = 8.19–8.12 (m, 2H, NH, CH$_{Ar}$), 8.11–8.07 (m, 1H, CH$_{Ar}$), 7.86 (dd, ^3J = 8.7 Hz, ^4J = 2.8 Hz, 1H, CH$_{Ar}$), 7.83–7.81 (m, 2H, CH$_{Ar}$), 7.78 (d, ^4J = 2.8 Hz, 1H, CH$_{Ar}$), 7.63 (d, ^3J = 8.7 Hz, 1H, CH$_{Ar}$), 7.27 (s, 1H, C=CH) ppm. – ^{13}C NMR (101 MHz, CDCl$_3$): δ = 191.9 (C$_q$, 1 × C=O), 184.9 (C$_q$, 1 × C=O), 182.8 (C$_q$,

1 × C=O), 145.3 (C_q, 1 × C_{qAr}), 140.0 (C_q, 1 × C_{qAr}), 138.0 (+, 1 × C=CH), 135.4 (C_q, 1 × C_{qAr}), 134.9 (+, 2 × CH$_{Ar}$), 134.7 (+, 1 × CH$_{Ar}$), 132.2 (C_q, 1 × C_{qAr}), 131.7 (C_q, 1 × C_{qAr}), 127.1 (+, 1 × CH$_{Ar}$), 126.7 (+, 1 × CH$_{Ar}$), 125.0 (+, 1 × CH$_{Ar}$), 122.5 (+, 1 × CH$_{Ar}$), 117.3 (C_q, 1 × C_{qAr}) ppm. Signals missing (C_q, 1 × COCF$_3$, 1 × CF$_3$). – IR (ATR): $\tilde{v}$ = 3350 (vw), 1727 (w), 1654 (w), 1583 (w), 1542 (w), 1469 (w), 1406 (w), 1349 (vw), 1299 (w), 1136 (m), 1047 (w), 881 (w), 830 (w), 777 (w), 760 (w), 724 (w), 670 (w), 632 (w), 580 (w), 495 (w), 458 (vw) cm^{-1}. – MS (APCI), *m/z* (%): 452/454 (97/100) [M+H]$^+$. – HRMS (APCI, C$_{19}$H$_{10}$^{79}BrF$_3$NO$_4$): calc. 451.9745; found 451.9729.

7.2.4 Syntheses of Dienes

(E)-(Buta-1,3-dien-1-yloxy)trimethylsilane (74d)

OTMS   To a solution of crotonaldehyde (7.01 g, 100 mmol, 1.00 equiv.) and triethylamine (11.1 g, 110 mmol, 1.10 equiv.) in benzene (20 mL) at 25 °C were added hydroquinone (200 mg, 1.82 mmol, 0.018 equiv.) and dry zinc chloride (160 mg, 1.17 mmol, 0.012 equiv.) rapidly with heavy stirring. Then freshly distilled trimethylsilyl chloride (11.4 mL, 9.78 g, 90.0 mmol, 0.90 equiv.) was added over 1–2 min, whereupon a white precipitate rapidly formed. After stirring the mixture for 30 min at 25 °C, a further 2.54 mL (2.17 g, 20.0 mmol, 0.20 equiv.) of trimethylsilyl chloride were added. The reaction was heated to 70 °C and stirred at this temperature for 12 h. The mixture was cooled to 0 °C, quenched by the addition of saturated aq. NaHCO$_3$ solution (20 mL), and the organic layer was separated. The aqueous layer was extracted with benzene (3 × 10 mL), the organic extracts were combined and washed with 10% aq. KHSO$_4$ solution. The organic phase was dried over Na$_2$SO$_4$, and benzene was removed *in vacuo* to leave a dark brown liquid. Distillation through a short Vigreux column (10 cm) yielded pure 1-trimethylsilyloxy-1,3-butadiene (**74d**) as a colorless liquid (bp 78–80 °C, 90 mbar). The yield was not determined.

– ^{1}H NMR (300 MHz, CDCl$_3$): δ = 6.54 (d, ^3J = 11.9 Hz, 1H, OC*H*), 6.29–6.16 (m, 1H, OCH=C*H*), 5.76–5.68 (m, 1H, CH$_2$=C*H*), 5.02–4.96 (m, 1H, C*H*H), 4.84–4.80 (m, 1H, C*H*H), 0.21 (s, 9H, Si(C*H*$_3$)$_3$) ppm. – Analytical data is in accordance with previously published literature.[195]

(E)-Trimethyl((2-methylbuta-1,3-dien-1-yl)oxy)silane (74e)

OTMS   Under argon atmosphere a solution of *trans*-2-methyl-2-butenal (**76**) (11.5 mL, 10.0 g, 119 mmol, 1.00 equiv.) in dry Et$_2$O (40 mL) was added to a suspension of ZnCl$_2$ (1.62 g, 11.9 mmol, 0.100 equiv.) in dry triethylamine (21.4 mL, 15.7 g, 155 mmol 1.30 equiv.). Freshly distilled trimethylsilyl chloride (18.2 mL, 15.5 g, 143 mmol, 1.20 equiv.) was added dropwise over 1 h and the mixture was stirred at 35 °C over night. The reaction was cooled to room temperature and diluted with Et$_2$O (40 mL). The precipitate was filtered and the resulting solution was concentrated under reduced pressure. Then the mixture was diluted with pentane (20 mL) and again filtered. The solvent was removed under reduced pressure and the diene **74e** was obtained as a colorless liquid (4.25 g, 23%) through distillation (bp 50 °C, 20 mbar).

– ^{1}H NMR (300 MHz, CDCl$_3$): δ = 6.40 (dd, ^4J = 1.2, 0.6 Hz, 1H, C=CH), 6.29 (dd, ^2J = 17.2 Hz, ^3J = 10.7 Hz, 1H, C=CH), 4.99 (ddd, ^2J = 17.2 Hz, ^4J = 1.3, 0.6 Hz, 1H, C=CH), 4.84 (dd, ^3J = 10.7 Hz, ^4J = 1.1 Hz, 1H, C=CH), 1.71 (d, ^4J = 1.3 Hz, 3H, CH_3), 0.21 (s, 9H, Si(CH_3)$_3$) ppm.

– Analytical data is in accordance with previously published literature.[117]

(E)-tert-Butyldimethyl((2-methylbuta-1,3-dien-1-yl)oxy)silane (74f)

OTBDMS   Under argon atomsphere a solution of potassium *tert*-butoxide (4.22 g, 37.6 mmol, 1.20 equiv.) in THF (35 mL) was added to a solution of the silylenolether **74e** (4.90 g, 31.6 mmol, 1.00 equiv.) in THF (50 mL) at −78 °C and the mixture was stirred at this temperature for 1 h. Then a solution of *tert*-butyldimethylchlorosilane (6.52 mL, 5.67 g, 37.6 mmol, 1.20 equiv.) in THF (15 mL) was added and the reaction mixture was stirred for 15 min at −78 °C. The reaction was quenched by the addition of saturated aq. NaHCO$_3$ solution (35 mL) and the aqueous phase was extracted with Et$_2$O (2 × 100 mL). The combined organic phases were dried over MgSO$_4$ and the solvent was removed under reduced pressure. After distillation (bp 80 °C, 20 mbar), the product **74f** was obtained as a colorless oil (3.86 g, 19.5 mmol, 62%).

– ^{1}H NMR (300 MHz, CDCl$_3$): δ = 6.40 (s, 1H, SiOCH), 6.28 (dd, J = 16.1, 12.2 Hz, 1H, C=CH), 4.97 (d, J = 16.1 Hz, 1H, C=CHH), 4.82 (d, J = 12.1 Hz, 1H, C=CHH), 1.70 (s, 3H, CH_3), 0.93 (s, 9H, SiC(CH_3)$_3$), 0.15 (s, 6H, Si(CH_3)$_2$) ppm. – Analytical data is in accordance with previously published literature.[196]

(Buta-1,3-dien-2-yloxy)(tert-butyl)diphenylsilane (74h)

OTBDPS   To a solution of sodium bis(trimethylsilyl)amide (1.5 – 2.0 M in THF, 19.5 mL, 7.15 g, 29.3 – 39.0 mmol, 0.98 – 1.30 equiv.) in dry THF (60 mL) at −78 °C was dropwise added but-3-en-2-one (**77**) (2.53 mL, 2.10 g, 30.0 mmol, 1.00 equiv.). The resulting pale yellow solution was stirred at −78 °C for 20 min after which *tert*-butyldiphenylsilylchloride (**78**) (8.58 mL, 9.07 g, 33.0 mmol, 1.10 equiv.) was added dropwise with a syringe. The reaction mixture was stirred at −78 °C for 30 min and then warmed to room temperature. After completion of the reaction (TLC control) the mixture was quenched by the dropwise addition of saturated aq. NaHCO$_3$ solution (45 mL). The product was extracted with Et$_2$O (2 × 150 mL), the combined organic phases were washed with brine (150 mL) and dried over MgSO$_4$. After filtration the solvent was removed under reduced pressure and the crude product was purified *via* column chromatography on silica gel (*c*Hex/EtOAc/Et$_3$N = 200:1:1). The product **74h** was obtained as a colorless oil (3.05 g, 9.90 mmol, 33%).

– ^{1}H NMR (500 MHz, CDCl$_3$): δ = 7.86–7.72 (m, 4H, CH$_{Ar}$), 7.52–7.38 (m, 6H, CH$_{Ar}$), 6.25 (dd, ^2J = 16.9 Hz, ^3J = 10.5 Hz, 1H, CHH), 5.85 (dd, ^2J = 17.0 Hz, ^4J = 1.9 Hz, 1H, CHH), 5.22 (dt, ^3J = 10.4 Hz, ^4J = 1.7 Hz, 1H, CHH), 4.20 (d, ^4J = 1.3 Hz, 1H, CHH), 4.03–3.93 (m, 1H, CH), 1.11 (s, 9H, Si(CH$_3$)$_3$) ppm. – ^{13}C NMR (126 MHz, CDCl$_3$): δ = 154.6 (C$_q$, 1 × C$_{qAr}$), 135.6 (+, 4 × CH$_{Ar}$), 134.9 (C$_q$, 1 × C$_{qAr}$), 132.8 (C$_q$, 1 × C$_q$), 130.0 (+, 2 × CH$_{Ar}$), 127.8 (+, 4 × CH$_{Ar}$), 114.5 (C$_q$, 1 × C$_q$), 97.4 (+, 1 × CH), 27.1 (–, 2 × CH$_2$), 26.7 (+, 3 × CH$_3$) ppm.– Analytical data is in accordance with previously published literature.[120]

(Buta-1,3-dien-2-yloxy)triisopropylsilane (74g)

Diisopropylamine (2.32 mL, 1.67 g, 16.5 mmol, 1.16 equiv.) was dissolved in THF (35 mL) and the resulting solution was cooled to –78 °C. Then n-BuLi (2.5 M in n-hexane, 6.60 mL, 1.06 g, 16.5 mmol, 1.16 equiv.) was added. The reaction mixture was warmed to 0 °C, stirred for 30 min and again cooled to –78 °C. In a separate reaction vessel methyl vinyl ketone 77 (1.75 mL, 1.45 g, 20.7 mmol, 1.00 equiv.) was dissolved in THF (4.5 mL) and the resulting mixture was added dropwise to the LDA-solution. After 1 h, TIPSOTf (4.45 mL, 5.06 g, 16.5 mmol, 1.16 equiv.) was slowly added and the mixture was allowed to warm to room temperature overnight (16 h). The reaction was quenched by the addition of ice water (15 mL) and the aqueous phase was extracted with cHex/Et$_2$O (1:1) (3 × 30 mL). The combined organic phases were dried over MgSO$_4$ and the solvent was removed under reduced pressure to yield the crude product as a yellow oil. After distillation (bp 70 °C, 0.1 mbar) the product 74g was obtained as a colorless oil (1.25 g, 5.53 mmol, 27%).

– ^{1}H NMR (300 MHz, CDCl$_3$): δ = 6.24–6.14 (m, 1H, C=CH), 5.62–5.55 (m, 1H, CHH), 5.11–5.05 (m, 1H, CHH), 4.31 (d, ^2J = 17.6 Hz, 2H, C=CH$_2$), 1.38–1.18 (m, 3H, SiCH), 1.11 (d, ^3J = 6.8 Hz, 18H, SiC(CH$_3$)$_6$) ppm. – Analytical data is in accordance with previously published literature.[119]

2-Methylbut-1-en-3-yne (148)

A three-necked flask was equipped with 2-methylbut-3-yn-2-ol 147 (58.7 mL, 50.5 g, 600 mmol, 1.00 equiv.). Acetic anhydride (70.9 mL, 76.6 g, 750 mmol, 1.25 equiv.) and sulfuric acid (1.60 mL, 2.94 g, 30.0 mmol, 0.05 equiv.) were added dropwise over 2 hours starting at 50 °C. After the addition of 15 mL, the temperature was increased to 70 °C and the product 148 was distilled from the mixture into a receiver flask cooled to –78 °C (bp 33 °C, 1 bar). After the addition was completed, the temperature was increased to 80 °C. The isolated product was washed with ice water to remove residues of acetic acid and alcohol, then dried over Na_2SO_4 to yield 2-methylbut-1-en-3-yne **148** (9.00 mL, 94.6 mmol, 16%) as a colorless liquid.

– ^1H NMR (300 MHz, CDCl$_3$): δ = 5.39 (s, 1H, C*H*), 5.30 (s, 1H, C*H*H), 2.87 (s, 1H, C*H*H), 1.90 (s, 3H, C*H*$_3$) ppm. – Analytical data is in accordance with previously published literature.[179]

2-Chloro-3-methylbuta-1,3-diene (149)

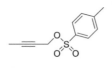

Copper(I) chloride (4.16 g, 42.1 mmol, 40 mol%), ammonium chloride (1.69 g, 31.5 mmol, 30 mol%) and conc. hydrochloric acid (26 mL) were stirred for 5 min at 0 °C. Subsequently, 2-methylbut-1-en-3-yne **148** (9.00 mL, 4.03 g, 61.0 mmol, 1.00 equiv.) was added dropwise. The dark-greenish reaction mixture was brought to room temperature and stirred for 1 h. Then, the reaction mixture was cooled to 0 °C to prevent the formation of by-products and stirred for one more hour. The organic layer of the reaction mixture was isolated, and the aqueous phase extracted with pentane (3 × 15 mL). The organic phases were combined and washed with water (3 × 15 mL) and brine (15 mL) to remove residues of acid and copper. After drying over Na_2SO_4, butylated hydroxytoluene (BHT) (~8.7 mg) was added and the solvent was removed under reduced pressure. The dark residue was distilled using a Vigreux column (receiver flasks were filled with 2 mL *m*-xylene and 8.75 mg BHT). The distillation flask was heated up to 80 °C within 20 min (bp 44 °C, 175 mbar). The concentration of the obtained product **149**/*m*-xylene solution (195 mg) was determined via ^1H NMR (4.52 µmol product per 1 mg solution, 881 µmol, 1.4%).

– ^1H NMR (300 MHz, CDCl$_3$): δ = 5.59 (s, 1H, C*H*H), 5.43 (d, J = 7.2 Hz, 2H, C*H*$_2$), 5.18 (s, 1H, C*H*H), 1.99 (s, 3H, C*H*$_3$) ppm. – Analytical data is in accordance with previously published literature.[179]

But-2-yn-1-yl 4-methylbenzenesulfonate (151)

Tosyl chloride (4.58 g, 24.0 mmol, 1.20 equiv.) followed by powdered KOH (13.5 g, 240 mmol, 12.0 equiv.) were added in portions to a stirred solution of but-2-yn-1-ol (**150**) (1.50 mL, 1.40 g, 20.0 mmol, 1.00 equiv.) in Et$_2$O (40 mL) at 0 °C. The reaction mixture was warmed to room temperature gradually and stirred overnight. Then water was added (50 mL), the organic layer was separated, and the aqueous layer was extracted with EtOAc (3 × 40 mL). The combined organic layers were washed with brine (60 mL), dried over anhydrous Na_2SO_4 and the solvent was removed under reduced pressure. The product **151** was obtained after column chromatography on silica gel (*n*Hex/EtOAc = 9:1) as an off-white solid (2.10 g, 9.36 mmol, 47%).

– R_f (*n*Hex/EtOAc = 9:1) = 0.25. – ^{1}H NMR (500 MHz, CDCl$_3$): δ = 7.81 (d, ^3J = 8.3 Hz, 2H, C*H*$_{Ar}$), 7.34 (d, ^3J = 8.3 Hz, 2H, C*H*$_{Ar}$), 4.66 (q, ^5J = 2.4 Hz, 2H, C*H*$_2$), 2.45 (s, 3H, C*H*$_3$), 1.73 (t, ^5J = 2.4 Hz, 3H, C*H*$_3$) ppm. – ^{13}C NMR (126 MHz, CDCl$_3$): δ = 145.0 (C$_q$, 1 × C$_{qAr}$), 133.5 (C$_q$, 1 × C$_{qAr}$), 129.9 (+, 2 × C*H*$_{Ar}$), 128.3 (+, 2 × C*H*$_{Ar}$), 86.3 (C$_q$, 1 × C$_q$), 71.2 (C$_q$, 1 × C$_q$), 58.8 (–, 1 × C*H*$_2$), 21.8 (+, 1 × C*H*$_3$), 3.7 (+, 1 × C*H*$_3$) ppm. – Analytical data is in accordance with previously published literature.[180]

Trimethyl(2-methylbuta-2,3-dien-1-yl)silane (153)

Me$_3$Si

To an ice-cooled suspension of CuCN (1.20 g, 13.4 mmol, 3.00 equiv.) and LiCl (1.11 g, 26.2 mmol, 5.88 equiv.) in Et$_2$O (18 mL) was added (trimethylsilyl)methylmagnesium chloride (1 M in Et$_2$O, 13.4 mL, 1.97 g, 13.4 mmol, 3.00 equiv.). After the mixture was stirred at 0 °C for 30 min, the reaction mixture was cooled to −78 °C and a solution of but-2-yn-1-yl-4-methylbenzolsulfonate (**151**) (1.00 g, 4.46 mmol, 1.00 equiv.) in Et$_2$O (4 mL) was added. After 1 h at −78 °C, the reaction was quenched by the addition of saturated aq. NH$_4$Cl solution (50 mL). The reaction mixture was diluted with Et$_2$O (20 mL), washed with water (3 × 40 mL), dried over MgSO$_4$ and concentrated *in vacuo*. The product **153** was obtained as yellow liquid (617 mg, 4.40 mmol, 99%).

– ^{1}H NMR (300 MHz, CDCl$_3$): δ = 4.59–4.48 (m, 2H, C=C*H*$_2$), 1.68 (t, J = 3.2 Hz, 3H, C*H*$_3$), 1.31 (t, J = 2.4 Hz, 2H, C*H*$_2$), 0.04 (s, 9H, Si(C*H*$_3$)$_3$) ppm. – Analytical data is in accordance with previously published literature.[181]

2-Iodo-3-methylbuta-1,3-diene (154)

To a stirred mixture of trimethyl(2-methylbuta-2,3-dien-1-yl)silane (**153**) (617 mg, 4.40 mmol, 1.00 equiv.) and NaHCO$_3$ (540 mg, 6.42 mmol, 1.46 equiv.) in CH$_2$Cl$_2$ (9 mL) at −78 °C under argon atmosphere was added a solution of iodine (1.63 g, 6.42 mmol, 1.46 equiv.) in CH$_2$Cl$_2$ (27 mL). After 15 min, TBAF (1 M in THF, 6.42 mL, 6.42 mmol, 1.46 equiv.) was added and the mixture was stirred at −78 °C for 1 h. The reaction mixture was diluted with CH$_2$Cl$_2$ (30 mL), washed with 5% aq. Na$_2$S$_2$O$_3$ solution (50 mL) and water (50 mL), dried over MgSO$_4$, and concentrated *in vacuo*. The residue was purified *via* column chromatography on silica gel (*n*-pentane) to give the pure diene **154** (0.12 mL, 195 mg, 1.01 mmol, 23%) as a light yellow liquid.

– ^{1}H NMR (300 MHz, CDCl$_3$): δ = 6.32 (d, ^4J = 1.7 Hz, 1H, C=CHH), 6.00 (s, 1H, C=CHH), 5.40 (s, 1H, C=CHH), 5.26 (d, ^4J = 1.4 Hz, 1H, C=CHH), 2.01 (s, 3H, CH_3) ppm. – Analytical data is in accordance with previously published literature.[181]

7.2.5   Synthesis of Anthraquinone Derivatives

(4a*R*,9a*R*)-4a-(2-Iodobenzoyl)-1,4,4a,9a-tetrahydroanthracene-9,10-dione (87aa)

 A suspension of 3-sulfolene (**75**) (5.00 g, 42.0 mmol, 16.4 equiv.) in *o*-xylene (15 mL) was heated to 125 °C for 0.5 h. The thereby developed gaseous 1,3-butadiene (**74a**) was led into a reaction vessel containing a solution of 2-(2-iodobenzoyl)naphthalene-1,4-dione   (**73a**)   (1.00 g,   3.00 mmol, 1.00 equiv.) in CH$_2$Cl$_2$ (5 mL) at –78 °C. After completion of the evolution of 1,3-butadiene gas (**74a**), the mixture of dienophile **73a** and diene **74a** in CH$_2$Cl$_2$ was slowly warmed to room temperature and stirred at this temperature for 2 h. The solvent was removed under reduced pressure. After column chromatography on silica gel (*c*Hex/EtOAc = 4:1) the product **87aa** was obtained as a colorless solid (925 mg, 2.09 mmol, 81%).

– R_f (*c*Hex/EtOAc = 4:1) = 0.45. – ^{1}H NMR (400 MHz, CDCl$_3$): δ = 8.17 (dd, ^3J = 7.6 Hz, ^4J = 1.5 Hz, 1H, C*H*$_{Ar}$), 8.03 (dd, ^3J = 7.8 Hz, ^4J = 1.4 Hz, 1H, C*H*$_{Ar}$), 7.87 (dd, ^3J = 7.8 Hz, ^4J = 1.1 Hz, 1H, C*H*$_{Ar}$), 7.76 (dtd, ^3J = 23.2, 7.5 Hz, ^4J = 1.5 Hz, 2H, C*H*$_{Ar}$), 7.31 (td, ^3J = 7.6 Hz, ^4J = 1.1 Hz, 1H, C*H*$_{Ar}$), 7.16–7.01 (m, 2H, C*H*$_{Ar}$), 5.70 (s, 2H, CH$_2$-C*H*=C*H*-CH$_2$), 3.68 (dd, ^3J = 9.9, 6.3 Hz, 1H, C=CH-CH$_2$-C*H*), 3.07–2.93 (m, 1H, C-C*H*H-CH), 2.50–2.45 (m, 1H, C-CH*H*-CH), 2.44–2.38 (m, 1H, CH-C*H*H-CH), 2.34–2.20 (m, 1H, CH-CH*H*-CH) ppm. – ^{13}C NMR (101 MHz, CDCl$_3$): δ = 200.6 (C$_q$, 1 × *C*=O), 196.1 (C$_q$, 1 × *C*=O), 194.0 (C$_q$, 1 × *C*=O), 142.6 (C$_q$, 1 × *C*$_{qAr}$), 141.0 (+, 1 × *C*H$_{Ar}$), 135.3 (+, 1 × *C*H$_{Ar}$), 134.3 (+, 1 × *C*H$_{Ar}$), 133.1 (C$_q$, 1 × *C*$_{qAr}$), 131.7 (+, 1 × *C*H$_{Ar}$), 127.5 (+, 1 × *C*H$_{Ar}$), 127.4 (+, 1 × *C*H$_{Ar}$), 127.2 (+, 1 × *C*H$_{Ar}$), 126.8 (+, 1 × *C*H$_{Ar}$), 124.4 (+, 1 × *C*=CH), 123.9 (+, 1 × *C*=CH), 93.1 (C$_q$, 1 × *C*$_{qAr}$), 68.1 (C$_q$, 1 × *C*$_{qAr}$), 50.0 (+, 1 × *C*H), 28.4 (–, 1 × *C*H$_2$), 26.4 (–, 1 × *C*H$_2$), 14.4 (C$_q$, 1 × *C*$_q$) ppm. – IR (ATR): $\tilde{v}$ = 2922 (vw), 1690 (w), 1672 (w), 1588 (w), 1424 (vw), 1290 (w), 1248 (w), 1219 (w), 1158 (vw), 1059 (w), 1016 (w), 984 (w), 941 (w), 920 (vw), 891 (vw), 807 (vw), 783 (vw), 761 (w), 746 (w), 728 (w), 687 (w), 660 (w), 635 (vw), 598 (w), 528 (vw), 442 (vw), 407 (w) cm^{-1}. – MS (EI, 70 eV), *m*/*z* (%): 443 (2) [M+H]$^+$, 442 (7) [M]$^+$, 231 (100) [C$_7$H$_4$IO]$^+$, 211 (26) [C$_{14}$H$_{11}$O$_2$]$^+$, 203 (19) [C$_6$H$_4$I]$^+$. – HRMS (EI, C$_{21}$H$_{15}$IO$_3$): calc. 442.0060; found 442.0062. – X-Ray: Crystallographic information on the product can be found in chapter 7.3.1.

(4a*R*,9a*R*)-4a-(2-Bromobenzoyl)-1,4,4a,9a-tetrahydroanthracene-9,10-dione (87ba)

A suspension of 3-sulfolene (**75**) (2.00 g, 16.9 mmol, 11.6 equiv.) in *o*-xylene (15 mL) was heated to 125 °C for 0.5 h. The thereby developed gaseous 1,3-butadiene (**74a**) was then led into a reaction vessel containing a solution of 2-(2-bromobenzoyl)naphthalene-1,4-dione (**73b**) (500 mg, 1.47 mmol, 1.00 equiv.) in CH$_2$Cl$_2$ (3 mL) at –78 °C. After completion of the evolution of 1,3-butadiene gas (**74a**), the mixture of dienophile **73b** and diene **74a** in CH$_2$Cl$_2$ was slowly warmed to room temperature and stirred at this temperature for 2 h. The solvent was removed under reduced pressure. After column chromatography on silica gel (*c*Hex/EtOAc = 8:1) the product **87ba** was obtained as a colorless solid (480 mg, 1.21 mmol, 82%).

– R_f (*c*Hex/EtOAc = 8:1) = 0.21. – ^{1}H NMR (400 MHz, CDCl$_3$): δ = 8.16 (dd, ^3J = 7.6 Hz, ^4J = 1.5 Hz, 1H, C*H*$_{Ar}$), 8.03 (dd, ^3J = 7.6 Hz, ^4J = 1.4 Hz, 1H, C*H*$_{Ar}$), 7.79 (td, ^3J = 7.6 Hz, ^4J = 1.5 Hz, 1H, C*H*$_{Ar}$), 7.74 (td, ^3J = 7.5 Hz, ^4J = 1.5 Hz, 1H, C*H*$_{Ar}$), 7.60–7.55 (m, 1H, C*H*$_{Ar}$), 7.31–7.22 (m, 2H, C*H*$_{Ar}$), 7.15–7.09 (m, 1H, C*H*$_{Ar}$), 5.72–5.64 (m, 2H, CH$_2$-C*H*=C*H*-CH$_2$), 3.68 (dd, ^3J = 9.6, 6.3 Hz, 1H, C=CH-CH$_2$-C*H*), 3.03–2.93 (m, 1H, C-C*H*H-CH), 2.51–2.40 (m, 2H, C*H*H, CH*H*), 2.33–2.24 (m, 1H, CH*H*) ppm. – ^{13}C NMR (101 MHz, CDCl$_3$): δ = 199.9 (C$_q$, 1 × *C*=O), 196.2 (C$_q$, 1 × *C*=O), 193.9 (C$_q$, 1 × *C*=O), 139.2 (C$_q$, 1 × *C*$_{qAr}$), 135.3 (+, 1 × *C*H$_{Ar}$), 134.3 (+, 1 × *C*H$_{Ar}$), 134.2 (C$_q$, 1 × *C*$_{qAr}$), 134.1 (+, 1 × *C*H$_{Ar}$), 133.1 (C$_q$, 1 × *C*$_{qAr}$), 131.6 (+, 1 × *C*H$_{Ar}$), 127.4 (+, 1 × *C*H$_{Ar}$), 127.2 (+, 2 × *C*H$_{Ar}$), 126.8 (+, 1 × *C*H$_{Ar}$), 124.5 (+, 1 × *C*=CH), 123.8 (+, 1 × *C*=CH), 119.8 (C$_q$, 1 × *C*$_{qAr}$), 68.3 (C$_q$, 1 × *C*$_q$), 49.7 (+, 1 × *C*H), 28.2 (–, 1 × *C*H$_2$), 26.2 (–, 1 × *C*H$_2$) ppm. – IR (ATR): $\tilde{v}$ = 2830 (vw), 2926 (vw), 2885 (vw), 1684 (w), 1589 (w), 1423 (w), 1252 (w), 1218 (w), 1064 (w), 1025 (w), 988 (w), 942 (w), 917 (w), 890 (w), 842 (vw), 798 (vw), 761 (w), 733 (w), 683 (w), 638 (w), 592 (w), 560 (w), 532 (vw), 442 (vw), 424 (vw), 401 (vw) cm^{-1}. – MS (FAB, 3-NBA), *m/z* (%): 395/397 (18/17) [M+H]$^+$. – HRMS (FAB, C$_{21}$H$_{16}$^{79}BrO$_3$): calc. 395.0283; found 395.0285. – X-Ray: Crystallographic information on the product can be found in chapter 7.3.2.

Cycloaddition product of 2-(2-iodobenzoyl)naphthalene-1,4-dione (73a) and isoprene (74b) (87/88ab)

87ab   **88ab**

According to **GP3** the cycloaddition was performed with 2-(2-iodobenzoyl)naphthalene-1,4-dione (**73a**) (388 mg, 1.00 mmol, 1.00 equiv.) and isoprene (**74b**) (0.50 mL, 341 mg, 5.00 mmol, 5.00 equiv.) in dry CH_2Cl_2 (2 mL). After 3 h the crude product was purified *via* column chromatography on silica gel (*c*Hex/EtOAc = 8:1) to obtain an off-white solid (289 mg, 633 µmol, 63%). The products **87/88ab** were isolated as a non-separable mixture of regioisomers in a 7.1 : 1 ratio as estimated by ^{1}H NMR.

Both possible regioisomers are drawn, as the exact structure of the products could not be resolved by analysis of the NMR spectra. Below the analytics of the major product are given.

– R_f (*c*Hex/EtOAc = 6:1) = 0.38. – ^{1}H NMR (400 MHz, CDCl$_3$): δ = 8.15 (dd, ^3J = 7.6 Hz, ^4J = 1.5 Hz, 1H, CH_{Ar}), 8.02 (dd, ^3J = 7.6 Hz, ^4J = 1.5 Hz, 1H, CH_{Ar}), 7.86 (dd, ^3J = 8.0 Hz, ^4J = 1.6 Hz, 1H, CH_{Ar}), 7.75 (dtd, J = 21.0 Hz, ^3J = 7.4 Hz, ^4J = 1.5 Hz, 2H, CH_{Ar}), 7.36–7.25 (m, 1H, CH_{Ar}), 7.16 (dd, ^3J = 7.7 Hz, ^4J = 1.7 Hz, 1H, CH_{Ar}), 7.07 (td, ^3J = 7.7 Hz, ^4J = 1.7 Hz, 1H, CH_{Ar}), 5.34 (ddd, ^3J = 4.8, 3.3 Hz, ^4J = 1.6 Hz, 1H, C=CH), 3.72 (dd, ^3J = 8.4, 7.1 Hz, 1H, CH), 2.90 (ddq, ^2J = 17.5 , ^3J = 3.9 Hz, ^4J = 1.9 Hz, 1H, CHH), 2.46 (dq, ^2J = 17.5 Hz, ^3J = 2.3 Hz, 1H, CHH), 2.25 (d, ^3J = 7.4 Hz, 2H, CH_2), 1.65 (s, 3H, CH_3) ppm. – ^{13}C NMR (101 MHz, CDCl$_3$): δ = 200.8 (C$_q$, 1 × C=O), 196.1 (C$_q$, 1 × C=O), 194.5 (C$_q$, 1 × C=O), 142.5 (C$_q$, 1 × C$_{qAr}$), 140.9 (+, 1 × CH_{Ar}), 135.2 (+, 1 × CH_{Ar}), 134.3 (+, 1 × CH_{Ar}), 133.0 (C$_q$, 1 × C$_{qAr}$), 132.0 (C$_q$, 1 × C$_{qAr}$), 131.6 (+, 1 × CH_{Ar}), 127.5 (+, 1 × CH_{Ar}), 127.4 (+, 1 × CH_{Ar}), 127.1 (+, 1 × CH_{Ar}), 126.9 (+, 1 × CH_{Ar}), 117.9 (+, 1 × C=CH), 93.2 (C$_q$, 1 × C$_{qAr}$), 67.6 (C$_q$, 1 × C$_q$), 50.3 (+, 1 × CH), 30.4 (–, 1 × CH_2), 28.9 (–, 1 × CH_2), 23.2 (+, 1 × CH_3) ppm. Signal missing (C$_q$, 1 × C$_q$). – IR (ATR): $\tilde{v}$ = 2910 (vw), 1676 (m), 1592 (w), 1423 (w), 1317 (vw), 1273 (w), 1253 (w), 1218 (w), 1161 (vw), 1061 (vw), 1010 (w), 967 (vw), 936 (w), 897 (vw), 794 (w), 761 (w), 738 (w), 677 (w), 637 (vw), 588 (w), 519 (vw), 430 (vw) cm^{-1}. – MS (FAB, 3-NBA), *m/z* (%): 457 (23) [M+H]$^+$, 231 (100), 225 (21), 154 (66). – HRMS (FAB, C$_{22}$H$_{18}$IO$_3$): calc. 457.0295; found 457.0294.

Cycloaddition product of 2-(2-bromobenzoyl)naphthalene-1,4-dione (73b) and isoprene (74b) (87/88bb)

87bb   **88bb**

According to **GP3** the cycloaddition ran with 2-(2-bromobenzoyl)naphthalene-1,4-dione (**73b**) (190 mg, 560 µmol, 1.00 equiv.) and isoprene (**74b**) (0.28 mL, 190 mg, 2.79 mmol, 5.00 equiv.) in dry CH_2Cl_2 (2 mL). After 3 h the crude product was purified *via* column chromatography on silica gel (cHex/EtOAc = 9:1 → 7:1) to obtain an off-white solid (121 mg, 296 µmol, 53%). The products **87/88bb** were isolated as a non-separable mixture of regioisomers in a 7.7 : 1 ratio as estimated by ^{1}H NMR.

Both possible regioisomers are drawn, as the exact structure of the products could not be resolved by analysis of the NMR spectra. Below the analytics of the major product are given.

– R_f (cHex/EtOAc = 9:1) = 0.32. – ^{1}H NMR (400 MHz, CDCl$_3$): δ = 8.14 (dd, ^3J = 7.6 Hz, ^4J = 1.5 Hz, 1H, CH_{Ar}), 8.03 (dd, ^3J = 7.6 Hz, ^4J = 1.5 Hz, 1H, CH_{Ar}), 7.83–7.70 (m, 2H, CH_{Ar}), 7.60–7.52 (m, 1H, CH_{Ar}), 7.34–7.13 (m, 3H, CH_{Ar}), 5.34–5.30 (m, 1H, C=CH), 3.74 (t, ^3J = 7.6 Hz, 1H, CH), 2.93–2.85 (m, 1H, CHH), 2.51–2.44 (m, 1H, CHH), 2.30–2.25 (m, 2H, CH_2), 1.64 (s, 3H, CH_3) ppm. – ^{13}C NMR (101 MHz, CDCl$_3$): δ = 200.1 (C$_q$, 1 × C=O), 196.2 (C$_q$, 1 × C=O), 194.5 (C$_q$, 1 × C=O), 139.1 (C$_q$, 1 × C_{qAr}), 135.3 (+, 1 × CH_{Ar}), 134.3 (+, 1 × CH_{Ar}), 134.2 (C$_q$, 1 × C_{qAr}), 134.0 (+, 1 × CH_{Ar}), 133.1 (C$_q$, 1 × C_{qAr}), 132.2 (C$_q$, 1 × C_{qAr}), 131.5 (+, 1 × CH_{Ar}), 127.4 (+, 1 × CH_{Ar}), 127.2 (+, 1 × CH_{Ar}), 127.1 (+, 1 × CH_{Ar}), 126.8 (+, 1 × CH_{Ar}), 120.0 (C$_q$, 1 × C_q), 117.8 (+, 1 × C=CH), 67.8 (C$_q$, 1 × C_q), 50.0 (+, 1 × CH), 30.2 (–, 1 × CH$_2$), 28.7 (–, 1 × CH$_2$), 23.14 (+, 1 × CH$_3$) ppm. – IR (ATR): $\tilde{v}$ = 2911 (w), 1679 (s), 1592 (m), 1427 (m), 1373 (w), 1274 (m), 1254 (s), 1218 (s), 1161 (w), 1052 (w), 1017 (m), 968 (w), 936 (m), 897 (w), 794 (m), 761 (m), 738 (s), 683 (m), 639 (w), 589 (m), 549 (w), 459 (w), 431 (w) cm^{-1}. – MS (FAB, 3-NBA), m/z (%): 409/411 (16/14) [M+H]$^+$, 408/410 (2/6) [M]$^+$, 225 (43), 183/185 (100/98). – HRMS (FAB, C$_{21}$H$_{16}$^{79}BrO$_3$N): calc. 409.0314; found 409.0314.

(4a*R*,9a*R*)-4a-(2-Iodobenzoyl)-2,3-dimethyl-1,4,4a,9a-tetrahydroanthracene-9,10-dione (87ac)

According to **GP3** the cycloaddition was performed with 2-(2-iodobenzoyl)naphthalene-1,4-dione (**73a**) (388 mg, 1.00 mmol, 1.00 equiv.) and 2,3-dimethylbuta-1,3-diene (**74c**) (0.34 mL, 246 mg, 3.00 mmol, 3.00 equiv.) in dry CH_2Cl_2 (5 mL). After 3 h the crude product was purified *via* column chromatography on silica gel (*c*Hex/EtOAc = 9:1) to obtain product **87ac** as a yellow solid (328 mg, 697 μmol, 70%).

– R_f (*c*Hex/EtOAc = 9:1) = 0.31. – ^{1}H NMR (400 MHz, CDCl$_3$): δ = 8.16 (dd, ^3J = 7.6 Hz, ^4J = 1.5 Hz, 1H, C*H*$_{Ar}$), 8.02 (dd, ^3J = 7.5 Hz, ^4J = 1.6 Hz, 1H, C*H*$_{Ar}$), 7.88 (dd, ^3J = 8.1 Hz, ^4J = 1.2 Hz, 1H, C*H*$_{Ar}$), 7.78 (td, ^3J = 7.5 Hz, ^4J = 1.5 Hz, 1H, C*H*$_{Ar}$), 7.73 (td, ^3J = 7.5 Hz, ^4J = 1.5 Hz, 1H, C*H*$_{Ar}$), 7.33 (td, ^3J = 7.6 Hz, ^4J = 1.1 Hz, 1H, C*H*$_{Ar}$), 7.22 (dd, ^3J = 7.8 Hz, ^4J = 1.7 Hz, 1H, C*H*$_{Ar}$), 7.08 (td, ^3J = 7.7 Hz, ^4J = 1.7 Hz, 1H, C*H*$_{Ar}$), 3.70 (t, ^3J = 7.7 Hz, 1H, C*H*), 2.81 (d, ^2J = 17.2 Hz, 1H, C*H*H), 2.38 (d, ^2J = 17.2 Hz, 1H, CH*H*), 2.29 (d, ^3J = 7.7 Hz, 2H, C*H*$_2$), 1.60 (s, 3H, C*H*$_3$), 1.55 (s, 3H, C*H*$_3$) ppm. – ^{13}C NMR (101 MHz, CDCl$_3$): δ = 200.9 (C$_q$, 1 × *C*=O), 196.4 (C$_q$, 1 × *C*=O), 194.8 (C$_q$, 1 × *C*=O), 142.6 (C$_q$, 1 × *C*$_{qAr}$), 141.0 (+, 1 × *C*H$_{Ar}$), 135.2 (+, 1 × *C*H$_{Ar}$), 134.4 (C$_q$, 1 × *C*$_{qAr}$), 134.3 (+, 1 × *C*H$_{Ar}$), 133.2 (C$_q$, 1 × *C*$_{qAr}$), 131.6 (+, 1 × *C*H$_{Ar}$), 127.4 (+, 1 × *C*H$_{Ar}$), 127.4 (+, 1 × *C*H$_{Ar}$), 127.2 (+, 1 × *C*H$_{Ar}$), 127.1 (+, 1 × *C*H$_{Ar}$), 123.9 (C$_q$, 1 × *C*$_q$), 123.1 (C$_q$, 1 × *C*$_q$), 93.2 (C$_q$, 1 × *C*$_{qAr}$), 68.5 (C$_q$, 1 × *C*$_q$), 50.5 (+, 1 × *C*H), 34.6 (–, 1 × *C*H$_2$), 31.9 (–, 1 × *C*H$_2$), 19.1 (+, 1 × *C*H$_3$), 18.9 (+, 1 × *C*H$_3$) ppm. – IR (ATR): $\tilde{v}$ = 3065 (vw), 2912 (vw), 1678 (w), 1592 (w), 1425 (w), 1252 (w), 1218 (w), 1055 (vw), 1005 (w), 933 (vw), 884 (vw), 836 (vw), 764 (w), 745 (w), 672 (vw), 638 (vw), 609 (vw), 554 (vw), 447 (vw), 386 (vw) cm^{-1}. – MS (FAB, 3-NBA), *m/z* (%): 470 (3) [M]$^+$, 471 (16) [M+H]$^+$, 307 (32), 231 (40), 154 (100). – HRMS (FAB, C$_{23}$H$_{20}$^{127}IO$_3$): calc. 471.0457; found 471.0456. – X-Ray: Crystallographic information on the product can be found in chapter 7.3.2.

(4a*R*,9a*S*)-4a-(2-Bromobenzoyl)-2,3-dimethyl-1,4,4a,9a-tetrahydroanthracene-9,10-dione (87bc)

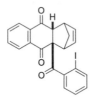

According to **GP3** the cycloaddition was performed with 2-(2-bromobenzoyl)naphthalene-1,4-dione (**73b**) (341 mg, 1.00 mmol, 1.00 equiv.) and 2,3-dimethylbuta-1,3-diene (**74c**) (0.34 mL, 246 mg, 3.00 mmol, 3.00 equiv.) in dry CH_2Cl_2 (5 mL). After 3 h the crude product was purified *via* column chromatography on silica gel (*c*Hex/EtOAc = 9:1) to obtain product **87bc** as a yellow solid (373 mg, 881 µmol, 88%).

– R_f (*c*Hex/EtOAc = 9:1) = 0.29. – ^{1}H NMR (500 MHz, CDCl$_3$): δ = 8.14 (dd, ^3J = 7.6 Hz, ^4J = 1.4 Hz, 1H, CH_{Ar}), 8.03 (dd, ^3J = 7.7 Hz, ^4J = 1.4 Hz, 1H, CH_{Ar}), 7.78 (td, ^3J = 7.5 Hz, ^4J = 1.5 Hz, 1H, CH_{Ar}), 7.74 (td, ^3J = 7.5 Hz, ^4J = 1.5 Hz, 1H, CH_{Ar}), 7.59–7.56 (m, 1H, CH_{Ar}), 7.31–7.23 (m, 3H, CH_{Ar}), 3.73 (dd, ^3J = 8.5, 6.7 Hz, 1H, CH), 2.79 (dt, ^2J = 16.8 Hz, ^3J = 2.2 Hz, 1H, CHH), 2.45–2.22 (d, 3H, CHH, CH_2), 1.60 (s, 3H, CH_3), 1.51 (s, 3H, CH_3) ppm. – ^{13}C NMR (126 MHz, CDCl$_3$): δ = 200.2 (C$_q$, 1 × C=O), 196.5 (C$_q$, 1 × C=O), 194.9 (C$_q$, 1 × C=O), 139.2 (C$_q$, 1 × C_{qAr}), 135.2 (+, 1 × CH$_{Ar}$), 134.4 (C$_q$, 1 × C_{qAr}), 134.3 (+, 1 × CH$_{Ar}$), 134.1 (+, 1 × CH$_{Ar}$), 133.2 (C$_q$, 1 × C_{qAr}), 131.4 (+, 1 × CH$_{Ar}$), 127.5 (+, 1 × CH$_{Ar}$), 127.4 (+, 1 × CH$_{Ar}$), 127.2 (+, 1 × CH$_{Ar}$), 126.7 (+, 1 × CH$_{Ar}$), 124.2 (C$_q$, 1 × C_{qAr}), 122.9 (C$_q$, 1 × C_q), 120.0 (C$_q$, 1 × C_q), 68.6 (C$_q$, 1 × C_q), 50.2 (+, 1 × CH), 34.4 (–, 1 × CH$_2$), 31.6 (–, 1 × CH$_2$), 19.1 (+, 1 × CH$_3$), 18.8 (+, 1 × CH$_3$) ppm. – IR (ATR): $\tilde{v}$ = 1694 (vs), 1679 (vs), 1589 (m), 1273 (m), 1256 (vs), 1247 (vs), 1214 (m), 1201 (s), 1057 (m), 1027 (m), 972 (m), 887 (m), 790 (m), 765 (s), 738 (vs), 711 (s), 688 (m), 674 (m), 635 (m), 586 (s), 551 (m), 439 (m), 422 (m), 408 (w), 394 (w) cm^{-1}. – MS (FAB, 3-NBA), *m/z* (%): 423/425 (24/21) [M+H]$^+$, 239 (100), 183/185 (92/89). – HRMS (FAB, C$_{23}$H$_{20}$O$_3$^{79}Br): calc. 423.0596; found 423.0597. – X-Ray: Crystallographic information on the product can be found in chapter 7.3.2.

(4a*R*,9a*R*)-4a-(2-Iodobenzoyl)-1,4,4a,9a-tetrahydro-1,4-methanoanthracene-9,10-dione (89)

According to **GP3** the cycloaddition was performed with 2-(2-iodobenzoyl)naphthalene-1,4-dione (**73a**) (77.6 mg, 200 µmol, 1.00 equiv.) and freshly distilled cyclopentadiene (**80**) (0.08 mL, 66.1 mg, 1.00 mmol, 5.00 equiv.) in dry CH_2Cl_2 (1 mL). After 2 h the crude product was purified *via* column chromatography on silica gel (*c*Hex/EtOAc = 9:1) to obtain **89** as an off-white solid (23.9 mg, 52.6 µmol, 26%).

– R_f (cHex/EtOAc = 9:1) = 0.28. – ^{1}H NMR (500 MHz, CDCl$_3$): δ = 8.05–7.97 (m, 2H, CH_{Ar}), 7.85 (dd, ^3J = 8.0 Hz, ^4J = 1.1 Hz, 1H, CH_{Ar}), 7.77–7.68 (m, 2H, CH_{Ar}), 7.21 (td, ^3J = 7.5 Hz, ^4J = 1.1 Hz, 1H, CH_{Ar}), 7.15 (dd, ^3J = 7.8 Hz, ^4J = 1.7 Hz, 1H, CH_{Ar}), 7.05 (ddd, ^3J = 7.9, 7.3 Hz, ^4J = 1.7 Hz, 1H, CH_{Ar}), 6.06 (t, ^3J = 1.8 Hz, 2H, 2 × C=CH), 4.11 (dq, ^3J = 3.1 Hz, ^4J = 1.6 Hz, 1H, CH), 3.84 (d, ^3J = 4.0 Hz, 1H, CH), 3.63 (dh, ^3J = 3.7 Hz, ^4J = 1.7 Hz, 1H, CH), 1.90 (dt, ^2J = 9.2 Hz, ^3J = 1.4 Hz, 1H, CHH), 1.59 (dt, ^2J = 9.2 Hz, ^3J = 1.8 Hz, 1H, CHH) ppm. – ^{13}C NMR (126 MHz, CDCl$_3$): δ = 200.7 (C$_q$, 1 × C=O), 196.1 (C$_q$, 1 × C=O), 194.1 (C$_q$, 1 × C=O), 144.1 (C$_q$, 1 × C_{qAr}), 140.6 (+, 1 × CH_{Ar}), 138.4 (+, 1 × C=CH), 136.7 (+, 1 × C=CH), 135.7 (C$_q$, 1 × C_{qAr}), 135.0 (C$_q$, 1 × C_{qAr}), 134.8 (+, 1 × CH_{Ar}), 134.6 (+, 1 × CH_{Ar}), 131.5 (+, 1 × CH_{Ar}), 127.7 (+, 1 × CH_{Ar}), 127.6 (+, 1 × CH_{Ar}), 127.2 (+, 1 × CH_{Ar}), 126.6 (+, 1 × CH_{Ar}), 92.6 (C$_q$, 1 × C_{qAr}), 72.0 (C$_q$, 1 × C_q), 55.1 (+, 1 × CH), 54.8 (+, 1 × CH), 49.9 (+, 1 × CH), 48.1 (–, 1 × CH$_2$) ppm. – Due to fast decomposition of the molecule no IR and MS data are available.

(1S,4aS,9aR)-9a-(2-Iodobenzoyl)-1-((trimethylsilyl)oxy)-1,4,4a,9a-tetrahydroanthracene-9,10-dione (87ad) and (1R,4aS,9aR)-9a-(2-iodobenzoyl)-1-((trimethylsilyl)oxy)-1,4,4a,9a-tetrahydroanthracene-9,10-dione (88ad)

87ad   88ad

According to **GP3** the cycloaddition was performed with 2-(2-iodobenzoyl)naphthalene-1,4-dione (**73a**) (97.0 mg, 250 µmol, 1.00 equiv.) and 1-(trimethylsiloxy)-1,3-butadiene (**74d**) (0.22 mL, 178 mg, 1.25 mmol, 5.00 equiv.) in dry CH$_2$Cl$_2$ (1 mL). After 3 h the crude product was purified *via* column chromatography on silica gel (cHex/EtOAc = 6:1) to obtain isomers **87ad** (26.5 mg, 50.0 µmol, 20%) and **88ad** (90.1 mg, 170 µmol, 68%) as yellow solids.

87ad: R_f (cHex/EtOAc = 9:1) = 0.29. – ^{1}H NMR (400 MHz, CDCl$_3$): δ = 8.08 (dd, ^3J = 7.8 Hz, ^4J = 1.2 Hz, 1H, CH_{Ar}), 7.94 (d, ^3J = 7.8 Hz, 1H, CH_{Ar}), 7.85 (dd, ^2J = 16.4, ^3J = 7.9 Hz, 2H, CH_{Ar}), 7.75 (t, ^3J = 7.5 Hz, 1H, CH_{Ar}), 7.66 (t, ^3J = 7.6 Hz, 1H, CH_{Ar}), 7.38 (t, ^3J = 7.7 Hz, 1H, CH_{Ar}), 7.08 (t, ^3J = 7.6 Hz, 1H, CH_{Ar}), 6.09–5.94 (m, 1H, C=CH), 5.82 (ddd, ^3J = 10.5, 5.1, ^4J = 2.5 Hz, 1H, C=CH), 5.38 (d, ^3J = 5.2 Hz, 1H, C=CH-CH), 3.88 (dd, ^3J = 11.9, 6.6 Hz, 1H, CH$_2$-CH), 2.53 (dt, ^2J = 18.8 Hz, ^3J = 5.7 Hz, 1H, CHH), 2.05 (dd, ^2J = 18.7 Hz, ^3J = 12.2 Hz, 1H, CHH), –0.03 (s, 9H, CH_3) ppm. – ^{13}C NMR (101 MHz, CDCl$_3$): δ = 197.0 (C$_q$, 1 × C=O), 195.5 (C$_q$, 1 × C=O), 193.1 (C$_q$, 1 × C=O), 141.8 (+, 1 × CH_{Ar}), 140.6 (C$_q$, 1 × C_{qAr}), 135.1 (+, 1 × CH_{Ar}), 134.3 (C$_q$, 1 × C_{qAr}), 133.9 (+, 1 × CH_{Ar}), 133.6 (C$_q$, 1 × C_{qAr}), 131.9 (+, 1 × CH_{Ar}), 128.0 (+, 1 × CH_{Ar}),

127.5 (+, 1 × C=CH), 127.2 (+, 1 × CH$_{Ar}$), 127.1 (+, 1 × CH$_{Ar}$), 127.0 (+, 1 × CH$_{Ar}$), 126.6 (+, 1 × C=CH), 94.9 (C$_q$, 1 × C_{qAr}), 73.3 (C$_q$, 1 × C_q), 65.9 (+, 1 × CH), 46.0 (+, 1 × CH), 27.8 (−, 1 × CH$_2$), 0.4 (+, 3 × SiCH$_3$) ppm. – IR (ATR): $\tilde{v}$ = 2952 (vw), 1700 (w), 1672 (w), 1593 (w), 1422 (w), 1322 (vw), 1248 (m), 1229 (w), 1076 (w), 1053 (w), 1013 (w), 934 (w), 887 (w), 838 (m), 750 (w), 735 (w), 693 (w), 637 (w), 593 (w), 482 (vw), 449 (vw) cm^{-1}. – MS (FAB, 3-NBA), m/z (%): 459 (39) [M–TMS+H]$^+$, 458 (71) [M–TMS]$^+$, 441 (8) [M–OTMS]$^+$, 415 (100). – HRMS (FAB, C$_{24}$H$_{23}$^{127}IO$_4$Si): calc. 530.0410; found. – X-Ray: Crystallographic information on the product can be found in chapter 7.3.2.

88ad: R_f (cHex/EtOAc = 9:1) = 0.31. – ^{1}H NMR (400 MHz, CDCl$_3$): δ = 8.19–8.12 (m, 1H, CH_{Ar}), 8.07–8.01 (m, 1H, CH_{Ar}), 7.97 (dd, ^3J = 7.9, ^4J = 1.2 Hz, 1H, CH_{Ar}), 7.82–7.64 (m, 3H, CH_{Ar}), 7.33 (td, ^3J = 7.6 Hz, ^4J = 1.2 Hz, 1H, CH_{Ar}), 7.09 (td, ^3J = 7.7 Hz, ^4J = 1.6 Hz, 1H, CH_{Ar}), 5.92 (ddd, ^3J = 10.2, 4.6 Hz, ^4J = 2.7 Hz, 1H, C=CH), 5.72 (ddt, ^3J = 10.0, 5.2 Hz, ^4J = 2.3 Hz, 1H, C=CH), 4.97 (d, ^3J = 5.2 Hz, 1H, C=CH-CH), 3.84 (d, ^3J = 7.0 Hz, 1H, CH$_2$-CH), 3.18 (ddt, ^2J = 19.1 Hz, ^3J = 4.2 Hz, ^4J = 1.5 Hz, 1H, CHH), 2.27–2.06 (m, 1H, CHH), −0.34 (s, 9H, SiCH_3) ppm. – ^{13}C NMR (101 MHz, CDCl$_3$): δ = 198.8 (C$_q$, 1 × C=O), 196.9 (C$_q$, 1 × C=O), 195.0 (C$_q$, 1 × C=O), 142.4 (+, 1 × CH$_{Ar}$), 141.3 (C$_q$, 1 × C_{qAr}), 138.0 (C$_q$, 1 × C_{qAr}), 135.5 (C$_q$, 1 × C_{qAr}), 134.8 (+, 1 × CH$_{Ar}$), 133.6 (+, 1 × CH$_{Ar}$), 132.2 (+, 1 × CH$_{Ar}$), 130.1 (+, 1 × C=CH), 128.1 (+, 1 × CH$_{Ar}$), 127.8 (+, 1 × CH$_{Ar}$), 127.3 (+, 1 × CH$_{Ar}$), 127.1 (+, 1 × C=CH), 126.1 (+, 1 × CH$_{Ar}$), 95.3 (C$_q$, 1 × C_{qAr}), 70.2 (C$_q$, 1 × C_q), 68.5 (+, 1 × CH), 46.4 (+, 1 × CH), 21.6 (−, 1 × CH$_2$), −0.2 (+, 3 × SiCH$_3$) ppm. – IR (ATR): $\tilde{v}$ = 2952 (vw), 1700 (w), 1672 (w), 1593 (w), 1422 (w), 1322 (vw), 1248 (m), 1229 (w), 1076 (w), 1053 (w), 1013 (w), 934 (w), 887 (w), 838 (m), 750 (w), 735 (w), 693 (w), 637 (w), 593 (w), 482 (vw), 449 (vw) cm^{-1}. – MS (FAB, 3-NBA), m/z (%): 531 (13) [M+H]$^+$, 389 (28), 231 (100). – HRMS (FAB, C$_{24}$H$_{24}$^{127}IO$_4$Si): calc. 531.0489; found 531.0488. – X-Ray: Crystallographic information on the product can be found in chapter 7.3.1.

(1*R*,4a*R*,9a*S*)-9a-(2-Bromobenzoyl)-1-((trimethylsilyl)oxy)-1,4,4a,9a-tetrahydroanthracene-9,10-dione (87bd) and (1*R*,4a*S*,9a*R*)-9a-(2-bromobenzoyl)-1-((trimethylsilyl)oxy)-1,4,4a,9a-tetrahydroanthracene-9,10-dione (88bd)

According to **GP3** the cycloaddition was performed with 2-(2-bromobenzoyl)naphthalene-1,4-dione **(73b)** (85.3 mg, 250 µmol, 1.00 equiv.) and 1-(trimethylsiloxy)-1,3-butadiene **(74d)** (0.22 mL, 178 mg, 1.25 mmol, 5.00 equiv.) in dry CH_2Cl_2 (1 mL). After 4 h the crude product was purified *via* column chromatography on silica gel (*c*Hex/EtOAc = 9:1) to obtain isomers **87bd** (24.1 mg, 50.0 µmol, 20%) and **88bd** (73.7 mg, 153 µmol, 61%) as yellow solids.

87bd: R_f (*c*Hex/EtOAc = 9:1) = 0.30. – ^{1}H NMR (400 MHz, CDCl$_3$): δ = 8.07 (dd, ^3J = 7.8 Hz, ^4J = 1.4 Hz, 1H, C*H*$_{Ar}$), 7.93 (dd, ^3J = 7.8 Hz, ^4J = 1.4 Hz, 1H, C*H*$_{Ar}$), 7.79 (dd, ^3J = 7.8 Hz, ^4J = 1.7 Hz, 1H, C*H*$_{Ar}$), 7.74 (td, ^3J = 7.6 Hz, ^4J = 1.4 Hz, 1H, C*H*$_{Ar}$), 7.66 (td, ^3J = 7.5 Hz, ^4J = 1.4 Hz, 1H, C*H*$_{Ar}$), 7.53 (dd, ^3J = 7.9 Hz, ^4J = 1.2 Hz, 1H, C*H*$_{Ar}$), 7.34 (td, ^3J = 7.6 Hz, ^4J = 1.3 Hz, 1H, C*H*$_{Ar}$), 7.29–7.22 (m, 1H, C*H*$_{Ar}$), 6.01 (ddd, ^3J = 9.8, 5.0 Hz, ^4J = 2.4 Hz, 1H, C=C*H*), 5.82 (ddd, ^3J = 10.0, 4.8 Hz, ^4J = 2.4 Hz, 1H, C=C*H*), 5.36 (dd, ^3J = 5.3 Hz, ^4J = 1.0 Hz, 1H, C=CH-C*H*), 3.87 (dd, ^3J = 11.8, 6.6 Hz, 1H, CH$_2$-C*H*), 2.58–2.50 (m, 1H, CH*H*), 2.13–1.97 (m, 1H, CH*H*), –0.03 (s, 9H, Si(C*H*$_3$)$_3$) ppm. – ^{13}C NMR (101 MHz, CDCl$_3$): δ = 196.9 (C$_q$, 1 × *C*=O), 196.1 (C$_q$, 1 × *C*=O), 193.6 (C$_q$, 1 × *C*=O), 138.5 (C$_q$, 1 × *C*$_{qAr}$), 135.6 (+, 1 × *C*H$_{Ar}$), 135.2 (+, 1 × *C*H$_{Ar}$), 134.8 (C$_q$, 1 × *C*$_{qAr}$), 134.4 (+, 1 × *C*H$_{Ar}$), 134.2 (C$_q$, 1 × *C*$_{qAr}$), 132.5 (+, 1 × *C*H$_{Ar}$), 128.8 (+, 1 × *C*=CH), 128.0 (+, 1 × *C*H$_{Ar}$), 127.7 (+, 1 × *C*H$_{Ar}$), 127.5 (+, 1 × *C*H$_{Ar}$), 127.4 (+, 1 × *C*=CH), 127.1 (+, 1 × *C*H$_{Ar}$), 122.1 (C$_q$, 1 × *C*$_{qAr}$), 74.2 (C$_q$, 1 × *C*$_q$), 66.3 (+, 1 × *C*H), 46.6 (+, 1 × *C*H), 28.4 (–, 1 × *C*H$_2$), 0.9 (+, 3 × Si*C*H$_3$) ppm. – IR (ATR): $\tilde{v}$ = 2922 (vw), 1698 (w), 1678 (w), 1593 (w), 1467 (vw), 1430 (vw), 1249 (w), 1230 (w), 1072 (w), 1052 (w), 1024 (w), 996 (vw), 945 (w), 890 (w), 841 (w), 763 (w), 683 (w), 641 (vw), 593 (vw), 515 (vw), 394 (vw) cm^{-1}. – MS (FAB, 3-NBA), *m/z* (%): 484/486 (38/32) [M+H]$^+$, 483/485 (100/88) [M]$^+$. – HRMS (FAB, C$_{24}$H$_{24}$^{79}BrO$_4$Si): calc. 483.0627; found 483.0625. – X-Ray: Crystallographic information on the product can be found in chapter 7.3.2.

88bd: R_f (*c*Hex/EtOAc = 9:1) = 0.36. – ^{1}H NMR (400 MHz, CDCl$_3$): δ = 8.17–8.12 (m, 1H, C*H*$_{Ar}$), 8.06–8.00 (m, 1H, C*H*$_{Ar}$), 7.81–7.69 (m, 3H, C*H*$_{Ar}$), 7.62 (dd, ^3J = 7.8 Hz, ^4J = 1.3 Hz, 1H, C*H*$_{Ar}$), 7.45–7.18 (m, 2H, C*H*$_{Ar}$), 5.90 (ddd, ^3J = 10.1, 4.5 Hz, ^4J = 2.8 Hz, 1H, C=C*H*), 5.67 (ddt, ^3J =

10.0, 5.0 Hz, ^4J = 2.3 Hz, 1H, C=CH), 4.94 (d, ^3J = 5.2 Hz, 1H, C=CH-CH), 3.89 (d, ^3J = 7.1 Hz, 1H, CH$_2$-CH), 3.22–3.16 (m, 1H, CHH), 2.30–2.06 (m, 1H, CHH), –0.35 (s, 9H, Si(CH_3)$_3$) ppm. – ^{13}C NMR (101 MHz, CDCl$_3$): δ = 197.9 (C$_q$, 1 × C=O), 196.6 (C$_q$, 1 × C=O), 194.8 (C$_q$, 1 × C=O), 138.6 (C$_q$, 1 × C$_{qAr}$), 137.6 (C$_q$, 1 × C$_{qAr}$), 135.1 (C$_q$, 1 × C$_{qAr}$), 134.8 (+, 1 × CH$_{Ar}$), 134.4 (+, 1 × CH$_{Ar}$), 133.3 (+, 1 × CH$_{Ar}$), 131.7 (+, 1 × CH$_{Ar}$), 130.1 (+, 1 × C=CH), 128.3 (+, 1 × CH$_{Ar}$), 126.9 (+, 1 × CH$_{Ar}$), 126.8 (+, 1 × CH$_{Ar}$), 126.5 (+, 1 × C=CH), 125.8 (+, 1 × CH$_{Ar}$), 121.2 (C$_q$, 1 × C$_{qAr}$), 70.2 (C$_q$, 1 × C$_q$), 68.2 (+, 1 × CH), 45.9 (+, 1 × CH), 21.1 (–, 1 × CH$_2$), –0.6 (+, 3 × SiCH$_3$) ppm. – IR (ATR): $\tilde{v}$ = 3066 (vw), 3033 (vw), 2955 (vw), 2898 (vw), 1701 (m), 1672 (w), 1593 (w), 1425 (w), 1324 (vw), 1249 (m), 1160 (vw), 1076 (w), 1056 (w), 1021 (w), 935 (w), 885 (w), 839 (m), 751 (w), 737 (w), 688 (w), 641 (w), 596 (vw), 542 (vw), 483 (vw), 456 (vw), 437 (vw) cm^{-1}. – MS (FAB, 3-NBA), m/z (%): 483 (42) [M+H]$^+$, 183 (100) [M–C$_{17}$H$_{19}$O$_3$Si]$^+$, 393 (13) [M–OTMS]$^+$. – HRMS (FAB, C$_{24}$H$_{24}$^{79}BrO$_4$Si): calc. 483.0627; found 483.0629. – X-Ray: Crystallographic information on the product can be found in chapter 7.3.2.

(1R,4aR,9aS)-9a-(2-Iodobenzoyl)-2-methyl-1-((trimethylsilyl)oxy)-1,4,4a,9a-tetrahydroanthracene-9,10-dione (87ae) and (1R,4aS,9aR)-9a-(2-iodobenzoyl)-2-methyl-1-((trimethylsilyl)oxy)-1,4,4a,9a-tetrahydroanthracene-9,10-dione (88ae)

87ae   **88ae**

According to **GP3** the cycloaddition was performed with 2-(2-iodobenzoyl)naphthalene-1,4-dione (**73a**) (233 mg, 600 µmol, 1.00 equiv.) and (E)-trimethyl((2-methylbuta-1,3-dien-1-yl)oxy)silane (**74e**) (0.34 mL, 281 mg, 1.80 mmol, 3.00 equiv.) in dry CH$_2$Cl$_2$ (2 mL). After 4 h the crude product was purified via column chromatography on silica gel (cHex/EtOAc = 9:1) to obtain a mixture of isomers **87ae** (84.9 mg, 156 µmol, 26%) and **88ae** (157 mg, 288 µmol, 48%) as yellow solids.

<u>**87ae**</u>: R_f (cHex/EtOAc = 9:1) = 0.26. – ^{1}H NMR (400 MHz, CDCl$_3$): δ = 7.96 (td, ^3J = 7.5 Hz, ^4J = 1.4 Hz, 2H, CH$_{Ar}$), 7.85 (dd, ^3J = 7.9 Hz, ^4J = 1.6 Hz, 1H, CH$_{Ar}$), 7.74 (dd, ^3J = 8.0 Hz, ^4J = 1.1 Hz, 1H, CH$_{Ar}$), 7.66 (td, ^3J = 7.5 Hz, ^4J = 1.3 Hz, 2H, CH$_{Ar}$), 7.42 (td, ^3J = 7.6 Hz, ^4J = 1.2 Hz, 1H, CH$_{Ar}$), 7.08 (ddd, ^3J = 8.0, 7.4 Hz, ^4J = 1.6 Hz, 1H, CH$_{Ar}$), 5.47 (ddt, ^3J = 4.4, 3.0 Hz, ^4J = 1.5 Hz, 1H, C=CH), 5.22 (s, 1H, C(OTMS)H), 3.93 (dd, ^3J = 10.9, 7.0 Hz, 1H, COCH), 2.63–2.39 (m, 1H, CHH), 2.12–1.99 (m, 1H, CHH), 1.95 (s, 3H, CH_3), 0.06 (s, 9H, Si(CH_3)$_3$) ppm.

– ^{13}C NMR (101 MHz, CDCl$_3$): δ = 198.7 (C$_q$, 1 × C=O), 196.2 (C$_q$, 1 × C=O), 192.2 (C$_q$, 1 × C=O), 141.5 (+, 1 × CH_{Ar}), 140.7 (C$_q$, 1 × C_{qAr}), 137.3 (C$_q$, 1 × C_{qAr}), 135.6 (C$_q$, 1 × C_{qAr}), 135.3 (+, 1 × CH_{Ar}), 134.9 (+, 1 × CH_{Ar}), 134.4 (+, 1 × CH_{Ar}), 133.7 (C$_q$, 1 × C_q), 132.1 (+, 1 × CH_{Ar}), 129.1 (+, 1 × CH_{Ar}), 127.7 (+, 1 × CH_{Ar}), 127.5 (+, 1 × CH_{Ar}), 122.0 (+, 1 × C=CH), 95.0 (C$_q$, 1 × C_{qAr}), 74.1 (C$_q$, 1 × C_q), 71.4 (+, 1 × CH), 45.9 (+, 1 × COCH), 28.4 (–, 1 × CH_2), 22.4 (+, 1 × CH_3), 1.0 (+, 3 × SiCH_3) ppm. – IR (ATR): $\tilde{v}$ = 2957 (w), 2857 (w), 1693 (m), 1586 (m), 1427 (w), 1299 (m), 1246 (m), 1220 (m), 1145 (w), 1112 (w), 1056 (s), 1006 (m), 955 (m), 928 (w), 895 (m), 839 (s), 771 (m), 744 (m), 699 (s), 639 (w), 602 (m), 570 (m), 503 (m), 395 (w) cm^{-1}. – MS (FAB, 3-NBA), m/z (%): 472 (3) [M–TMS]$^+$, 309 (25), 251 (26), 197 (43), 135 (100). – HRMS (FAB, C$_{25}$H$_{26}$O$_4$^{127}ISi): calc. 545.0645; found 545.0647. – X-Ray: Crystallographic information on the product can be found in chapter 7.3.1.

88ae: R_f (cHex/EtOAc = 9:1) = 0.34. – ^{1}H NMR (400 MHz, CDCl$_3$): δ = 8.16–8.14 (m, 1H, CH_{Ar}), 8.07–8.05 (m, 2H, CH_{Ar}), 7.96 (dd, ^3J = 7.9 Hz, ^4J = 1.1 Hz, 1H, CH_{Ar}), 7.70–7.80 (m, 2H, CH_{Ar}), 7.41 (td, ^3J = 7.6 Hz, ^4J = 1.2 Hz, 1H, CH_{Ar}), 7.08 (td, ^3J = 7.6 Hz, ^4J = 1.6 Hz, 1H, CH_{Ar}), 5.63–5.62 (m, 1H, C=CH), 4.68 (s, 1H, CH), 3.95 (d, ^3J = 7.8 Hz, 1H, COCH), 3.32–3.28 (m, 1H, CHH), 3.28–3.24 (m, 1H, CHH), 1.29 (d, ^4J = 1.9 Hz, 3H, CH_3), –0.38 (s, 9H, Si(CH_3)$_3$) ppm. – ^{13}C NMR (101 MHz, CDCl$_3$): δ = 198.4 (C$_q$, 1 × C=O), 197.4 (C$_q$, 1 × C=O), 194.9 (C$_q$, 1 × C=O), 141.5 (+, 1 × CH_{Ar}), 141.4 (C$_q$, 1 × C_{qAr}), 137.3 (C$_q$, 1 × C_{qAr}), 134.5 (C$_q$, 1 × C_{qAr}), 134.4 (+, 1 × CH_{Ar}), 133.0 (+, 1 × CH_{Ar}), 132.8 (C$_q$, 1 × C_q), 131.0 (+, 1 × CH_{Ar}), 127.9 (+, 1 × CH_{Ar}), 126.9 (+, 1 × CH_{Ar}), 126.8 (+, 1 × CH_{Ar}), 126.2 (+, 1 × CH_{Ar}), 125.8 (+, 1 × C=CH), 95.1 (C$_q$, 1 × C_{qAr}), 72.6 (+, 1 × CH), 71.5 (C$_q$, 1 × C_q), 45.0 (+, 1 × CO-CH), 21.6 (+, 1 × CH_3), 20.7 (–, 1 × CH_2), –0.7 (+, 3 × SiCH_3) ppm. – IR (ATR): $\tilde{v}$ = 3059 (vw), 2955 (vw), 2927 (vw), 2881 (vw), 1692 (m), 1667 (m), 1590 (w), 1422 (w), 1247 (m), 1084 (m), 1011 (w), 872 (m), 841 (m), 802 (w), 768 (m), 751 (m), 735 (m), 715 (m), 607 (m), 541 (w), 453 (w), 390 (w) cm^{-1}. – MS (FAB, 3-NBA), m/z (%): 545 (34) [M+H]$^+$, 455 (100) [M–OTMS]$^+$. – HRMS (FAB, C$_{25}$H$_{26}$^{127}IO$_4$Si): calc. 545.0645; found 545.0643. – X-Ray: Crystallographic information on the product can be found in chapter 7.3.2.

(1S,4aS,9aR)-1-((tert-Butyldimethylsilyl)oxy)-9a-(2-iodobenzoyl)-2-methyl-1,4,4a,9a-tetrahydroanthracene-9,10-dione (88af) and (1S,4aR,9aS)-1-((tert-butyldimethylsilyl)oxy)-9a-(2-iodobenzoyl)-2-methyl-1,4,4a,9a-tetrahydroanthracene-9,10-dione (87af)

87af   88af

According to **GP3** the cycloaddition was performed with 2-(2-iodobenzoyl)naphthalene-1,4-dione (**73a**) (200 mg, 515 µmol, 1.00 equiv.) and (E)-tert-butyldimethyl((2-methylbuta-1,3-dien-1-yl)oxy)silane (**74f**) (512 mg, 2.58 mmol, 5.00 equiv.) in dry CH_2Cl_2 (7 mL). After 6 h the crude product was purified via column chromatography on silica gel (cHex/EtOAc = 9:1) to obtain **87af** (86.8 mg, 148 µmol, 29%) and **88af** (114 mg, 194 µmol, 38%) both as yellow solids.

87af: R_f (cHex/EtOAc = 9:1) = 0.33. – ^1H NMR (500 MHz, CDCl$_3$): δ = 7.99 (d, ^3J = 7.6 Hz, 1H, CH$_{Ar}$), 7.88 (d, ^3J = 7.6 Hz, 1H, CH$_{Ar}$), 7.83 (dd, ^3J = 7.6 Hz, ^4J = 1.6 Hz, 1H, CH$_{Ar}$), 7.76–7.70 (m, 2H, CH$_{Ar}$), 7.65 (td, ^3J = 7.5 Hz, ^4J = 1.3 Hz, 1H, CH$_{Ar}$), 7.40 (td, ^3J = 7.6 Hz, ^4J = 1.2 Hz, 1H, CH$_{Ar}$), 7.06 (td, ^3J = 7.7 Hz, ^4J = 1.6 Hz, 1H, CH$_{Ar}$), 5.46 (s, 1H, C=CH), 5.32 (s, 1H, C(OTBDMS)H), 4.03 (dd, ^3J = 10.8, 7.7 Hz, 1H, COCH), 2.53–2.61 (m, 1H, CHH), 2.10–1.99 (m, 1H, CHH), 1.96 (s, 3H, CH$_3$), 0.86 (s, 9H, SiC(CH$_3$)$_3$), 0.19 (s, 3H, SiCH$_3$), –0.08 (s, 3H, SiCH$_3$) ppm. – ^{13}C NMR (126 MHz, CDCl$_3$): δ = 198.4 (C$_q$, 1 × C=O), 196.4 (C$_q$, 1 × C=O), 192.1 (C$_q$, 1 × C=O), 141.4 (+, 1 × CH$_{Ar}$), 141.0 (C$_q$, 1 × C$_{qAr}$), 135.5 (C$_q$, 1 × C$_{qAr}$), 135.3 (C$_q$, 1 × C$_q$), 134.9 (+, 1 × CH$_{Ar}$), 134.4 (+, 1 × CH$_{Ar}$), 133.6 (1 × CH=C$_q$), 132.0 (+, 1 × CH$_{Ar}$), 128.7 (+, 1 × CH$_{Ar}$), 127.8 (+, 1 × CH$_{Ar}$), 127.5 (+, 1 × CH$_{Ar}$), 127.2 (+, 1 × C=CH), 122.5 (+, 1 × CH$_{Ar}$), 94.7 (C$_q$, 1 × C$_{qAr}$), 74.4 (+, 1 × CH), 71.3 (C$_q$, 1 × C$_q$), 45.0 (+, 1 × CH), 28.2 (+, 3 × CH$_3$), 26.1 (+, 1 × CH$_3$), 23.0 (–, 1 × CH$_2$), 18.7 (C$_q$, 1 × C$_q$), –3.5 (+, 1 × SiCH$_3$), –3.8 (+, 1 × SiCH$_3$) ppm. – IR (ATR): $\tilde{v}$ = 2924 (w), 2850 (w), 1697 (m), 1672 (w), 1593 (w), 1456 (w), 1423 (w), 1248 (m), 1220 (w), 1074 (m), 1050 (m), 1006 (w), 930 (w), 885 (w), 836 (m), 782 (m), 765 (m), 735 (m), 716 (w), 701 (w), 679 (w), 637 (w), 605 (w), 540 (w), 453 (w), 414 (w) cm^{-1}. – MS (FAB, 3-NBA), m/z (%): 587 (7) [M+H]$^+$, 529 (16) [M–C(CH$_3$)$_3$]$^+$, 455 (4) [M–OTBDMS]$^+$, 389 (11), 231 (100). – HRMS (EI, C$_{28}$H$_{31}$IO$_4$Si): calc. 586.1109; found 586.1111. – X-Ray: Crystallographic information on the product can be found in chapter 7.3.

88af: R_f (cHex/EtOAc = 9:1) = 0.42. – ^1H NMR (500 MHz, CDCl$_3$): δ = 8.16–8.08 (m, 2H, CH$_{Ar}$), 8.03 (dd, ^3J = 7.5 Hz, ^4J = 1.7 Hz, 1H, CH$_{Ar}$), 7.96 (dd, ^3J = 7.9 Hz, ^4J = 1.1 Hz, 1H, CH$_{Ar}$), 7.78–

7.71 (m, 2H, CH_{Ar}), 7.41 (td, ^3J = 7.6 Hz, ^4J = 1.2 Hz, 1H, CH_{Ar}), 7.07 (td, ^3J = 7.6 Hz, ^4J = 1.6 Hz, 1H, CH_{Ar}), 5.64–5.61 (m, 1H, C=CH), 4.76 (s, 1H, C(OTBDMS)H), 3.95 (d, ^3J = 8.0 Hz, 1H, COCH), 3.39–3.26 (m, 1H, CHH), 2.30–2.22 (m, 1H, CHH), 1.28 (s, 3H, CH_3), 0.50 (s, 9H, SiC(CH_3)$_3$), –0.12 (s, 3H, SiCH_3), –0.70 (s, 3H, SiCH_3) ppm. – ^{13}C NMR (126 MHz, CDCl$_3$): δ = 198.8 (C$_q$, 1 × C=O), 197.4 (C$_q$, 1 × C=O), 195.0 (C$_q$, 1 × C=O), 142.0 (+, 1 × CH_{Ar}), 141.9 (C$_q$, 1 × C_{qAr}), 137.4 (C$_q$, 1 × C_{qAr}), 134.9 (C$_q$, 1 × C_{qAr}), 134.7 (+, 1 × CH_{Ar}), 133.4 (+, 1 × CH_{Ar}), 133.3 (C$_q$, 1 × C_q), 131.3 (+, 1 × CH_{Ar}), 128.1 (+, 1 × CH_{Ar}), 127.7 (+, 1 × CH_{Ar}), 127.1 (+, 1 × CH_{Ar}), 127.0 (+, C=CH), 126.4 (+, 1 × CH_{Ar}), 95.5 (C$_q$, 1 × C_{qAr}), 73.6 (+, 1 × CH), 71.9 (C$_q$, 1 × C_q), 45.3 (+, 1 × CH), 25.4 (+, 3 × CH_3), 22.2 (+, 1 × CH_3), 21.0 (–, 1 × CH_2), 18.2 (C$_q$, 1 × C_q), –4.7 (+, SiCH_3), –4.8 (+, SiCH_3) ppm. – IR (ATR): $\tilde{v}$ = 2921 (w), 2849 (w), 1697 (w), 1672 (w), 1593 (vw), 1457 (vw), 1424 (vw), 1248 (w), 1219 (w), 1074 (w), 1050 (w), 1009 (w), 930 (vw), 885 (vw), 871 (vw), 836 (w), 781 (w), 765 (w), 735 (w), 717 (w), 701 (w), 679 (w), 637 (vw), 605 (w), 541 (vw), 452 (vw), 416 (vw) cm^{-1}. – MS (FAB, 70 eV), m/z (%): 587 (62) [M+H]$^+$, 586 (100) [M]$^+$, 529 (82) [M–CMe$_3$]$^+$, 455 (35) [M–OTBDMS]$^+$. – HRMS (EI, C$_{28}$H$_{31}$IO$_4$Si): calc. 586.1031; found 586.1033. – X-Ray: Crystallographic information on the product can be found in chapter 7.3.1.

Cycloaddition products (87/88bf) of 2-(2-bromobenzoyl)naphthalene-1,4-dione (73b) and (E)-*tert*-butyldimethyl((2-methylbuta-1,3-dien-1-yl)oxy)silane (74f)

87bf   88bf

According to **GP3** the cycloaddition was performed with 2-(2-bromobenzoyl)naphthalene-1,4-dione (**73b**) (500 mg, 1.47 mmol, 1.00 equiv.) and (E)-*tert*-butyldimethyl((2-methylbuta-1,3-dien-1-yl)oxy)silane (**74f**) (1.16 g, 5.86 mmol, 4.00 equiv.) in dry CH$_2$Cl$_2$ (7 mL). After 4 h the crude product was purified *via* column chromatography on silica gel (cHex/EtOAc = 12:1) to obtain **87bf** (215 mg, 399 µmol, 27%) and **88bf** (419 mg, 776 µmol, 53%) both as yellow solids.

The exact structure of isomer **88bf** could be verified by X-ray analysis while the exact structure of the other isolated isomer (**87bf**) could not be resolved by analysis of the NMR spectra.

87bf: R_f (cHex/EtOAc = 12:1) = 0.26. – ^{1}H NMR (500 MHz, CDCl$_3$): δ = 8.00 (dd, ^3J = 7.7 Hz, ^4J = 1.3 Hz, 1H, CH_{Ar}), 7.87 (dd, ^3J = 7.7 Hz, ^4J = 1.3 Hz, 1H, CH_{Ar}), 7.78 (dd, ^3J = 7.8 Hz, ^4J = 1.7 Hz, 1H, CH_{Ar}), 7.72 (td, ^3J = 7.6 Hz, ^4J = 1.4 Hz, 1H, CH_{Ar}), 7.65 (td, ^3J = 7.5 Hz, ^4J = 1.3 Hz,

1H, CH_{Ar}), 7.41 (dd, ^3J = 8.1 Hz, ^4J = 1.2 Hz, 1H, CH_{Ar}), 7.37 (td, ^3J = 7.6 Hz, ^4J = 1.2 Hz, 1H, CH_{Ar}), 7.26–7.22 (m, 1H, CH_{Ar}), 5.48–5.44 (m, 1H, C=CH), 5.31 (s, 1H, C(OTBDMS)H), 4.02 (dd, ^3J = 10.7, 7.8 Hz, 1H, COCH), 2.63–2.56 (m, 1H, CHH), 2.08–2.00 (m, 1H, CHH), 1.98–1.95 (m, 3H, CH_3), 0.87 (s, 9H, SiC(CH_3)$_3$), 0.18 (s, 3H, SiCH_3), –0.02 (s, 3H, SiCH_3) ppm. – ^{13}C NMR (126 MHz, CDCl$_3$): δ = 197.6 (C$_q$, 1 × C=O), 196.4 (C$_q$, 1 × C=O), 192.1 (C$_q$, 1 × C=O), 138.0 (C$_q$, 1 × C_{qAr}), 135.3 (C$_q$, 1 × C_{qAr}), 135.2 (+, 1 × CH$_{Ar}$), 134.9 (C$_q$, 1 × C_{qAr}), 134.3 (+, 1 × CH$_{Ar}$), 134.2 (+, 1 × CH$_{Ar}$), 133.5 (C$_q$, 1 × C_q), 131.9 (+, 1 × CH$_{Ar}$), 128.8 (+, 1 × CH$_{Ar}$), 127.6 (+, 1 × CH$_{Ar}$), 127.3 (+, 1 × CH$_{Ar}$), 126.6 (+, 1 × CH$_{Ar}$), 122.7 (+, 1 × C=CH), 121.0 (C$_q$, 1 × C_{qAr}), 74.8 (C$_q$, 1 × C_q), 71.0 (+, 1 × C(OTBDMS)H), 44.6 (+, 1 × COCH), 28.1 (–, 1 × CH$_2$), 26.1 (+, 3 × CH$_3$), 23.0 (+, 1 × CH$_3$), 18.8 (C$_q$, 1 × C_q), –3.5 (+, 1 × SiCH$_3$), –3.8 (+, 1 × SiCH$_3$) ppm. – IR (ATR): $\tilde{v}$ = 2925 (w), 2851 (w), 1708 (s), 1693 (vs), 1591 (m), 1462 (w), 1428 (w), 1265 (s), 1231 (vs), 1071 (vs), 1057 (vs), 1006 (s), 953 (m), 932 (m), 888 (vs), 858 (m), 834 (vs), 810 (vs), 768 (vs), 722 (s), 673 (s), 646 (m), 602 (s), 568 (m), 533 (m), 395 (w) cm^{-1}. – MS (FAB, 3-NBA), m/z (%): 539/541 (7/6) [M+H]$^+$, 483 (15), 281 (19), 185 (100). – HRMS (FAB, C$_{28}$H$_{32}$^{79}BrO$_4$Si): calc. 539.1253; found 539.1252.

88bf: R_f (cHex/EtOAc = 12:1) = 0.34. – ^{1}H NMR (500 MHz, CDCl$_3$): δ = 8.16–8.09 (m, 2H, CH_{Ar}), 8.04–8.00 (m, 1H, CH_{Ar}), 7.74 (pd, ^3J = 7.3 Hz, ^4J = 1.6 Hz, 2H, CH_{Ar}), 7.62 (dd, ^3J = 8.0 Hz, ^4J = 1.2 Hz, 1H, CH_{Ar}), 7.39 (td, ^3J = 7.6 Hz, ^4J = 1.2 Hz, 1H, CH_{Ar}), 7.28–7.23 (m, 1H, CH_{Ar}), 5.62–5.58 (m, 1H, C=CH), 4.74 (s, 1H, C(OTBDMS)H), 3.97 (d, ^3J = 8.0 Hz, 1H, COCH), 3.39–3.24 (m, 1H, CHH), 2.23–2.14 (m, 1H, CHH), 1.28–1.26 (m, 3H, CH_3), 0.50 (s, 9H, SiC(CH_3)$_3$), –0.13 (s, 3H, SiCH_3), –0.70 (s, 3H, SiCH_3) ppm. – ^{13}C NMR (126 MHz, CDCl$_3$): δ = 198.2 (C$_q$, 1 × C=O), 197.3 (C$_q$, 1 × C=O), 195.1 (C$_q$, 1 × C=O), 139.3 (C$_q$, 1 × C_{qAr}), 137.3 (C$_q$, 1 × C_{qAr}), 134.9 (C$_q$, 1 × C_{qAr}), 134.6 (+, 1 × CH$_{Ar}$), 134.5 (+, 1 × CH$_{Ar}$), 133.3 (+, 1 × CH$_{Ar}$), 133.0 (C$_q$, 1 × C_q), 131.3 (+, 1 × CH$_{Ar}$), 128.5 (+, 1 × CH$_{Ar}$), 127.6 (+, 1 × CH$_{Ar}$), 127.0 (+, 1 × C=CH), 126.5 (+, 1 × CH$_{Ar}$), 126.3 (+, 1 × CH$_{Ar}$), 121.4 (C$_q$, 1 × C_{qAr}), 73.6 (+, 1 × C(OTBDMS)H), 72.0 (C$_q$, 1 × C_q), 45.1 (+, 1 × COCH), 25.4 (+, 3 × CH$_3$), 22.2 (+, 1 × CH$_3$), 20.9 (–, 1 × CH$_2$), 18.1 (C$_q$, 1 × C_q), –4.8 (+, 1 × SiCH$_3$), –4.9 (+, 1 × SiCH$_3$) ppm. – IR (ATR): $\tilde{v}$ = 1701 (vs), 1672 (s), 1594 (w), 1425 (w), 1252 (vs), 1220 (m), 1075 (vs), 1051 (vs), 1014 (m), 870 (m), 858 (m), 837 (vs), 806 (m), 783 (vs), 764 (vs), 751 (m), 737 (vs), 718 (s), 703 (s), 684 (s), 639 (w), 606 (m), 582 (w), 541 (m), 460 (m), 439 (w), 415 (m), 392 (m) cm^{-1}. – MS (FAB, 3-NBA), m/z (%): 539/541 (13/11) [M+H]$^+$, 281 (27), 207 (30), 185 (100). – HRMS (FAB, C$_{28}$H$_{32}$^{79}BrO$_4$Si): calc. 539.1253; found 539.1252. – X-Ray: Crystallographic information on the product can be found in chapter 7.3.2.

(4a*R*,9a*R*)-4a-(2-Iodobenzoyl)-2-((triisopropylsilyl)oxy)-1,4,4a,9a-tetrahydroanthracene-9,10-dione (87ag)

According to **GP3** the cycloaddition was performed with 2-(2-iodobenzoyl)naphthalene-1,4-dione (**73a**) (233 mg, 600 µmol, 1.00 equiv.) and (buta-1,3-dien-2-yloxy)triisopropylsilane (**74g**) (0.33 mL, 272 mg, 1.20 mmol, 2.00 equiv.) in dry CH_2Cl_2 (3 mL) at 40 °C. After 3.5 h the crude product was purified *via* column chromatography on silica gel (*c*Hex/EtOAc = 10:1) to obtain product **87ag** as a yellow solid (274 mg, 446 µmol, 75%).

– R_f (cylohexane/EtOAc = 14:1) = 0.29. – ^{1}H NMR (400 MHz, CDCl$_3$): δ = 8.17 (dd, ^3J = 7.5 Hz, ^4J = 1.6 Hz, 1H, C*H*$_{Ar}$), 8.03 (dd, ^3J = 7.5 Hz, ^4J = 1.6 Hz, 1H, C*H*$_{Ar}$), 7.84 (dd, ^3J = 7.9 Hz, ^4J = 1.1 Hz, 1H, C*H*$_{Ar}$), 7.82–7.68 (m, 2H, C*H*$_{Ar}$), 7.33–7.26 (m, 1H, C*H*$_{Ar}$), 7.14 (dd, ^3J = 7.8 Hz, ^4J = 1.7 Hz, 1H, C*H*$_{Ar}$), 7.06 (td, ^3J = 7.7 Hz, ^4J = 1.7 Hz, 1H, C*H*$_{Ar}$), 4.83–4.71 (m, 1H, C=C*H*), 3.80 (dd, ^3J = 9.8, 6.6 Hz, 1H, C*H*), 3.03–2.96 (m, 1H, C*H*H), 2.56–2.22 (m, 3H, CH*H*, C*H*$_2$), 1.15–0.93 (m, 21H, 3 × C*H*, 6 × C*H*$_3$) ppm. – ^{13}C NMR (101 MHz, CDCl$_3$): δ = 200.7 (C$_q$, 1 × *C*=O), 195.4 (C$_q$, 1 × *C*=O), 193.8 (C$_q$, 1 × *C*=O), 148.0 (C$_q$, 1 × C_{qAr}), 142.7 (C$_q$, 1 × C_{qAr}), 140.8 (+, 1 × *C*H$_{Ar}$), 135.3 (+, 1 × *C*H$_{Ar}$), 134.3 (+, 1 × *C*H$_{Ar}$), 134.2 (C$_q$, 1 × C_{qAr}), 133.1 (C$_q$, 1 × C_q), 131.6 (+, 1 × *C*H$_{Ar}$), 127.4 (+, 1 × *C*H$_{Ar}$), 127.4 (+, 1 × *C*H$_{Ar}$), 127.3 (+, 1 × *C*H$_{Ar}$), 126.9 (+, 1 × *C*H$_{Ar}$), 99.4 (+, 1 × *C*=CH), 93.0 (C$_q$, 1 × C_{qAr}), 67.9 (C$_q$, 1 × C_q), 50.7 (+, 1 × *C*H), 30.3 (–, 1 × *C*H$_2$), 27.8 (–, 1 × *C*H$_2$), 18.0 (+, 6 × *C*H$_3$), 12.6 (+, 3 × Si*C*H) ppm. – IR (ATR): $\tilde{v}$ = 2890 (w), 2862 (w), 1689 (w), 1668 (m), 1590 (w), 1460 (w), 1376 (w), 1287 (w), 1222 (m), 1201 (m), 1061 (w), 1014 (w), 881 (w), 853 (m), 797 (w), 767 (w), 753 (w), 736 (m), 692 (m), 633 (m), 501 (w), 431 (w), 402 (w) cm^{-1}. – MS (FAB, 3-NBA), *m/z* (%): 614 (10) [M]$^+$, 615 (31) [M+H]$^+$, 231 (100) [M–C$_{23}$H$_{31}$O$_3$Si]$^+$, 383 (99) [M–C$_7$H$_4$IO]$^+$. – HRMS (FAB, C$_{30}$H$_{36}$^{127}IO$_4$Si): calc. 615.1422; found 615.1424. – X-Ray: Crystallographic information on the product can be found in chapter 7.3.2.

(4a*R*,9a*R*)-2-((*tert*-Butyldiphenylsilyl)oxy)-4a-(2-iodobenzoyl)-1,4,4a,9a-tetrahydroanthracene-9,10-dione (87ah)

According to **GP3** the cycloaddition was performed with 2-(2-iodobenzoyl)naphthalene-1,4-dione (**73a**) (311 mg, 800 μmol, 1.00 equiv.) and (buta-1,3-dien-2-yloxy)(*tert*-butyl)diphenylsilane (**74h**) (1.26 mL, 1.23 g, 4.00 mmol, 5.00 equiv.) in dry CH_2Cl_2 (4 mL). After 5 h the crude product was purified *via* column chromatography on silica gel (*c*Hex/EtOAc = 9:1) to obtain product **87ah** as a yellow solid (401 mg, 576 mmol, 72%).

– R_f (*c*Hex/EtOAc = 8:1) = 0.34. – ^{1}H NMR (400 MHz, CDCl$_3$): δ = 8.17–8.11 (m, 1H, C*H*$_{Ar}$), 8.05–7.99 (m, 1H, C*H*$_{Ar}$), 7.84 (d, ^3J = 7.9 Hz, 1H, C*H*$_{Ar}$), 7.82–7.68 (m, 3H, C*H*$_{Ar}$), 7.68–7.58 (m, 3H, C*H*$_{Ar}$), 7.41–7.32 (m, 6H, C*H*$_{Ar}$), 7.24–7.22 (m, 1H, C*H*$_{Ar}$), 7.10 (dd, ^3J = 7.8 Hz, ^4J = 1.6 Hz, 1H, C*H*$_{Ar}$), 7.04 (td, ^3J = 7.7 Hz, ^4J = 1.6 Hz, 1H, C*H*$_{Ar}$), 4.60–4.50 (m, 1H, C=C*H*), 3.76 (dd, ^3J = 9.4, 6.6 Hz, 1H, C*H*), 2.77 (dd, ^2J = 16.5 Hz, ^3J = 4.8 Hz, 1H, C*HH*), 2.56–2.44 (m, 1H, C*HH*), 2.40–2.22 (m, 2H, C*H*$_2$), 0.99 (s, 9H, 3 × C*H*$_3$) ppm. – ^{13}C NMR (101 MHz, CDCl$_3$): δ = 200.8 (C$_q$, 1 × *C*=O), 195.2 (C$_q$, 1 × *C*=O), 193.7 (C$_q$, 1 × *C*=O), 147.7 (C$_q$, 1 × *C*$_{qAr}$), 142.5 (+, 1 × *C*H$_{Ar}$), 140.8 (+, 1 × *C*H$_{Ar}$), 135.5 (+, 3 × *C*H$_{Ar}$), 135.2 (C$_q$, 1 × *C*$_{qAr}$), 134.9 (+, 1 × *C*H$_{Ar}$), 134.2 (C$_q$, 1 × *C*$_{qAr}$), 134.2 (C$_q$, 1 × *C*$_{qAr}$), 133.2 (C$_q$, 1 × *C*$_{qAr}$), 133.0 (C$_q$, 1 × *C*$_{qAr}$), 131.5 (+, 1 × *C*H$_{Ar}$), 129.9 (+, 2 × *C*H$_{Ar}$), 129.8 (C$_q$, 1 × *C*$_q$), 127.9 (+, 1 × *C*H$_{Ar}$), 127.8 (+, 2 × *C*H$_{Ar}$), 127.8 (+, 2 × *C*H$_{Ar}$), 127.5 (+, 1 × *C*H$_{Ar}$), 127.4 (+, 1 × *C*H$_{Ar}$), 127.2 (+, 1 × *C*H$_{Ar}$), 126.9 (+, 1 × *C*H$_{Ar}$), 101.4 (+, 1 × *C*=CH), 67.6 (C$_q$, 1 × *C*$_q$), 50.4 (+, 1 × *C*H), 30.0 (–, 1 × *C*H$_2$), 27.7 (–, 1 × *C*H$_2$), 26.6 (+, 3 × *C*H$_3$), 19.3 (C$_q$, 1 × *C*$_q$) ppm. – IR (ATR): $\tilde{v}$ = 2890 (w), 2862 (w), 1689 (w), 1668 (m), 1590 (w), 1460 (w), 1376 (w), 1287 (w), 1222 (m), 1201 (m), 1061 (w), 1014 (w), 881 (w), 853 (m), 797 (w), 767 (w), 753 (w), 736 (m), 692 (m), 633 (m), 501 (w), 431 (w), 402 (w) cm^{-1}. – MS (FAB, 3-NBA), *m/z* (%): 697 (20) [M+H]$^+$, 696 (11) [M]$^+$, 466 (28), 465 (67), 231 (79), 197 (46), 135 (100). – HRMS (FAB, C$_{37}$H$_{34}$O$_4$^{127}ISi): calc. 697.1271; found 697.1272.

(4a*S*,9a*S*)-4a-(2-Bromobenzoyl)-2-((*tert*-butyldiphenylsilyl)oxy)-1,4,4a,9a-tetrahydroanthracene-9,10-dione (87bh)

According to **GP3** the cycloaddition was performed with 2-(2-bromobenzoyl)naphthalene-1,4-dione (**73b**) (273 mg, 800 µmol, 1.00 equiv.) and (buta-1,3-dien-2-yloxy)(*tert*-butyl)diphenylsilane (**74h**) (1.26 mL, 1.23 g, 4.00 mmol, 5.00 equiv.) in dry CH_2Cl_2 (4 mL). After 5 h the crude product was purified *via* column chromatography on silica gel (*c*Hex/EtOAc = 9:1) to obtain product **87bh** as a yellow solid (411 mg, 632 mmol, 79%).

$- R_f$ (*c*Hex/EtOAc = 9:1) = 0.37. $-$ ^{1}H NMR (500 MHz, CDCl$_3$): δ = 8.13 (dd, ^3J = 7.5 Hz, ^4J = 1.4 Hz, 1H, C*H*$_{Ar}$), 8.02 (dd, ^3J = 7.5 Hz, ^4J = 1.4 Hz, 1H, C*H*$_{Ar}$), 7.80–7.72 (m, 2H, C*H*$_{Ar}$), 7.65–7.61 (m, 4H, C*H*$_{Ar}$), 7.57–7.53 (m, 1H, C*H*$_{Ar}$), 7.40–7.31 (m, 6H, C*H*$_{Ar}$), 7.21 (dd, ^3J = 5.8 Hz, ^4J = 3.5 Hz, 2H, C*H*$_{Ar}$), 7.14 (dd, ^3J = 5.8 Hz, ^4J = 3.5 Hz, 1H, C*H*$_{Ar}$), 4.52–4.50 (m, 1H, C=C*H*), 3.78 (dd, ^3J = 8.8, 6.6 Hz, 1H, C*H*), 2.76–2.71 (m, 1H, C*H*H), 2.52–2.47 (m, 1H, C*H*H), 2.41–2.31 (m, 2H, CH*H*, CH*H*), 0.99 (s, 9H, 3 × C*H*$_3$) ppm. $-$ ^{13}C NMR (126 MHz, CDCl$_3$): δ = 200.0 (C$_q$, 1 × C=O), 195.2 (C$_q$, 1 × C=O), 193.7 (C$_q$, 1 × C=O), 147.8 (C$_q$, 1 × C$_{qAr}$), 138.9 (C$_q$, 1 × C$_{qAr}$), 135.4 (+, 2 × CH$_{Ar}$), 135.4 (+, 2 × CH$_{Ar}$), 135.1 (+, 1 × CH$_{Ar}$), 134.2 (+, 1 × CH$_{Ar}$), 134.0 (C$_q$, 1 × C$_{qAr}$), 133.8 (+, 1 × CH$_{Ar}$), 133.0 (C$_q$, 1 × C$_{qAr}$), 132.8 (C$_q$, 1 × C$_{qAr}$), 132.5 (C$_q$, 1 × C$_q$), 131.3 (+, 1 × CH$_{Ar}$), 129.8 (+, 2 × CH$_{Ar}$), 127.7 (+, 2 × CH$_{Ar}$), 127.7 (+, 2 × CH$_{Ar}$), 127.4 (+, 1 × CH$_{Ar}$), 127.3 (+, 1 × CH$_{Ar}$), 127.1 (+, 1 × CH$_{Ar}$), 126.6 (+, 1 × CH$_{Ar}$), 119.8 (C$_q$, 1 × C$_{qAr}$), 101.0 (+, 1 × C=CH), 67.5 (C$_q$, 1 × C$_q$), 50.0 (+, 1 × CH), 29.5 (–, 1 × CH$_2$), 27.4 (–, 1 × CH$_2$), 26.4 (+, 3 × CH$_3$), 19.2 (C$_q$, 1 × C$_q$) ppm. $-$ IR (ATR): $\tilde{v}$ = 3070 (vw), 2930 (vw), 2856 (vw), 1679 (w), 1591 (w), 1470 (vw), 1427 (w), 1372 (vw), 1255 (w), 1241 (w), 1219 (w), 1194 (w), 1111 (w), 1062 (w), 1019 (w), 939 (w), 893 (vw), 851 (w), 821 (w), 796 (vw), 761 (vw), 736 (w), 699 (m), 646 (w), 612 (w), 550 (vw), 492 (w) cm^{-1}. $-$ MS (FAB, 3-NBA), *m/z* (%): 649/651 (20/20) [M+H]$^+$, 648/650 (7/14) [M]$^+$, 466 (41), 465 (100), 185 (48), 183 (51). $-$ HRMS (FAB, C$_{37}$H$_{34}$O$_4$^{79}BrSi): calc. 649.1410; found 649.1408. $-$ X-Ray: Crystallographic information on the product can be found in chapter 7.3.1.

Cycloaddition products (87/88bb) of 2-(2-bromo-5-methoxybenzoyl)naphthalene-1,4-dione (73c) and isoprene (74b)

87bb   88bb

According to **GP3** the cycloaddition was performed with 2-(2-bromo-5-methoxybenzoyl)naphthalene-1,4-dione (**73c**) (148 mg, 400 μmol, 1.00 equiv.) and isoprene (**74b**) (0.20 mL, 136 mg, 2.00 mmol, 5.00 equiv.) in dry CH_2Cl_2 (2 mL). After 3 h the crude product was purified *via* column chromatography on silica gel (*c*Hex/EtOAc = 6:1) to obtain a light yellow solid (89.6 mg, 204 μmol, 51%). The products **87/88bb** were isolated as a non-separable mixture of regioisomers in a $\sim$ 10 : 1 ratio as estimated by ^{1}H NMR.

Both possible regioisomers are drawn, as the exact structure of the products could not be resolved by analysis of the NMR spectra. Below the analytics of the major product are given.

– R_f (*c*Hex/EtOAc = 6:1) = 0.27. – ^{1}H NMR (500 MHz, $CDCl_3$): δ = 8.13 (dd, ^3J = 7.6 Hz, ^4J = 1.5 Hz, 1H, CH_{Ar}), 8.02 (dd, ^3J = 7.6 Hz, ^4J = 1.4 Hz, 1H, CH_{Ar}), 7.80–7.70 (m, 2H, CH_{Ar}), 7.41 (d, ^3J = 8.8 Hz, 1H, CH_{Ar}), 6.78 (dd, ^3J = 8.8 Hz, ^4J = 3.0 Hz, 1H, CH_{Ar}), 6.72 (d, ^4J = 3.0 Hz, 1H, CH_{Ar}), 5.33–5.31 (m, 1H, C=CH), 3.75–3.72 (m, 4H, CH, OCH_3), 2.92–2.87 (m, 1H, CHH), 2.51–2.45 (m, 1H, CHH), 2.30–2.23 (m, 2H, CH_2), 1.63 (d, ^4J = 1.9 Hz, 3H, CH_3) ppm. – ^{13}C NMR (126 MHz, $CDCl_3$): δ = 199.9 (C_q, 1 × C=O), 196.2 (C_q, 1 × C=O), 194.5 (C_q, 1 × C=O), 158.1 (C_q, 1 × C_{qAr}), 139.7 (C_q, 1 × C_{qAr}), 135.2 (+, 1 × CH_{Ar}), 134.7 (+, 1 × CH_{Ar}), 134.3 (+, 1 × CH_{Ar}), 134.2 (C_q, 1 × C_{qAr}), 133.1 (C_q, 1 × C_{qAr}), 132.2 (C_q, 1 × C_{qAr}), 127.3 (+, 1 × CH_{Ar}), 127.1 (+, 1 × CH_{Ar}), 117.8 (+, 1 × C=CH), 117.2 (+, 1 × CH_{Ar}), 113.3 (+, 1 × CH_{Ar}), 110.0 (C_q, 1 × C_q), 67.7 (C_q, 1 × C_q), 55.7 (+, 1 × OCH$_3$), 49.9 (+, 1 × CH), 30.2 (–, 1 × CH$_2$), 28.6 (–, 1 × CH$_2$), 23.1 (+, 1 × CH$_3$) ppm. – IR (ATR): $\tilde{v}$ = 2932 (w), 2836 (w), 1679 (vs), 1591 (s), 1568 (s), 1465 (s), 1275 (vs), 1242 (vs), 1200 (vs), 1176 (vs), 1065 (m), 1023 (vs), 969 (m), 936 (s), 807 (vs), 739 (vs), 687 (m), 599 (s), 561 (m), 483 (w), 431 (m) cm^{-1}. – MS (FAB, 3-NBA), *m/z* (%): 439/441 (9/8) [M+H]$^+$, 225 (41), 215 (100). – HRMS (FAB, $C_{23}H_{20}^{79}BrO_4Si$): calc. 439.0545; found 439.0544.

(4a*R*,9a*R*)-4a-(2-Bromo-5-methoxybenzoyl)-2,3-dimethyl-1,4,4a,9a-tetrahydroanthracene-9,10-dione (87cc)

According to **GP3** the cycloaddition was performed with 2-(2-bromo-5-methoxybenzoyl)naphthalene-1,4-dione (**73c**) (74.2 mg, 200 µmol, 1.00 equiv.) and 2,3-dimethylbuta-1,3-diene (**74c**) (0.12 mL, 82.2 mg, 1.00 mmol, 5.00 equiv.) in dry CH_2Cl_2 (1 mL). After 3 h the crude product was purified *via* column chromatography on silica gel (*c*Hex/EtOAc = 6:1) to obtain product **87cc** as a yellow solid (58.9 mg, 130 µmol, 65%).

– R_f (*c*Hex/EtOAc = 6:1) = 0.43. – ^{1}H NMR (500 MHz, CDCl$_3$): δ = 8.17–8.11 (m, 1H, C*H*$_{Ar}$), 8.07–8.01 (m, 1H, C*H*$_{Ar}$), 7.78 (td, ^3J = 7.5 Hz, ^4J = 1.5 Hz, 1H, C*H*$_{Ar}$), 7.74 (td, ^3J = 7.5 Hz, ^4J = 1.5 Hz, 1H, C*H*$_{Ar}$), 7.47–7.41 (m, 1H, C*H*$_{Ar}$), 6.83–6.74 (m, 2H, C*H*$_{Ar}$), 3.77 (s, 3H, OC*H*$_3$), 3.72 (dd, ^3J = 8.4, 6.7 Hz, 1H, C*H*), 2.86–2.74 (m, 1H, C*HH*), 2.41 (d, ^2J = 16.7 Hz, 1H, CH*H*), 2.35–2.23 (m, 2H, C*H*$_2$), 1.59 (s, 3H, C*H*$_3$), 1.54 (s, 3H, C*H*$_3$) ppm. – ^{13}C NMR (126 MHz, CDCl$_3$): δ = 200.0 (C$_q$, 1 × *C*=O), 196.5 (C$_q$, 1 × *C*=O), 194.8 (C$_q$, 1 × *C*=O), 158.1 (C$_q$, 1 × *C*$_{qAr}$), 139.8 (C$_q$, 1 × *C*$_{qAr}$), 135.2 (+, 1 × *C*H$_{Ar}$), 134.7 (+, 1 × *C*H$_{Ar}$), 134.3 (C$_q$, 1 × *C*$_{qAr}$), 134.3 (+, 1 × *C*H$_{Ar}$), 133.2 (C$_q$, 1 × *C*$_{qAr}$), 127.3 (+, 1 × *C*H$_{Ar}$), 127.2 (+, 1 × *C*H$_{Ar}$), 124.2 (C$_q$, 1 × *C*$_{qAr}$), 122.9 (C$_q$, 1 × *C*$_q$), 117.1 (+, 1 × *C*H$_{Ar}$), 113.4 (+, 1 × *C*H$_{Ar}$), 110.1 (C$_q$, 1 × *C*$_q$), 68.6 (C$_q$, 1 × *C*$_q$), 55.7 (+, 1 × O*C*H$_3$), 50.2 (+, 1 × *C*H), 34.3 (–, 1 × *C*H$_2$), 31.6 (–, 1 × *C*H$_2$), 19.1 (+, 1 × *C*H$_3$), 18.8 (+, 1 × *C*H$_3$) ppm. – IR (ATR): $\tilde{v}$ = 3070 (w), 3003 (w), 2912 (w), 2837 (w), 1681 (vs), 1591 (vs), 1568 (s), 1463 (s), 1442 (m), 1402 (w), 1390 (w), 1310 (m), 1276 (vs), 1242 (vs), 1220 (s), 1201 (vs), 1176 (s), 1123 (w), 1089 (w), 1062 (m), 1021 (vs), 1009 (s), 976 (w), 932 (w), 914 (w), 849 (w), 823 (m), 779 (m), 751 (vs), 728 (vs), 715 (s), 683 (w), 667 (m), 647 (w), 601 (m), 560 (w), 540 (w), 477 (w), 439 (w), 391 (w) cm^{-1}. – MS (FAB, 3-NBA), *m/z* (%): 453/455 (20/17) [M+H]$^+$, 239 (80), 215 (98) [C$_8$H$_8$BrO$_2$]$^+$, 213 (100). – HRMS (FAB, C$_{24}$H$_{22}$O$_4$^{79}Br): calc. 453.0701; found 453.0702.

Cycloaddition products (87/88cd) of 2-(2-bromo-5-methoxybenzoyl)naphthalene-1,4-dione (73c) and 1-(trimethylsiloxy)-1,3-butadiene (74d)

According to **GP3** the cycloaddition was performed with 2-(2-bromo-5-methoxybenzoyl)naphthalene-1,4-dione **(73c)** (260 mg, 700 μmol, 1.00 equiv.) and 1-(trimethylsiloxy)-1,3-butadiene **(74d)** (0.61 mL, 498 mg, 3.50 mmol, 5.00 equiv.) in dry CH_2Cl_2 (4 mL). After 4 h the crude product was purified *via* column chromatography on silica gel (cHex/EtOAc = 6:1) to obtain isomers **87/88cd F1** (191,8 mg, 374 μmol, 53%) and **87/88cd F2** (34.2 mg, 66.6 μmol, 10%) as yellow solids.

Both possible regioisomers are drawn, as the exact structure of the products could not be resolved by analysis of the NMR spectra.

87/88cd F1: R_f (cHex/EtOAc = 6:1) = 0.34. – ^{1}H NMR (500 MHz, CDCl$_3$): δ = 8.18–8.12 (m, 1H, CH_{Ar}), 8.06–8.00 (m, 1H, CH_{Ar}), 7.78–7.68 (m, 2H, CH_{Ar}), 7.48 (d, ^3J = 8.8 Hz, 1H, CH_{Ar}), 7.42 (d, ^4J = 3.0 Hz, 1H, CH_{Ar}), 6.82 (dd, ^3J = 8.8 Hz, ^4J = 3.0 Hz, 1H, CH_{Ar}), 5.91 (ddd, ^3J = 10.1, 4.5 Hz, ^4J = 2.8 Hz, 1H, C=CH), 5.69–5.63 (m, 1H, C=CH), 4.94 (d, ^3J = 5.2 Hz, 1H, CH), 3.93 (d, ^3J = 7.1 Hz, 1H, CH), 3.76 (s, 3H, OCH_3), 3.23–3.16 (m, 1H, CHH), 2.32–2.12 (m, 1H, CHH), –0.34 (s, 9H, Si(CH_3)$_3$) ppm. – ^{13}C NMR (126 MHz, CDCl$_3$): δ = 197.8 (C$_q$, 1 × C=O), 196.7 (C$_q$, 1 × C=O), 194.9 (C$_q$, 1 × C=O), 158.2 (C$_q$, 1 × C_{qAr}), 139.3 (C$_q$, 1 × C_{qAr}), 137.5 (C$_q$, 1 × C_{qAr}), 135.3 (+, 1 × CH$_{Ar}$), 135.1 (C$_q$, 1 × C_{qAr}), 134.4 (+, 1 × CH$_{Ar}$), 133.3 (+, 1 × CH$_{Ar}$), 130.3 (+, 1 × =CH), 126.8 (+, 1 × CH$_{Ar}$), 126.4 (+, 1 × =CH), 125.8 (+, 1 × CH$_{Ar}$), 117.8 (+, 1 × CH$_{Ar}$), 113.8 (+, 1 × CH$_{Ar}$), 111.3 (C$_q$, 1 × C_{qAr}), 70.3 (C$_q$, 1 × C_q), 68.2 (+, 1 × CH), 55.7 (+, 1 × OCH$_3$), 45.8 (+, 1 × CH), 21.1 (–, 1 × CH$_2$), –0.6 (+, 3 × SiCH$_3$) ppm. – IR (ATR): $\tilde{v}$ = 2956 (w), 1701 (vs), 1673 (s), 1591 (m), 1568 (w), 1463 (m), 1421 (w), 1391 (w), 1309 (w), 1273 (s), 1247 (vs), 1218 (s), 1194 (s), 1176 (m), 1077 (vs), 1043 (vs), 1017 (s), 969 (w), 943 (m), 864 (vs), 840 (vs), 783 (s), 751 (vs), 737 (s), 722 (s), 693 (vs), 601 (m), 472 (w), 438 (w), 380 (w) cm^{-1}. – MS (FAB, 3-NBA), m/z (%): 513/515 (17/16) [M+H]$^+$, 371/373 (27/27), 299 (56), 213/215 (100/98). – HRMS (FAB, C$_{25}$H$_{26}$O$_5$^{79}BrSi): calc. 513.0733; found 513.0735.

87/88cd F2: R_f (cHex/EtOAc = 6:1) = 0.24. – ^{1}H NMR (500 MHz, CDCl$_3$): δ = 8.10–8.03 (m, 1H, CH_{Ar}), 7.97–7.90 (m, 1H, CH_{Ar}), 7.74 (td, ^3J = 7.5 Hz, ^4J = 1.3 Hz, 1H, CH_{Ar}), 7.66 (td, ^3J = 7.5 Hz, ^4J = 1.3 Hz, 1H, CH_{Ar}), 7.42–7.34 (m, 2H, CH_{Ar}), 6.81 (dd, ^3J = 8.8 Hz, ^4J = 3.0 Hz, 1H,

CH_{Ar}), 6.05–5.99 (m, 1H, C=CH), 5.83 (ddd, ^3J = 10.0, 4.8 Hz, ^4J = 2.4 Hz, 1H, C=CH), 5.44–5.37 (m, 1H, CH), 3.87 (dd, ^3J = 11.8, 6.6 Hz, 1H, CH), 3.83 (s, 3H, OCH_3), 2.58–2.50 (m, 1H, CHH), 2.11–2.01 (m, 1H, CHH), 0.02 (s, 9H, Si(CH_3)$_3$) ppm. – ^{13}C NMR (126 MHz, CDCl$_3$): δ = 196.4 (C$_q$, 1 × C=O), 195.8 (C$_q$, 1 × C=O), 193.2 (C$_q$, 1 × C=O), 158.4 (C$_q$, 1 × C_{qAr}), 138.5 (C$_q$, 1 × C_{qAr}), 135.4 (+, 1 × CH_{Ar}), 135.3 (+, 1 × CH_{Ar}), 134.6 (C$_q$, 1 × C_{qAr}), 134.2 (+, 1 × CH_{Ar}), 133.9 (C$_q$, 1 × C_{qAr}), 127.7 (+, 1 × C=CH), 127.4 (+, 1 × CH_{Ar}), 127.2 (+, 1 × CH_{Ar}), 127.1 (+, 1 × C=CH), 117.9 (+, 1 × CH_{Ar}), 114.5 (+, 1 × CH_{Ar}), 111.8 (C$_q$, 1 × C_{qAr}), 73.8 (C$_q$, 1 × C_q), 66.1 (+, 1 × CH), 55.9 (+, 1 × OCH$_3$), 46.1 (+, 1 × CH), 28.1 (–, 1 × CH$_2$), 0.7 (+, 3 × SiCH$_3$) ppm. – IR (ATR): $\tilde{v}$ = 2955 (w), 1698 (s), 1677 (vs), 1592 (m), 1568 (m), 1463 (m), 1402 (w), 1288 (s), 1266 (vs), 1244 (vs), 1218 (vs), 1197 (s), 1174 (s), 1072 (s), 1051 (vs), 1017 (s), 943 (s), 894 (vs), 839 (vs), 782 (s), 742 (vs), 714 (s), 686 (s), 585 (s), 492 (m) cm^{-1}. – MS (FAB, 3-NBA), m/z (%): 513/515 (9/8) [M+H]$^+$, 371/373 (27/27), 213/215 (100/97). – HRMS (FAB, C$_{25}$H$_{26}$O$_5$^{79}BrSi): calc. 513.0733; found 513.0731.

Cycloaddition products (87/88db) of 2-(2-bromo-4,5-dimethoxybenzoyl)naphthalene-1,4-dione (73d) and isoprene (74b)

87db   **88db**

According to **GP3** the cycloaddition was performed with 2-(2-bromo-4,5-dimethoxybenzoyl)naphthalene-1,4-dione (**73d**) (120 mg, 300 μmol, 1.00 equiv.) and isoprene (**74b**) (0.15 mL, 102 mg, 1.50 mmol, 5.00 equiv.) in dry CH$_2$Cl$_2$ (1 mL). After 4 h the crude product was purified via column chromatography on silica gel (cHex/EtOAc = 6:1) to obtain a yellow solid (98.6 mg, 210 μmol, 70%). The products **87/88db** were isolated as a non-separable mixture of regioisomers in a ~ 6.7 : 1 ratio as estimated by ^{1}H NMR.

Both possible regioisomers are drawn, as the exact structure of the products could not be resolved by analysis of the NMR spectra. Below the analytics of the major product are given.

– R_f (cHex/EtOAc = 6:1) = 0.33. – ^{1}H NMR (400 MHz, CDCl$_3$): δ = 8.15 (dd, ^3J = 7.6 Hz, ^4J = 1.4 Hz, 1H, CH_{Ar}), 8.00 (dd, ^3J = 7.5 Hz, ^4J = 1.4 Hz, 1H, CH_{Ar}), 7.78 (td, ^3J = 7.6 Hz, ^4J = 1.6 Hz, 1H, CH_{Ar}), 7.72 (td, ^3J = 7.6 Hz, ^4J = 1.6 Hz, 1H, CH_{Ar}), 6.99 (s, 1H, CH_{Ar}), 6.88 (s, 1H, CH_{Ar}), 5.36–5.33 (m, 1H, C=CH), 3.87 (s, 3H, OCH_3), 3.83 (s, 3H, OCH_3), 3.75 (t, ^3J = 7.7 Hz, 1H, CH), 3.01–2.95 (m, 1H, CHH), 2.47–2.39 (m, 1H, CHH), 2.27 (d, ^3J = 7.5 Hz, 2H, CH_2), 1.65 (s, 3H,

CH_3) ppm. – ^{13}C NMR (101 MHz, CDCl$_3$): δ = 198.8 (C$_q$, 1 × C=O), 196.3 (C$_q$, 1 × C=O), 195.5 (C$_q$, 1 × C=O), 151.0 (C$_q$, 1 × C$_{qAr}$), 147.5 (C$_q$, 1 × C$_{qAr}$), 135.3 (+, 1 × CH$_{Ar}$), 134.3 (+, 1 × CH$_{Ar}$), 134.2 (C$_q$, 1 × C$_{qAr}$), 133.3 (C$_q$, 1 × C$_{qAr}$), 132.3 (C$_q$, 1 × C$_{qAr}$), 130.8 (C$_q$, 1 × C$_q$), 127.2 (+, 1 × CH$_{Ar}$), 127.2 (+, 1 × CH$_{Ar}$), 117.8 (+, 1 × C=CH), 116.7 (+, 1 × CH$_{Ar}$), 111.9 (C$_q$, 1 × C$_{qAr}$), 110.9 (+, 1 × CH$_{Ar}$), 67.9 (C$_q$, 1 × C$_q$), 56.4 (+, 1 × OCH$_3$), 56.3 (+, 1 × OCH$_3$), 50.3 (+, 1 × CH), 30.3 (–, 1 × CH$_2$), 29.3 (–, 1 × CH$_2$), 23.1 (+, 1 × CH$_3$) ppm. – IR (ATR): $\tilde{v}$ = 2842 (vw), 2931 (vw), 1676 (w), 1593 (w), 1505 (w), 1439 (w), 1370 (w), 1329 (w), 1256 (m), 1201 (w), 1170 (w), 1050 (w), 1023 (w), 937 (w), 851 (w), 786 (w), 742 (w), 681 (vw), 593 (vw), 429 (vw) cm^{-1}. – MS (FAB, 3-NBA), m/z (%): 468/470 (3/5) [M]$^+$, 469/471 (10/9) [M+H]$^+$, 245 (37). – HRMS (FAB, C$_{24}$H$_{22}$O$_5$^{79}Br): calc. 469.0651; found 469.0650.

(4aR,9aR)-4a-(2-Bromo-4,5-dimethoxybenzoyl)-2,3-dimethyl-1,4,4a,9a-tetrahydroanthracene-9,10-dione (87dc)

According to **GP3** the cycloaddition was performed with 2-(2-bromo-4,5-dimethoxybenzoyl)naphthalene-1,4-dione (**73d**) (100 mg, 250 µmol, 1.00 equiv) and 2,3-dimethylbuta-1,3-diene (**74c**) (0.15 mL, 103 mg, 1.25 mmol, 5.00 equiv.) in dry CH$_2$Cl$_2$ (1.5 mL). After 3 h the crude product was purified *via* column chromatography on silica gel (cHex/EtOAc = 3:1) to obtain product **87dc** as a yellow solid (47.1 mg, 97.5 µmol, 39%).

– R_f (cHex/EtOAc = 3:1) = 0.31. – ^{1}H NMR (500 MHz, CDCl$_3$): δ = 8.13 (dd, ^3J = 7.6 Hz, ^4J = 1.4 Hz, 1H, CH$_{Ar}$), 8.00 (dd, ^3J = 7.7 Hz, ^4J = 1.3 Hz, 1H, CH$_{Ar}$), 7.77 (td, ^3J = 7.5 Hz, ^4J = 1.4 Hz, 1H, CH$_{Ar}$), 7.72 (td, ^3J = 7.5 Hz, ^4J = 1.4 Hz, 1H, CH$_{Ar}$), 7.00 (s, 1H, CH$_{Ar}$), 6.91 (s, 1H, CH$_{Ar}$), 3.87 (s, 3H, OCH$_3$), 3.83 (s, 3H, OCH$_3$), 3.74–3.69 (m, 1H, CH), 2.94–2.77 (m, 1H, CHH), 2.43–2.23 (m, 3H, CHH + CH$_2$), 1.59 (s, 3H, CH$_3$), 1.55 (s, 3H, CH$_3$) ppm. – ^{13}C NMR (126 MHz, CDCl$_3$): δ = 198.9 (C$_q$, 1 × C=O), 196.5 (C$_q$, 1 × C=O), 195.6 (C$_q$, 1 × C=O), 150.8 (C$_q$, 1 × C$_{qAr}$), 147.4 (C$_q$, 1 × C$_{qAr}$), 135.2 (C$_q$, 1 × C$_{qAr}$), 134.3 (+, 2 × CH$_{Ar}$), 133.3 (C$_q$, 1 × C$_{qAr}$), 130.9 (C$_q$, 1 × C$_{qAr}$), 127.2 (+, 1 × CH$_{Ar}$), 127.1 (+, 1 × CH$_{Ar}$), 124.1 (C$_q$, 1 × C$_{qAr}$), 122.9 (C$_q$, 1 × C$_q$), 116.7 (+, 1 × CH$_{Ar}$), 111.8 (C$_q$, 1 × C$_q$), 110.9 (+, 1 × CH$_{Ar}$), 68.7 (C$_q$, 1 × C$_q$), 56.4 (+, 1 × OCH$_3$), 56.3 (+, 1 × OCH$_3$), 50.4 (+, 1 × CH), 34.9 (–, 1 × CH$_2$), 31.6 (–, 1 × CH$_2$), 19.1 (+, 1 × CH$_3$), 18.8 (+, 1 × CH$_3$) ppm. – IR (ATR): $\tilde{v}$ = 2911 (w), 1676 (vs), 1592 (s), 1504 (vs), 1439 (m), 1370 (m), 1329 (m), 1256 (vs), 1200 (vs), 1169 (vs), 1057 (s), 1024 (s), 858 (w), 788 (m), 749 (vs) cm^{-1}.

– MS (FAB, 3-NBA), m/z (%): 483/485 (9/7) [M+H]$^+$, 243/245 (80/80) [C$_9$H$_9$BrO$_3$]. – HRMS (FAB, C$_{25}$H$_{24}$O$_5$^{79}Br): calc. 483.0807; found 483.0808.

(4aR,9aR)-4a-(2-Bromo-4,5-dihydroxybenzoyl)-2,3-dimethyl-1,4,4a,9a-tetrahydroanthracene-9,10-dione (87ec)

According to **GP3** the cycloaddition was performed with 2-(2-bromo-4,5-dihydroxybenzoyl)naphthalene-1,4-dione **(73e)** (56.0 mg, 150 µmol, 1.00 equiv.) and 2,3-dimethylbuta-1,3-diene **(74c)** (0.09 mL, 61.6 mg, 750 µmol, 5.00 equiv.) in dry CH$_2$Cl$_2$ (2 mL). After 4 h the crude product was purified *via* column chromatography on silica gel (*c*Hex/EtOAc = 3:1) to obtain product **87ec** as a yellow solid (36.9 mg, 81.0 µmol, 54%).

– R_f (*c*Hex/EtOAc = 3:1) = 0.08. – ^{1}H NMR (500 MHz, CDCl$_3$): δ = 8.13 (d, ^3J = 7.7 Hz, 1H, CH_{Ar}), 7.97 (d, ^3J = 7.7 Hz, 1H, CH_{Ar}), 7.77 (t, ^3J = 7.5 Hz, 1H, CH_{Ar}), 7.71 (t, ^3J = 7.5 Hz, 1H, CH_{Ar}), 7.06 (s, 1H, CH_{Ar}), 6.93 (s, 1H, CH_{Ar}), 3.68 (dd, ^3J = 8.6, 6.6 Hz, 1H, CH), 2.83 (d, ^2J = 17.1 Hz, 1H, CHH), 2.37–2.22 (m, 3H, CHH, CH_2), 1.59 (s, 3H, CH_3), 1.56 (s, 3H, CH_3) ppm. Signals missing (2H, 2 × OH). – ^{13}C NMR (126 MHz, CDCl$_3$): δ = 199.0 (C$_q$, 1 × C=O), 196.9 (C$_q$, 1 × C=O), 195.8 (C$_q$, 1 × C=O), 146.6 (C$_q$, 1 × C_{qAr}), 142.1 (C$_q$, 1 × C_{qAr}), 135.3 (+, 1 × CH_{Ar}), 134.4 (+, 1 × CH_{Ar}), 134.3 (C$_q$, 1 × C_{qAr}), 133.3 (C$_q$, 1 × C_{qAr}), 130.9 (C$_q$, 1 × C_{qAr}), 127.3 (+, 1 × CH_{Ar}), 127.2 (+, 1 × CH_{Ar}), 124.0 (C$_q$, 1 × C_{qAr}), 123.0 (C$_q$, 1 × C_q), 120.8 (+, 1 × CH_{Ar}), 114.8 (+, 1 × CH_{Ar}), 111.4 (C$_q$, 1 × C_q), 68.7 (C$_q$, 1 × C_q), 50.5 (+, 1 × CH), 34.9 (–, 1 × CH_2), 31.6 (–, 1 × CH_2), 19.1 (+, 1 × CH_3), 18.8 (+, 1 × CH_3) ppm. – IR (ATR): $\tilde{v}$ = 3338 (vw), 2923 (w), 1678 (w), 1590 (w), 1501 (w), 1415 (w), 1257 (w), 1173 (w), 1093 (w), 1043 (w), 1021 (w), 944 (vw), 863 (vw), 797 (w), 753 (w), 728 (vw), 675 (vw), 634 (vw), 608 (vw), 560 (vw), 448 (vw), 399 (vw) cm^{-1}. – MS (APCI), m/z (%): 455/457 (97/100) [M+H]$^+$. – HRMS (APCI, C$_{23}$H$_{20}$^{79}BrO$_5$): calc. 455.0494; found 455.0476.

Cycloaddition products (87/88fb) of *N*-(4-bromo-3-(1,4-dioxo-1,4-dihydronaphthalene-2-carbonyl)phenyl)acetamide (73f) and isoprene (74b)

87fb 88fb

According to **GP3** the cycloaddition was performed with *N*-(4-bromo-3-(1,4-dioxo-1,4-dihydronaphthalene-2-carbonyl)phenyl)acetamide (**73f**) (79.6 mg, 200 μmol, 1.00 equiv.) and isoprene (**74b**) (0.06 mL, 40.9 mg, 600 μmol, 3.00 equiv.) in dry CH_2Cl_2 (1 mL). After 4 h the crude product was purified *via* column chromatography on silica gel (*c*Hex/EtOAc = 1:2) to obtain a yellow solid (35.4 mg, 76.0 μmol, 38%). The products **87/88fb** were isolated as a non-separable mixture of regioisomers in a ~ 7.7 : 1 ratio as estimated by ^{1}H NMR.

Both possible regioisomers are drawn, as the exact structure of the products could not be resolved by analysis of the NMR spectra. Below the analytics of the major product are given.

– R_f (*c*Hex/EtOAc = 1:2) = 0.30. – ^{1}H NMR (500 MHz, $CDCl_3$): δ = 8.09 (dd, ^3J = 7.7 Hz, ^4J = 1.3 Hz, 1H, CH_{Ar}), 8.03 (brs, 1H, NH), 7.99 (dd, ^3J = 7.7 Hz, ^4J = 1.4 Hz, 1H, CH_{Ar}), 7.74 (td, ^3J = 7.5 Hz, ^4J = 1.4 Hz, 1H, CH_{Ar}), 7.67 (td, ^3J = 7.5 Hz, ^4J = 1.4 Hz, 1H, CH_{Ar}), 7.54 (t, ^4J = 1.5 Hz, 1H, CH_{Ar}), 7.42 (d, ^4J = 1.5 Hz, 2H, CH_{Ar}), 5.27 (tq, ^4J = 3.5, 1.7 Hz, 1H, C=CH), 3.74 (dd, ^3J = 8.2, 6.6 Hz, 1H, CH), 2.84 (ddq, ^2J = 17.4 Hz, ^3J = 4.1 Hz, ^4J = 1.9 Hz, 1H, CHH), 2.49 (ddq, ^2J = 17.4 Hz, ^3J = 4.2 Hz, ^4J = 2.1 Hz, 1H, CHH), 2.40–2.19 (m, 2H, CH_2), 2.13 (s, 3H, COCH_3), 1.62 (d, ^4J = 2.0 Hz, 3H, CH=CCH_3) ppm. – ^{13}C NMR (126 MHz, $CDCl_3$): δ = 200.2 (C_q, 1 × C=O), 196.4 (C_q, 1 × C=O), 194.6 (C_q, 1 × C=O), 169.0 (C_q, 1 × C_{qAr}), 139.1 (C_q, 1 × C_{qAr}), 137.1 (C_q, 1 × C_{qAr}), 135.3 (+, 1 × CH_{Ar}), 134.4 (+, 1 × CH_{Ar}), 134.3 (+, 1 × CH_{Ar}), 134.1 (C_q, 1 × C_{qAr}), 133.0 (C_q, 1 × C_{qAr}), 132.3 (C_q, 1 × C_q), 127.6 (+, 1 × CH_{Ar}), 127.0 (+, 1 × CH_{Ar}), 122.4 (+, 1 × CH_{Ar}), 118.2 (+, 1 × CH_{Ar}), 117.7 (+, 1 × C=CH), 113.7 (C_q, 1 × C_q), 67.7 (C_q, 1 × C_q), 49.9 (+, 1 × CH), 29.9 (–, 1 × CH$_2$), 28.6 (–, 1 × CH$_2$), 24.6 (+, 1 × COCH$_3$), 23.2 (+, 1 × C=CCH$_3$) ppm. – IR (ATR): $\tilde{v}$ = 3306 (w), 2912 (w), 1677 (vs), 1581 (vs), 1530 (vs), 1466 (vs), 1391 (vs), 1373 (s), 1252 (vs), 1187 (vs), 1047 (s), 1016 (vs), 936 (s), 909 (m), 802 (vs), 773 (s), 728 (vs), 688 (s), 647 (s), 591 (s), 561 (s), 484 (s), 418 (s) cm^{-1}. – MS (FAB, 3-NBA), *m/z* (%): 466/468 (19/20) [M+H]$^+$, 242 (97), 240 (100). – HRMS (FAB, $C_{24}H_{21}O_4N^{79}Br$): calc. 466.0654; found 466.0652.

N-(4-Bromo-3-((4aR,9aR)-2,3-dimethyl-9,10-dioxo-1,4,4a,9,9a,10-hexahydroanthracene-4a-carbonyl)phenyl)acetamide (87fc)

According to **GP3** the cycloaddition was performed with N-(4-bromo-3-(1,4-dioxo-1,4-dihydronaphthalene-2-carbonyl)phenyl)acetamide **(73f)** (119 mg, 300 µmol, 1.00 equiv.) and 2,3-dimethylbuta-1,3-diene **(74c)** (0.10 mL, 73.9 mg, 900 µmol, 3.00 equiv.) in dry CH_2Cl_2 (1 mL). After 5 h the crude product was purified *via* column chromatography on silica gel (cHex/EtOAc = 1:1) to obtain product **87fc** as a yellow solid (98.0 mg, 204 µmol, 68%).

$- R_f$ (cHex/EtOAc = 1:1) = 0.30. $-$ ^{1}H NMR (500 MHz, CDCl$_3$): δ = 8.13 (d, ^3J = 7.6 Hz, 1H, CH_{Ar}), 8.06 (d, ^3J = 7.5 Hz, 1H, CH_{Ar}), 7.76 (dt, ^3J = 21.3, 7.4 Hz, 2H, CH_{Ar}), 7.52–7.47 (m, 3H, CH_{Ar}), 7.40 (brs, 1H, NH), 3.74 (t, ^3J = 7.3 Hz, 1H, CH), 2.77 (d, ^2J = 17.1 Hz, 1H, CHH), 2.42–2.36 (m, 2H, CHH, CHH), 2.33–2.23 (m, 1H, CHH), 2.18 (s, 3H, COCH_3), 1.59 (s, 3H, CH_3), 1.49 (s, 3H, CH_3) ppm. $-$ ^{13}C NMR (126 MHz, CDCl$_3$): δ = 200.0 (C$_q$, 1 × C=O), 196.6 (C$_q$, 1 × C=O), 195.2 (C$_q$, 1 × C=O), 168.6 (C$_q$, 1 × C=O), 139.3 (C$_q$, 1 × C_{qAr}), 136.8 (C$_q$, 1 × C_{qAr}), 135.3 (+, 1 × CH_{Ar}), 134.4 (+, 1 × CH_{Ar}), 134.3 (+, 1 × CH_{Ar}), 133.1 (C$_q$, 1 × C_{qAr}), 127.6 (+, 1 × CH_{Ar}), 127.1 (+, 1 × CH_{Ar}), 124.4 (C$_q$, 1 × C_{qAr}), 122.8 (C$_q$, 1 × C_{qAr}), 122.3 (+, 1 × CH_{Ar}), 118.3 (+, 1 × CH_{Ar}), 114.0 (C$_q$, 1 × C_q), 68.5 (C$_q$, 1 × C_q), 50.1 (+, 1 × CH), 34.3 (−, 1 × CH_2), 31.3 (−, 1 × CH_2), 24.8 (+, 1 × CH_3), 19.1 (+, 1 × CH_3), 18.9 (+, 1 × CH_3), 14.3 (C$_q$, 1 × C_q) ppm. $-$ IR (ATR): $\tilde{v}$ = 3310 (vw), 3265 (vw), 3186 (vw), 3111 (vw), 3065 (vw), 2928 (vw), 2858 (vw), 1678 (m), 1591 (w), 1538 (m), 1471 (w), 1370 (w), 1317 (w), 1247 (m), 1199 (w), 1012 (w), 886 (w), 856 (w), 835 (w), 779 (w), 755 (w), 718 (w), 592 (w), 558 (w), 486 (w), 421 (w), 388 (w) cm^{-1}. $-$ MS (FAB, 3-NBA), m/z (%): 480 (13) [M+H]$^+$, 240/242 (100/95). $-$ HRMS (FAB, $C_{25}H_{23}O_4N^{79}$Br): calc. 480.0810; found 480.0811. $-$ X-Ray: Crystallographic information on the product can be found in chapter 7.3.2.

Cycloaddition products (87/88gb) of *N*-(4-bromo-3-(1,4-dioxo-1,4-dihydronaphthalene-2-carbonyl)phenyl)-2,2,2-trifluoroacetamide (73g) and isoprene (74b)

87gb   88gb

According to **GP3** the cycloaddition was performed with *N*-(4-bromo-3-(1,4-dioxo-1,4-dihydronaphthalene-2-carbonyl)phenyl)-2,2,2-trifluoroacetamide (**73g**) (90.4 mg, 200 µmol, 1.00 equiv.) and isoprene (**74b**) (0.06 mL, 40.9 mg, 600 µmol, 3.00 equiv.) in dry CH_2Cl_2 (1 mL). After 3 h the crude product was purified *via* column chromatography on silica gel (*c*Hex/EtOAc = 4:1) to obtain a light yellow solid (82.2 mg, 158 µmol, 79%). The products **87/88gb** were isolated as a non-separable mixture of regioisomers in a ~ 7.7 : 1 ratio as estimated by ^{1}H NMR.

Both possible regioisomers are drawn, as the exact structure of the products could not be resolved by analysis of the NMR spectra. Below the analytics of the major product are given.

– R_f (*c*Hex/EtOAc = 4:1) = 0.32. – ^{1}H NMR (500 MHz, CDCl$_3$): δ = 8.73 (brs, 1H, N*H*), 8.08 (dd, ^3J = 7.6 Hz, ^4J = 1.4 Hz, 1H, C*H*$_{Ar}$), 7.99 (dd, ^3J = 7.6 Hz, ^4J = 1.4 Hz, 1H, C*H*$_{Ar}$), 7.76 (td, ^3J = 7.5 Hz, ^4J = 1.5 Hz, 1H, C*H*$_{Ar}$), 7.72 (td, ^3J = 7.5 Hz, ^4J = 1.5 Hz, 1H, C*H*$_{Ar}$), 7.67 (d, ^3J = 2.4 Hz, 1H, C*H*$_{Ar}$), 7.57–7.48 (m, 2H, C*H*$_{Ar}$), 5.25 (app. s, 1H, C=C*H*), 3.79 (dd, ^3J = 7.7, 6.4 Hz, 1H, C*H*), 2.84–2.78 (m, 1H, C*H*H), 2.53–2.48 (m, 1H, CH*H*), 2.42–2.30 (m, 1H, C*H*H), 2.25 (dd, ^3J = 18.2, 6.4 Hz, 1H, CH*H*), 1.63 (s, 3H, C*H*$_3$) ppm. – ^{13}C NMR (126 MHz, CDCl$_3$): δ = 200.1 (C$_q$, 1 × *C*=O), 196.6 (C$_q$, 1 × *C*=O), 194.9 (C$_q$, 1 × *C*=O), 155.4 (q, ^2J = 38.1 Hz, 1 × *C*OCF$_3$), 139.5 (C$_q$, 1 × *C*$_{qAr}$), 135.4 (+, 1 × *C*H$_{Ar}$), 134.8 (+, 1 × *C*H$_{Ar}$), 134.6 (+, 1 × *C*H$_{Ar}$), 134.3 (C$_q$, 1 × *C*$_{qAr}$), 134.1 (C$_q$, 1 × *C*$_{qAr}$), 132.8 (C$_q$, 1 × *C*$_{qAr}$), 132.8 (C$_q$, 1 × *C*$_{qAr}$), 127.7 (+, 1 × *C*H$_{Ar}$), 127.0 (+, 1 × *C*H$_{Ar}$), 123.3 (+, 1 × *C*H$_{Ar}$), 119.4 (+, 1 × *C*H$_{Ar}$), 117.4 (+, 1 × C=*C*H), 115.6 (q, ^1J = 288.0 Hz, 1 × *C*F$_3$), 67.6 (C$_q$, 2 × C$_q$), 49.8 (+, 1 × *C*H), 29.6 (–, 1 × *C*H$_2$), 28.7 (–, 1 × *C*H$_2$), 23.2 (+, 1 × *C*H$_3$) ppm. – IR (ATR): $\tilde{v}$ = 3298 (w), 2917 (w), 1684 (vs), 1589 (s), 1545 (m), 1468 (m), 1276 (s), 1252 (vs), 1210 (vs), 1150 (vs), 1047 (s), 936 (w), 899 (w), 824 (m), 800 (w), 738 (s), 725 (m), 686 (w), 591 (w), 562 (w), 484 (w), 435 (w) cm^{-1}. – MS (FAB, 3-NBA), *m/z* (%): 520/522 (27/23) [M+H]$^+$, 294/296 (100/98), 225 (71), 209 (26). – HRMS (FAB, C$_{24}$H$_{18}$O$_4$N^{79}BrF$_3$): calc. 520.0371; found 520.0370.

N-(4-Bromo-3-((4aS,9aS)-2,3-dimethyl-9,10-dioxo-1,4,4a,9,9a,10-hexahydroanthracene-4a-carbonyl)phenyl)-2,2,2-trifluoroacetamide (87gc)

According to **GP3** the cycloaddition was performed with *N*-(4-bromo-3-(1,4-dioxo-1,4-dihydronaphthalene-2-carbonyl)phenyl)-2,2,2-trifluoroacetamide (**73g**) (181 mg, 400 µmol, 1.00 equiv.) and 2,3-dimethylbuta-1,3-diene (**74c**) (0.14 mL, 98.6 mg, 1.20 mmol, 3.00 equiv.) in dry CH$_2$Cl$_2$ (4 mL). After 5 h the crude product was purified *via* column chromatography on silica gel (*c*Hex/EtOAc = 4:1) to obtain product **87gc** as a yellow solid (180 mg, 336 µmol, 84%).

– R_f (*c*Hex/EtOAc = 4:1) = 0.37. – ^{1}H NMR (500 MHz, CDCl$_3$): δ = 8.15 (brs, 1H, N*H*), 8.12 (dd, ^3J = 7.5 Hz, ^4J = 1.5 Hz, 1H, C*H*$_{Ar}$), 8.07 (dd, ^3J = 7.5 Hz, ^4J = 1.5 Hz, 1H, C*H*$_{Ar}$), 7.78 (dtd, ^3J = 17.0, 7.5 Hz, ^4J = 1.5 Hz, 2H, C*H*$_{Ar}$), 7.65 (d, ^3J = 2.5 Hz, 1H, C*H*$_{Ar}$), 7.60 (d, ^3J = 8.7 Hz, 1H, C*H*$_{Ar}$), 7.55 (dd, ^3J = 8.7 Hz, ^4J = 2.5 Hz, 1H, C*H*$_{Ar}$), 3.80 (t, ^3J = 6.9 Hz, 1H, C*H*), 2.73 (d, ^2J = 17.1 Hz, 1H, C*H*H), 2.49–2.40 (m, 2H, 2 × C*H*H), 2.34–2.24 (m, 1H, C*H*H), 1.61 (s, 3H, C*H*$_3$), 1.45 (s, 3H, C*H*$_3$) ppm. – ^{13}C NMR (126 MHz, CDCl$_3$): δ = 199.9 (C$_q$, 1 × *C*=O), 196.5 (C$_q$, 1 × *C*=O), 195.5 (C$_q$, 1 × *C*=O), 155.1 (q, ^2J = 38.1 Hz, 1 × *C*OCF$_3$), 139.8 (C$_q$, 1 × *C*$_{qAr}$), 135.4 (+, 1 × *C*H$_{Ar}$), 134.9 (+, 1 × *C*H$_{Ar}$), 134.5 (+, 1 × *C*H$_{Ar}$), 134.3 (C$_q$, 1 × *C*$_{qAr}$), 134.0 (C$_q$, 1 × *C*$_{qAr}$), 132.9 (C$_q$, 1 × *C*$_{qAr}$), 127.7 (+, 1 × *C*H$_{Ar}$), 127.1 (+, 1 × *C*H$_{Ar}$), 125.1 (C$_q$, 1 × *C*$_{qAr}$), 123.0 (+, 1 × *C*H$_{Ar}$), 122.5 (C$_q$, 1 × *C*$_q$), 119.4 (+, 1 × *C*H$_{Ar}$), 116.8 (C$_q$, 1 × *C*$_q$), 115.6 (q, ^1J = 289.0 Hz, 1 × *C*F$_3$), 68.2 (C$_q$, 1 × *C*$_q$), 49.8 (+, 1 × *C*H), 34.4 (–, 1 × *C*H$_2$), 30.8 (–, 1 × *C*H$_2$), 19.1 (+, 1 × *C*H$_3$), 19.0 (+, 1 × *C*H$_3$) ppm. – IR (ATR): $\tilde{v}$ = 3361 (vw), 2911 (vw), 1699 (w), 1673 (w), 1588 (w), 1543 (vw), 1466 (vw), 1402 (vw), 1282 (w), 1258 (w), 1239 (w), 1135 (w), 1057 (w), 875 (vw), 831 (w), 793 (vw), 757 (vw), 741 (w), 728 (vw), 671 (vw), 645 (vw), 597 (vw), 566 (vw), 477 (vw), 434 (vw), 402 (vw) cm^{-1}. – MS (FAB, 3-NBA), *m/z* (%): 534/536 (25/21) [M+H]$^+$, 294 (68), 239 (87). – HRMS (FAB, C$_{25}$H$_{20}$O$_4$N^{79}BrF$_3$): calc. 534.0528; found 534.0530. – X-Ray: Crystallographic information on the product can be found in chapter 7.3.2.

Cycloaddition products (87/88gd) of *N*-(4-bromo-3-(1,4-dioxo-1,4-dihydronaphthalene-2-carbonyl)phenyl)-2,2,2-trifluoroacetamide (73g) and 1-(trimethylsiloxy)-1,3-butadiene (74d)

87gd   88gd

According to **GP3** the cycloaddition was performed with *N*-(4-bromo-3-(1,4-dioxo-1,4-dihydronaphthalene-2-carbonyl)phenyl)-2,2,2-trifluoroacetamide (**73g**) (136 mg, 300 µmol, 1.00 equiv.) and 1-(trimethylsiloxy)-1,3-butadiene (**74d**) (0.26 mL, 213 mg, 1.50 mmol, 5.00 equiv.) in dry CH_2Cl_2 (2 mL). After 4 h the crude product was purified *via* column chromatography on silica gel (*c*Hex/EtOAc = 5:1) to obtain isomers **87gd** (28.9 mg, 48.8 µmol, 16%) and **88gd** (107 mg, 180 µmol, 60%) as yellow solids.

The exact structure of **88gd** could be verified by X-ray analysis. For **87gd** two possible isomers are drawn as the exact structure of the product could not be resolved by analysis of the NMR spectra.

87gd: R_f (*c*Hex/EtOAc = 5:1) = 0.31. – ^1H NMR (500 MHz, CDCl$_3$): δ = 8.26 (brs, 1H, N*H*), 8.08 (s, 1H, C*H*$_{Ar}$), 8.05 (d, ^3J = 7.8 Hz, 1H, C*H*$_{Ar}$), 7.97 (d, ^3J = 7.8 Hz, 1H, C*H*$_{Ar}$), 7.76 (td, ^3J = 7.6 Hz, ^4J = 1.3 Hz, 1H, C*H*$_{Ar}$), 7.68 (td, ^3J = 7.6 Hz, ^4J = 1.3 Hz, 1H, C*H*$_{Ar}$), 7.54 (s, 2H, C*H*$_{Ar}$), 6.04–6.00 (m, 1H, C=C*H*), 5.85–5.81 (m, 1H, C=C*H*), 5.38 (d, ^3J = 5.2 Hz, 1H, C(OTMS)*H*), 3.90 (dd, ^3J = 11.6, 6.6 Hz, 1H, COC*H*), 2.58–2.52 (m, 1H, C*H*H), 2.12–2.05 (m, 1H, CH*H*), 0.01 (s, 9H, Si(C*H*$_3$)$_3$) ppm. – ^{13}C NMR (126 MHz, CDCl$_3$): δ = 196.2 (C$_q$, 1 × *C*=O), 196.0 (C$_q$, 1 × *C*=O), 192.8 (C$_q$, 1 × *C*=O), 155.2 (q, ^2J = 38.0 Hz, 1 × *C*OCF$_3$), 138.9 (C$_q$, 1 × *C*$_{qAr}$), 135.5 (+, 1 × *C*H$_{Ar}$), 135.5 (+, 1 × *C*H$_{Ar}$), 134.6 (C$_q$, 1 × *C*$_{qAr}$), 134.4 (+, 1 × *C*H$_{Ar}$), 134.1 (C$_q$, 1 × *C*$_{qAr}$), 133.8 (C$_q$, 1 × *C*$_{qAr}$), 127.6 (+, 1 × *C*=CH), 127.5 (+, 1 × *C*H$_{Ar}$), 127.4 (+, 1 × *C*H$_{Ar}$), 127.2 (+, 1 × *C*=CH), 123.7 (+, 1 × *C*H$_{Ar}$), 120.4 (+, 1 × *C*H$_{Ar}$), 118.4 (C$_q$, 1 × *C*$_{qAr}$), 115.7 (q, ^1J = 291.0 Hz, 1 × *C*F$_3$), 73.7 (C$_q$, 1 × *C*$_q$), 66.1 (+, 1 × *C*H), 46.1 (+, 1 × *C*H), 28.0 (–, 1 × *C*H$_2$), 0.6 (+, 3 × Si*C*H$_3$) ppm. – IR (ATR): $\tilde{v}$ = 3289 (w), 1730 (s), 1713 (s), 1683 (vs), 1587 (s), 1545 (w), 1468 (m), 1404 (w), 1290 (s), 1248 (vs), 1218 (vs), 1203 (vs), 1154 (vs), 1137 (vs), 1072 (vs), 1051 (vs), 1021 (s), 945 (m), 894 (vs), 836 (vs), 782 (s), 745 (vs), 718 (vs), 701 (s), 645 (m), 608 (s), 578 (s), 492 (s), 401 (s) cm^{-1}. – MS (FAB, 3-NBA) *m/z* (%): 594/596 (15/13) [M+H]$^+$, 452/454

(20/22), 299 (40), 294/296 (100/96). – HRMS (FAB, $C_{26}H_{24}O_5N^{79}BrF_3Si$): calc. 594.0559; found 594.0561.

88gd: R_f (cHex/EtOAc = 5:1) = 0.39. – ^{1}H NMR (500 MHz, CDCl$_3$): δ = 8.21 (d, ^3J = 2.6 Hz, 1H, CH_{Ar}), 8.15–8.12 (m, 1H, CH_{Ar}), 8.10 (brs, 1H, NH), 8.07–8.02 (m, 1H, CH_{Ar}), 7.80–7.70 (m, 2H, CH_{Ar}), 7.63 (d, ^3J = 8.7 Hz, 1H, CH_{Ar}), 7.56 (dd, ^3J = 8.7 Hz, ^4J = 2.7 Hz, 1H, CH_{Ar}), 5.95 (ddd, ^3J = 10.1, 4.5 Hz, ^4J = 2.9 Hz, 1H, C=CH), 5.62 (ddt, ^3J = 10.0, 5.0 Hz, ^4J = 2.3 Hz, 1H, C=CH), 4.90 (d, ^3J = 5.3 Hz, 1H, C(OTMS)H), 4.01 (d, ^3J = 7.2 Hz, 1H, COCH), 3.28–3.23 (m, 1H, CHH), 2.32–2.17 (m, 1H, CHH), –0.34 (s, 9H, Si(CH_3)$_3$) ppm. – ^{13}C NMR (126 MHz, CDCl$_3$): δ = 197.8 (C$_q$, 1 × C=O), 197.4 (C$_q$, 2 × C=O), 194.8 (C$_q$, 1 × C=O), 155.0 (q, ^2J = 38.0 Hz, 1 × COCF$_3$), 139.8 (C$_q$, 1 × C_{qAr}), 137.5 (C$_q$, 1 × C_{qAr}), 135.4 (+, 1 × CH$_{Ar}$), 135.0 (C$_q$, 1 × C_{qAr}), 134.7 (+, 1 × CH$_{Ar}$), 134.0 (C$_q$, 1 × C_{qAr}), 133.4 (+, 1 × CH$_{Ar}$), 131.2 (+, 1 × C=CH), 127.0 (+, 1 × CH$_{Ar}$), 126.2 (+, 1 × C=CH), 125.8 (+, 1 × CH$_{Ar}$), 123.4 (+, 1 × CH$_{Ar}$), 120.5 (+, 1 × CH$_{Ar}$), 118.1 (C$_q$, 1 × C_{qAr}), 115.6 (q, ^1J = 290.0 Hz, 1 × CF$_3$), 70.7 (C$_q$, 1 × C_q), 68.0 (+, 1 × CH), 45.7 (+, 1 × CH), 21.2 (–, 1 × CH$_2$), –0.6 (+, 3 × SiCH$_3$) ppm. – IR (ATR): $\tilde{v}$ = 1732 (s), 1701 (m), 1687 (vs), 1664 (s), 1592 (w), 1545 (s), 1472 (w), 1279 (s), 1254 (vs), 1187 (vs), 1157 (vs), 1079 (s), 1061 (vs), 1044 (s), 909 (w), 867 (vs), 841 (vs), 765 (m), 738 (s), 691 (vs), 670 (m), 611 (m), 497 (m), 475 (m), 416 (m) cm^{-1}. – MS (FAB, 3-NBA) m/z (%): 594/596 (28/24) [M+H]$^+$, 299 (36), 294/296 (91/87). – HRMS (FAB, $C_{26}H_{24}O_5N^{79}BrF_3Si$): calc. 594.0559; found 594.0561. – X-Ray: Crystallographic information on the product can be found in chapter 7.3.2.

3-(Furan-2-yl)-4-hydroxy-2-(2-iodobenzoyl)naphthalen-1(2H)-one (92)

2-(2-Iodobenzoyl)naphthalene-1,4-dione (**73a**) (77.6 mg, 200 µmol, 1.00 equiv.) in furane (0.73 mL, 681 mg, 10.0 mmol, 50.0 equiv.) was stirred at room temperature for 80 h. The reaction mixture was concentrated under reduced pressure. After column chromatography on silica gel (cHex/EtOAc = 5:1) the product was isolated as a brown solid (21.4 mg, 46.8 µmol, 39%).

– R_f (cHex/EtOAc = 5:1) = 0.39. – ^{1}H NMR (300 MHz, CDCl$_3$): δ = 12.66 (d, ^5J = 1.4 Hz, 1H, C=COH), 8.54 (d, ^3J = 8.2 Hz, 1H, CH_{Ar}), 8.28 (d, ^3J = 8.3 Hz, 1H, CH_{Ar}), 7.84–7.60 (m, 3H, CH_{Ar}), 7.30–7.26 (m, 1H, CH_{Ar}), 7.18–7.08 (m, 1H, CH_{Ar}), 7.04 (dd, ^3J = 7.7 Hz, ^4J = 1.8 Hz, 1H, CH_{Ar}), 6.87 (td, ^3J = 7.6 Hz, ^4J = 1.7 Hz, 1H, CH_{Ar}), 6.33–6.25 (m, 1H, CH_{Ar}), 6.16–6.09 (m, 1H, CH_{Ar}), 6.07 (s, 1H, CH) ppm. – IR (ATR): $\tilde{v}$ = 3151 (vw), 3061 (w), 2922 (w), 2852 (w), 1671 (m), 1640 (m), 1571 (m), 1459 (m), 1427 (m), 1388 (m), 1321 (w), 1267 (s), 1161 (m), 1051 (m), 1016 (m), 986 (m), 913 (w), 883 (w), 828 (w), 780 (m), 720 (m), 677 (m), 636 (m),

607 (m), 589 (w) cm^{-1}. – MS (FAB, 3-NBA), m/z (%): 457 (36) [M+H]$^+$, 456 (69) [M]$^+$. – HRMS (FAB, C$_{21}$H$_{13}$O$_4$^{127}I): calc. 455.9859; found 455.9860. – Due to the instability of the product, a proper analysis of the ^{13}C NMR spectrum was not possible.

Cycloaddition products (161a/b) of 2-(2-iodobenzoyl)naphthalene-1,4-dione (73a) and 2-chloro-3-methylbuta-1,3-diene (149)

According to **GP3** the cycloaddition was performed with 2-(2-iodobenzoyl)naphthalene-1,4-dione (**73a**) (155 mg, 400 µmol, 1.00 equiv.) and 2-chloro-3-methylbuta-1,3-diene (**149**) (0.19 mL, 180 mg, 880 µmol, 2.20 equiv.) in dry CH$_2$Cl$_2$ (2 mL). After 3 h the crude product was purified *via* column chromatography on silica gel (*c*Hex/EtOAc = 8:1) to obtain an off-white solid (91.0 mg, 190 µmol, 48%). The products **161a** and **161b** were isolated as a non-separable mixture in a ~ 1.8 : 1 ratio as estimated by ^{1}H NMR.

Both possible regioisomers are drawn, as the exact structure of the products could not be resolved by analysis of the NMR spectra. Below the analytics of the major product are given.

– R_f (*c*Hex/EtOAc = 8:1) = 0.32. – ^{1}H NMR (500 MHz, CDCl$_3$): δ = 8.20–8.14 (m, 1H, CH_{Ar}), 8.07–8.02 (m, 1H, CH_{Ar}), 7.92–7.87 (m, 1H, CH_{Ar}), 7.84–7.79 (m, 1H, CH_{Ar}), 7.79–7.73 (m, 1H, CH_{Ar}), 7.37–7.31 (m, 1H, CH_{Ar}), 7.20–7.15 (m, 1H, CH_{Ar}), 7.14–7.08 (m, 1H, CH_{Ar}), 3.84 (t, ^3J = 7.7 Hz, 1H, CH), 2.98–2.89 (m, 1H, CHH), 2.67 (dq, ^3J = 7.7, 2.0 Hz, 2H, CH_2), 2.61–2.50 (m, 1H, CH-H), 1.73 (s, 3H, CH_3) ppm. – ^{13}C NMR (126 MHz, CDCl$_3$): δ = 200.0 (C$_q$, 1 × C=O), 194.4 (C$_q$, 1 × C=O), 193.7 (C$_q$, 1 × C=O), 142.2 (C$_q$, 1 × C_{qAr}), 141.1 (+, 1 × CH_{Ar}), 135.6 (+, 1 × CH_{Ar}), 134.6 (+, 1 × CH_{Ar}), 134.1 (C$_q$, 1 × C_{qAr}), 132.8 (C$_q$, 1 × C_{qAr}), 131.8 (+, 1 × CH_{Ar}), 127.6 (+, 1 × CH_{Ar}), 127.6 (+, 1 × CH_{Ar}), 127.4 (+, 1 × CH_{Ar}), 127.0 (+, 1 × CH_{Ar}), 126.8 (C$_q$, 1 × C_{qAr}), 123.5 (C$_q$, 1 × C_{qAr}), 93.1 (C$_q$, 1 × C_{qAr}), 67.9 (C$_q$, 1 × C_q), 50.9 (+, 1 × CH), 34.8 (–, 1 × CH_2), 33.1 (–, 1 × CH_2), 20.0 (+, 1 × CH_3) ppm. – IR (ATR): $\tilde{v}$ = 2917 (w), 1690 (vs), 1679 (vs), 1589 (s), 1439 (w), 1422 (m), 1377 (w), 1344 (w), 1272 (vs), 1254 (vs), 1244 (vs), 1213 (s), 1200 (vs), 1163 (m), 1119 (m), 1062 (m), 1054 (m), 1033 (m), 1006 (vs), 975 (s), 946 (m), 897 (s), 880 (m), 841 (w), 812 (w), 790 (s), 765 (s), 752 (s), 742 (vs), 734 (vs), 707 (vs), 690 (m), 670 (s), 636 (s), 584 (m), 547 (s), 500 (m), 470 (m), 450 (m), 432 (s), 421 (s), 387 (s) cm^{-1}. – MS (FAB, 3-NBA) m/z (%): 491/493 (62/29) [M+H]$^+$. – HRMS (FAB, C$_{22}$H$_{17}$O$_3$^{35}Cl^{127}I): calc. 490.9911; found 490.9912. – X-Ray: Crystallographic information on the product can be found in chapter 7.3.2.

Cycloaddition products (162a/b) of 2-(2-bromobenzoyl)naphthalene-1,4-dione (73b) and 2-iodo-3-methylbuta-1,3-diene (154)

According to **GP3** the cycloaddition was performed with 2-(2-bromobenzoyl)naphthalene-1,4-dione (**73b**) (188 mg, 550 μmol, 1.00 equiv.) and 2-iodo-3-methylbuta-1,3-diene (**154**) (0.26 mL, 427 mg, 2.20 mmol, 4.00 equiv.) in dry CH_2Cl_2 (1 mL) at room temperature under the exclusion of light. After 20 h the crude product was purified *via* column chromatography on silica gel (*c*Hex/EtOAc = 9:1) to obtain a yellow solid (212 mg, 395 μmol, 72%). The products **162a** and **162b** were isolated as a non-separable mixture of regioisomers in a ~ 1 : 1 ratio as estimated by ^{1}H NMR.

Both possible regioisomers are drawn, as the exact structure of the products could not be resolved by analysis of the NMR spectra. Due to the 1 : 1 ratio of the isomers NMR signals could not be assigned.

– R_f (*c*Hex/EtOAc = 9:1) = 0.26. – IR (ATR): $\tilde{v}$ = 3058 (w), 2898 (w), 1691 (vs), 1680 (vs), 1650 (w), 1588 (m), 1439 (w), 1419 (m), 1273 (s), 1254 (vs), 1244 (vs), 1213 (m), 1200 (vs), 1160 (w), 1123 (w), 1082 (w), 1060 (w), 1027 (m), 996 (s), 970 (m), 942 (m), 912 (w), 885 (m), 874 (m), 790 (s), 764 (s), 754 (s), 737 (vs), 705 (s), 694 (m), 676 (m), 657 (w), 636 (m), 613 (m), 584 (m), 550 (m), 524 (w), 469 (w), 446 (m), 439 (m), 425 (m), 407 (m), 384 (m) cm^{-1}. – MS (FAB, 3-NBA) *m/z* (%): 535/537 (16/15) [M+H]$^+$, 351 (25), 183/185 (100/98) [C_7H_4BrO]$^+$. – HRMS (ESI+, $C_{22}H_{17}BrIO_3$): calc. 534.9406; found 534.9407.

7.2.6 Synthesis of the Heck Products

(5*S*,13a*R*)-6*H*-5,13a-Methanobenzo[4,5]cycloocta[1,2-b]naphthalene-8,13,14(5*H*)-trione (111a)

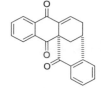

According to **GP4** the coupling was performed with (4a*R*,9a*R*)-4a-(2-iodobenzoyl)-1,4,4a,9a-tetrahydroanthracene-9,10-dione **(87aa)** (53.1 mg, 120 μmol, 1.00 equiv.), Pd(OAc)₂ (5.39 mg, 24.0 μmol, 20 mol%), PPh₃ (12.6 mg, 48.0 μmol, 40 mol%) and PMP (45 μL, 37.2 mg, 240 μmol, 2.00 equiv.) in dry DMA (0.5 mL) at 70 °C. After 22 h the workup was done and the crude product was purified *via* column chromatography on silica gel (*c*Hex/EtOAc = 6:1) to obtain **111a** as a yellow solid (27.0 mg, 85.9 μmol, 72%). Other procedures on the synthesis of this compound are explained in Table 4, chapter 5.5.2.

– R_f (*c*Hex/EtOAc = 6:1) = 0.22. – ^{1}H NMR (400 MHz, CDCl₃): δ = 8.25–8.22 (m, 1H, C*H*$_{Ar}$), 8.18–8.15 (m, 1H, C*H*$_{Ar}$), 7.88–7.84 (m, 1H, C*H*$_{Ar}$), 7.82–7.78 (m, 1H, C*H*$_{Ar}$), 7.56 (td, ^3J = 7.5 Hz, ^4J = 1.5 Hz, 1H, C*H*$_{Ar}$), 7.34–7.29 (m, 2H, C*H*$_{Ar}$), 7.24 (m, 2H, C=C*H*, C*H*$_{Ar}$), 3.58 (m, 1H, C*H*), 3.16 (ddd, ^2J = 13.5 Hz, ^3J = 3.8 Hz, ^4J = 1.7 Hz, 1H, CH*H*), 3.01 (ddd, ^2J = 20.6 Hz, ^3J = 6.1, 2.7 Hz, 1H, C*H*H), 2.52 (ddt, ^2J = 20.6 Hz, ^3J = 5.2 Hz, ^4J = 1.6 Hz, 1H, CH*H*), 2.28 (dd, ^2J = 13.5 Hz, ^3J = 2.6 Hz, 1H, C*H*H) ppm. – ^{13}C NMR (101 MHz, CDCl₃): δ = 196.2 (C$_q$, 1 × *C*=O), 195.1 (C$_q$, 1 × *C*=O), 183.0 (C$_q$, 1 × *C*=O), 147.1 (C$_q$, 1 × C$_{qAr}$), 140.2 (+, 1 × CH$_{Ar}$), 137.3 (C$_q$, 1 × C$_{qAr}$), 135.4 (C$_q$, 1 × C$_{qAr}$), 135.1 (+, 1 × CH$_{Ar}$), 134.9 (+, 1 × CH$_{Ar}$), 134.8 (+, 1 × CH$_{Ar}$), 132.7 (C$_q$, 1 × C$_{qAr}$), 129.4 (C$_q$, 1 × C$_{qAr}$), 128.8 (+, 1 × CH$_{Ar}$), 128.7 (+, 1 × CH$_{Ar}$), 128.1 (+, 2 × CH$_{Ar}$), 127.0 (+, 1 × C=CH), 59.8 (C$_q$, 1 × C$_q$), 34.3 (–, 1 × CH₂), 32.5 (+, 1 × CH), 31.1 (–, 1 × CH₂) ppm. – IR (ATR): $\tilde{v}$ = 2923 (w), 1731 (m), 1697 (m), 1669 (m), 1589 (m), 1453 (w), 1411 (w), 1269 (s), 1250 (s), 1156 (m), 969 (w), 946 (w), 927 (m), 890 (w), 858 (m), 825 (w), 780 (w), 754 (m), 718 (m), 694 (m), 635 (w), 578 (w), 541 (w), 519 (w), 445 (w) cm^{-1}. – MS (EI, 70 eV), *m/z* (%): 314 (91) [M+H]⁺. – HRMS (EI, C₂₁H₁₄O₃): calc. 314.0937; found 314.0937. – X-Ray: Crystallographic information on the product can be found in chapter 7.3.2.

(4a*R*,9a*S*)-4a-(2-Iodobenzoyl)-1,4,4a,9a-tetrahydroanthracene-9,10-dione (87aa-epi)

Under argon atmosphere a mixture of the anthraquinone derivative **87aa** (88.5 mg, 200 μmol, 1.00 equiv.), Pd(OAc)$_2$ (8.98 mg, 40.0 μmol, 0.200 equiv.) and BINAP (49.8 mg, 80.0 μmol, 0.400 equiv.) was placed into a high-pressure glass tube. The mixture was dissolved in DMA (1 mL), PMP (0.07 mL, 62.1 mg, 400 μmol, 2.00 equiv.) was added and the mixture was stirred at 70 °C for 14 h. After completion of the reaction as indicated by TLC, it was quenched by the addition of water (2 mL). The aqueous phase was extracted with EtOAc (2 × 5 mL) and the combined organic phases were dried over Na$_2$SO$_4$. The solvents were removed under reduced pressure and the remaining crude product was purified *via* column chromatography on silica gel (*n*Hex/EtOAc = 6:1). The product was obtained as a mixture of starting material **87aa** and epimer of the starting material **87aa-epi** in a ratio of 1 : 3.0 as estimated by ^{1}H NMR as a colorless solid (36.0 mg, 115 μmol, 58%). The signals of epimer **87aa-epi** were extracted from the spectra of the mixture.

– R_f (*n*Hex/EtOAc = 6:1) = 0.29. – ^{1}H NMR (500 MHz, CDCl$_3$): δ = 8.26 (dd, ${}^3J$ = 7.8 Hz, ${}^4J$ = 1.3 Hz, 1H, C*H*$_{Ar}$), 8.07 (dd, ${}^3J$ = 7.7 Hz, ${}^4J$ = 1.4 Hz, 1H, C*H*$_{Ar}$), 7.85 (dd, ${}^3J$ = 7.9 Hz, ${}^4J$ = 1.2 Hz, 1H, C*H*$_{Ar}$), 7.79 (td, ${}^3J$ = 7.5 Hz, ${}^4J$ = 1.3 Hz, 1H, C*H*$_{Ar}$), 7.71 (td, ${}^3J$ = 7.5 Hz, ${}^4J$ = 1.4 Hz, 1H, C*H*$_{Ar}$), 7.45 (dd, ${}^3J$ = 7.8 Hz, ${}^4J$ = 1.7 Hz, 1H, C*H*$_{Ar}$), 7.32 (td, ${}^3J$ = 7.6 Hz, ${}^4J$ = 1.2 Hz, 1H, C*H*$_{Ar}$), 7.07 (td, ${}^3J$ = 7.7 Hz, ${}^4J$ = 1.6 Hz, 1H, C*H*$_{Ar}$), 5.95–5.82 (m, 1H, C=C*H*), 5.64–5.51 (m, 1H, C=C*H*), 3.23–3.06 (m, 2H, C*H*, C*H*H), 2.96–2.70 (m, 3H, CH*H*, C*H*$_2$) ppm. – ^{13}C NMR (126 MHz, CDCl$_3$): δ = 200.3 (C$_q$, 1 × *C*=O), 194.5 (C$_q$, 1 × *C*=O), 194.1 (C$_q$, 1 × *C*=O), 141.3 (C$_q$, 1 × *C*$_{qAr}$), 141.3 (+, 1 × *C*H$_{Ar}$), 136.1 (C$_q$, 1 × *C*$_{qAr}$), 135.0 (+, 1 × *C*H$_{Ar}$), 133.5 (+, 1 × *C*H$_{Ar}$), 133.1 (C$_q$, 1 × *C*$_{qAr}$), 131.7 (+, 1 × *C*H$_{Ar}$), 128.0 (+, 1 × *C*=CH), 127.8 (+, 1 × *C*H$_{Ar}$), 127.2 (+, 1 × *C*H$_{Ar}$), 127.0 (+, 1 × *C*H$_{Ar}$), 126.3 (+, 1 × *C*H$_{Ar}$), 123.2 (+, 1 × C=*C*H), 93.2 (C$_q$, 1 × *C*$_{qAr}$), 69.8 (C$_q$, 1 × *C*$_q$), 50.0 (+, 1 × *C*H), 30.8 (–, 1 × *C*H$_2$), 24.5 (–, 1 × *C*H$_2$) ppm. – MS (FAB, 3-NBA), *m/z* (%): 443 (12) [M+H]$^+$, 231 (100) [C$_7$H$_4$IO]$^+$. – HRMS (FAB, C$_{21}$H$_{16}$^{127}IO$_3$): calc. 443.0144; found 443.0142. – X-Ray: Crystallographic information on the product can be found in chapter 7.3.1.

(5*S*,7a*S*,13a*S*)-6-Methylene-7,7a-dihydro-6*H*-5,13a-methanobenzo[4,5]cycloocta[1,2-b]naphthalene-8,13,14(5*H*)-trione (113ab)

According to **GP4** the coupling was performed with (4a*R*,9a*R*)-4a-(2-iodobenzoyl)-2-methyl-1,4,4a,9a-tetrahydroanthracene-9,10-dione **(87ab)** (137 mg, 300 μmol, 1.00 equiv.), Pd(OAc)$_2$ (13.5 mg, 60.0 μmol, 20 mol%), PPh$_3$ (31.5 mg, 120 μmol, 40 mol%) and PMP (0.11 mL, 93.2 mg, 600 μmol, 2.00 equiv.) in dry DMA (3 mL). After 72 h the workup was done, and the crude product was purified *via* column chromatography on silica gel (*n*Hex/EtOAc = 8:1) to obtain isomer **113ab** as a colorless solid. The yield was not determined.

– R_f (*c*Hex/EtOAc = 8:1) = 0.29. – ^{1}H NMR (500 MHz, CDCl$_3$): δ = 8.14–8.11 (m, 1H, C*H*$_{Ar}$), 8.06–8.03 (m, 1H, C*H*$_{Ar}$), 7.86 (d, ^3J = 7.8 Hz, 1H, C*H*$_{Ar}$), 7.78–7.74 (m, 2H, C*H*$_{Ar}$), 7.54 (td, ^3J = 7.5 Hz, ^4J = 1.5 Hz, 1H, C*H*$_{Ar}$), 7.32 (t, ^3J = 7.8 Hz, 2H, C*H*$_{Ar}$), 5.17 (s, 1H, C=C*HH*), 4.86 (s, 1H, C=CH*H*), 3.92 (t, ^3J = 3.1 Hz, 1H, COC*H*), 3.29–3.24 (m, 2H, C*H*, C*HH*), 2.69 (dd, 1H, ^2J = 15.2 Hz, ^3J = 4.7 Hz, 1H, C*HH*), 2.36 (ddt, ^2J = 14.9 Hz, ^3J = 12.6, 2.1 Hz, 1H, CH*H*), 2.25 (dd, ^2J = 13.3 Hz, ^3J = 3.2 Hz, 1H, C*HH*) ppm. – ^{13}C NMR (126 MHz, CDCl$_3$): δ = 195.9 (C$_q$, 1 × *C*=O), 195.8 (C$_q$, 1 × *C*=O), 195.3 (C$_q$, 1 × *C*=O), 145.6 (C$_q$, 1 × *C*$_{qAr}$), 145.3 (C$_q$, 1 × *C*$_{qAr}$), 137.0 (C$_q$, 1 × *C*$_{qAr}$), 136.1 (C$_q$, 1 × *C*$_{qAr}$), 135.0 (+, 1 × *C*H$_{Ar}$), 134.3 (+, 1 × *C*H$_{Ar}$), 134.0 (+, 1 × *C*H$_{Ar}$), 132.2 (C$_q$, 1 × *C*$_q$), 128.3 (+, 1 × *C*H$_{Ar}$), 127.7 (+, 1 × *C*H$_{Ar}$), 127.5 (+, 1 × *C*H$_{Ar}$), 127.2 (+, 1 × *C*H$_{Ar}$), 126.2 (+, 1 × *C*H$_{Ar}$), 111.1 (–, 1 × *C*H$_2$), 61.1 (C$_q$, 1 × *C*$_q$), 53.6 (+, 1 × *C*H), 44.7 (+, 1 × *C*H), 36.2 (–, 1 × *C*H$_2$), 27.9 (–, 1 × *C*H$_2$) ppm. – IR (ATR): $\tilde{v}$ = 2883 (w), 1704 (vs), 1683 (s), 1664 (vs), 1645 (s), 1594 (vs), 1476 (w), 1452 (m), 1432 (w), 1341 (m), 1319 (w), 1278 (vs), 1262 (s), 1237 (vs), 1210 (vs), 1173 (m), 1140 (w), 1109 (m), 1068 (m), 1058 (w), 1034 (w), 1021 (m), 1007 (w), 973 (w), 962 (w), 945 (m), 926 (s), 914 (vs), 898 (m), 878 (m), 839 (w), 820 (w), 793 (m), 785 (vs), 766 (s), 748 (vs), 724 (vs), 691 (s), 676 (m), 632 (m), 606 (vs), 569 (s), 526 (s), 466 (s), 458 (m), 431 (s), 416 (m), 394 (m), 382 (m) cm^{-1}. – MS (FAB, 3-NBA), *m/z* (%): 329 (6) [M+H]$^+$, 307 (24), 279 (37), 154 (100). – HRMS (FAB, C$_{22}$H$_{17}$O$_3$): calc. 329.1178; found 329.1178. – X-Ray: Crystallographic information on the product can be found in chapter 7.3.2.

(5*R*,13a*S*)-5-Methyl-6*H*-5,13a-methanobenzo[4,5]cycloocta[1,2-b]naphthalene-8,13,14(5*H*)-trione (115ab)

As starting material the inseparable product mixture (137 mg, 300 μmol, 1.00 equiv.) of the cycloaddtion between 2-(2-iodobenzoyl)naphthalene-1,4-dione (**73a**) and isoprene (**74b**) was used. Together with Pd(OAc)$_2$ (13.5 mg, 60.0 μmol, 20 mol%) and PPh$_3$ (63.0 mg, 240 μmol, 80 mol%) it was placed into a high-pressure glass vial under argon atmosphere. Then dry DMA (3 mL) and *i*-Pr$_2$NEt (0.61 mL, 465 mg, 3.60 mmol, 12.0 equiv.) were added and the mixture was stirred for 72 h at 80 °C. After completion the reaction mixture was quenched by the addition of water (3 mL), and the aqueous phase was extracted with EtOAc (3 × 10 mL). The crude product was purified *via* column chromatography on silica gel (*n*Hex/EtOAc = 8:1) to obtain isomer **115ab** as a colorless solid. The yield was not determined.

– R_f (*c*Hex/EtOAc = 8:1) = 0.20. – ^{1}H NMR (500 MHz, CDCl$_3$): δ = 8.25–8.20 (m, 1H, C*H*$_{Ar}$), 8.19–8.15 (m, 1H, C*H*$_{Ar}$), 7.86 (dd, ^3J = 7.9 Hz, ^4J = 1.5 Hz, 1H, C*H*$_{Ar}$), 7.84–7.75 (m, 2H, C*H*$_{Ar}$), 7.59 (td, ^3J = 7.6 Hz, ^4J = 1.5 Hz, 1H, C*H*$_{Ar}$), 7.45 (dd, ^3J = 8.0 Hz, ^4J = 1.1 Hz, 1H, C*H*$_{Ar}$), 7.30 (td, ^3J = 7.5 Hz, ^4J = 1.2 Hz, 1H, C*H*$_{Ar}$), 7.21 (dd, ^3J = 5.1, 2.7 Hz, 1H, C=C*H*), 2.98 (dd, ^2J = 13.6 Hz, ^3J = 2.0 Hz, 1H, CH*H*), 2.65 (dd, ^2J = 20.4 Hz, ^3J = 2.7 Hz, 1H, CH*H*), 2.54 (ddd, ^2J = 20.4 Hz, ^3J = 5.1, 2.0 Hz, 1H, CH*H*), 2.12 (d, ^2J = 13.5 Hz, 1H, CH*H*), 1.64 (s, 3H, C*H*$_3$) ppm. – ^{13}C NMR (126 MHz, CDCl$_3$): δ = 195.8 (C$_q$, 1 × *C*=O), 195.1 (C$_q$, 1 × *C*=O), 182.8 (C$_q$, 1 × *C*=O), 149.7 (C$_q$, 1 × *C*$_{qAr}$), 141.0 (+, 1 × *C*=CH), 136.9 (C$_q$, 1 × *C*$_{qAr}$), 135.0 (C$_q$, 1 × *C*$_{qAr}$), 134.9 (+, 1 × *C*H$_{Ar}$), 134.6 (+, 1 × *C*H$_{Ar}$), 134.4 (+, 1 × *C*H$_{Ar}$), 132.1 (C$_q$, 1 × *C*$_{qAr}$), 129.0 (C$_q$, 1 × *C*$_{qAr}$), 128.2 (+, 1 × *C*H$_{Ar}$), 127.8 (+, 1 × *C*H$_{Ar}$), 127.4 (+, 1 × *C*H$_{Ar}$), 126.7 (+, 1 × *C*H$_{Ar}$), 125.4 (+, 1 × *C*H$_{Ar}$), 61.0 (C$_q$, 1 × *C*$_q$), 41.7 (–, 1 × *C*H$_2$), 38.7 (–, 1 × *C*H$_2$), 33.5 (C$_q$, 1 × *C*$_q$), 28.8 (+, 1 × *C*H$_3$) ppm. – IR (ATR): $\tilde{v}$ = 1698 (s), 1666 (vs), 1613 (s), 1588 (vs), 1574 (s), 1265 (vs), 1244 (vs), 1222 (vs), 1190 (s), 1021 (s), 919 (s), 822 (s), 759 (vs), 725 (s), 714 (vs), 697 (vs), 671 (s), 639 (vs), 577 (s), 545 (s), 404 (s), 378 (s) cm^{-1}. – MS (FAB, 3-NBA), *m/z* (%): 329 (57) [M+H]$^+$, 307 (21), 154 (100), 138 (37). – HRMS (FAB, C$_{22}$H$_{17}$O$_3$): calc. 329.1178; found 329.1178. – X-Ray: Crystallographic information on the product can be found in chapter 7.3.

(4aR,9aS)-4a-(2-Iodobenzoyl)-2-methyl-1,4,4a,9a-tetrahydroanthracene-9,10-dione (87ab-epi)

Under argon atmosphere a mixture of anthraquinone derivative **87ab** (137 mg, 300 μmol, 1.00 equiv.), Pd(OAc)$_2$ (6.73 mg, 30.0 μmol, 10 mol%) and BINAP (37.4 mg, 60.0 μmol, 20 mol%) was placed into a high-pressure glass tube. The mixture was dissolved in DMA (3 mL), PMP (0.11 mL, 93.2 mg, 600 μmol, 2.00 equiv.) was added and the mixture was stirred at 70 °C for 63 h. Then the mixture was quenched by the addition of water (5 mL). The aqueous phase was extracted with EtOAc (2 × 10 mL) and the combined organic phases were dried over Na$_2$SO$_4$. The solvents were removed under reduced pressure and the remaining crude product was purified *via* column chromatography on silica gel (*c*Hex/EtOAc = 8:1). The compound was isolated as a non-separable mixture of isomers. The yield was not determined. The signals of epimer **87ab-epi** were extracted from the spectra of the mixture.

Colorless solid. – R_f (*c*Hex/EtOAc = 8:1) = 0.20. – ^{1}H NMR (500 MHz, CDCl$_3$): δ = 8.29–8.23 (m, 1H, C*H*$_{Ar}$), 8.10–8.04 (m, 1H, C*H*$_{Ar}$), 7.88–7.83 (m, 1H, C*H*$_{Ar}$), 7.81–7.75 (m, 1H, C*H*$_{Ar}$), 7.74–7.67 (m, 1H, C*H*$_{Ar}$), 7.45–7.39 (m, 1H, C*H*$_{Ar}$), 7.37–7.28 (m, 1H, C*H*$_{Ar}$), 7.10–7.03 (m, 1H, C*H*$_{Ar}$), 5.33–5.22 (m, 1H, C=C*H*), 3.20–3.00 (m, 2H, C*H*, C*H*H), 2.88–2.68 (m, 2H, C*H*H, C*H*H), 2.67–2.54 (m, 1H, C*H*H) ppm. – ^{13}C NMR (126 MHz, CDCl$_3$): δ = 200.6 (C$_q$, 1 × *C*=O), 194.9 (C$_q$, 1 × *C*=O), 194.5 (C$_q$, 1 × *C*=O), 141.6 (C$_q$, 1 × *C*$_{qAr}$), 141.3 (+, 1 × *C*H$_{Ar}$), 135.6 (C$_q$, 1 × *C*$_{qAr}$), 135.2 (+, 1 × *C*H$_{Ar}$), 133.8 (+, 1 × *C*H$_{Ar}$), 131.8 (+, 1 × *C*H$_{Ar}$), 128.0 (C$_q$, 1 × *C*$_{qAr}$), 127.9 (+, 1 × *C*H$_{Ar}$), 127.4 (+, 1 × *C*H$_{Ar}$), 127.2 (+, 1 × *C*H$_{Ar}$), 126.5 (+, 1 × *C*H$_{Ar}$), 117.9 (+, 1 × *C*=*C*H), 93.4 (C$_q$, 1 × *C*$_{qAr}$), 69.8 (C$_q$, 1 × *C*$_q$), 50.4 (+, 1 × *C*H), 31.3 (–, 1 × *C*H$_2$), 29.4 (–, 1 × *C*H$_2$), 23.8 (+, 1 × *C*H$_3$) ppm. Signal missing (1 × C$_q$). – X-Ray: Crystallographic information on the product can be found in chapter 7.3.2.

Side product: **2-Methylanthracene-9,10-dione (116)**

Crystallized from the crude product. Pale yellow solid. – R_f (*n*Hex/EtOAc = 12:1) = 0.31. – ^{1}H NMR (300 MHz, CDCl$_3$): δ = 8.34–8.28 (m, 2H, C*H*$_{Ar}$), 8.21 (d, ^3J = 7.9 Hz, 1H, C*H*$_{Ar}$), 8.13–8.08 (m, 1H, C*H*$_{Ar}$), 7.82–7.75 (m, 2H, C*H*$_{Ar}$), 7.60 (ddd, ^3J = 7.9, 1.8 Hz, ^4J = 0.8 Hz, 1H, C*H*$_{Ar}$), 2.54 (s, 3H, C*H*$_3$) ppm. – Analytical data is in accordance with previously published literature.[197]

(5*S*,7a*S*,13a*S*)-5-Methyl-6-methylene-7,7a-dihydro-6*H*-5,13a-methanobenzo[4,5]cycloocta[1,2-b]naphthalene-8,13,14(5*H*)-trione (113ac)

According to **GP4** the coupling was performed with (4a*R*,9a*R*)-4a-(2-iodobenzoyl)-2,3-dimethyl-1,4,4a,9a-tetrahydroanthracene-9,10-dione (**87ac**) (56.4 mg, 120 μmol, 1.00 equiv.), Pd(OAc)$_2$ (5.39 mg, 20.0 μmol, 20 mol%), PPh$_3$ (12.6 mg, 50.0 μmol, 40 mol%) and PMP (0.04 mL, 37.3 mg, 240 μmol, 2.00 equiv.) in dry DMA (1 mL). After 40 h the workup was done and the crude product was purified *via* column chromatography on silica gel (*n*Hex/EtOAc = 8:1) to obtain a non-separable mixture of isomers. The yield was not determined.

– R_f (*c*Hex/EtOAc = 8:1) = 0.27. – ^{1}H NMR (400 MHz, CDCl$_3$): δ = 8.15–8.11 (m, 1H, C*H*$_{Ar}$), 8.06–8.01 (m, 1H, C*H*$_{Ar}$), 7.90 (dd, ^3J = 7.9 Hz, ^4J = 1.5 Hz, 1H, C*H*$_{Ar}$), 7.80–7.71 (m, 2H, C*H*$_{Ar}$), 7.57 (ddd, ^3J = 8.5, 7.4 Hz, ^4J = 1.5 Hz, 1H, C*H*$_{Ar}$), 7.42–7.37 (m, 1H, C*H*$_{Ar}$), 7.35–7.28 (m, 1H, C*H*$_{Ar}$), 5.14 (d, ^4J = 1.3 Hz, 1H, C=C*H*H), 4.95 (d, ^4J = 1.9 Hz, 1H, C=C*H*H), 3.27 (dd, ^2J = 12.7 Hz, ^3J = 4.6 Hz, 1H, C*H*), 3.11 (d, ^2J = 13.5 Hz, 1H, C*H*H), 2.71 (dd, ^2J = 15.1 Hz, ^3J = 4.7 Hz, 1H, C*H*H), 2.39–2.24 (m, 1H, C*H*H), 2.04 (d, ^2J = 13.5 Hz, 1H, C*H*H), 1.74 (s, 3H, C*H*$_3$) ppm. – ^{13}C NMR (101 MHz, CDCl$_3$): δ = 196.1 (C$_q$, 1 × *C*=O), 196.0 (C$_q$, 1 × *C*=O), 195.3 (C$_q$, 1 × *C*=O), 148.6 (C$_q$, 1 × *C*$_{qAr}$), 136.8 (C$_q$, 1 × *C*$_{qAr}$), 136.2 (C$_q$, 1 × *C*$_{qAr}$), 135.1 (+, 1 × *C*H$_{Ar}$), 134.3 (+, 1 × *C*H$_{Ar}$), 134.1 (+, 1 × *C*H$_{Ar}$), 132.6 (C$_q$, 1 × *C*$_{qAr}$), 127.5 (+, 1 × *C*H$_{Ar}$), 127.2 (+, 1 × *C*H$_{Ar}$), 127.1 (+, 1 × *C*H$_{Ar}$), 126.2 (+, 1 × *C*H$_{Ar}$), 126.0 (+, 1 × *C*H$_{Ar}$), 109.5 (–, 1 × *C*=CH$_2$), 61.7 (C$_q$, 1 × *C*$_q$), 54.0 (+, 1 × *C*H), 43.7 (–, 1 × *C*H$_2$), 40.3 (C$_q$, 1 × *C*$_q$), 30.2 (–, 1 × *C*H$_2$), 24.8 (+, 1 × *C*H$_3$), 22.5 (C$_q$, 1 × *C*$_q$) ppm. – IR (ATR): ṽ = 3371 (w), 2917 (w), 2846 (w), 1676 (vs), 1673 (vs), 1589 (s), 1468 (w), 1429 (m), 1380 (w), 1312 (w), 1269 (vs), 1247 (vs), 1222 (vs), 1207 (vs), 1160 (m), 1122 (m), 1095 (w), 1054 (m), 1027 (m), 1013 (s), 956 (s), 907 (vs), 887 (s), 819 (w), 802 (w), 752 (vs), 731 (vs), 691 (vs), 646 (s), 598 (m), 554 (m), 540 (m), 499 (m), 455 (s), 426 (m), 399 (m), 380 (s) cm^{-1}. – MS (FAB, 3-NBA), *m/z* (%): 343 (23) [M+H]$^+$, 154 (100). – HRMS (FAB, C$_{23}$H$_{19}$O$_3$): calc. 343.1334; found 343.1333. – X-Ray: Crystallographic information on the product can be found in chapter 7.3.1.

Side product: **(4a*R*,9a*S*)-4a-(2-Bromobenzoyl)-2,3-dimethyl-1,4,4a,9a-tetrahydroanthracene-9,10-dione (87bc-epi)**

Crystallized from the crude product. Pale yellow solid. – R_f (*c*Hex/EtOAc = 8:1) = 0.27. – ¹H NMR (400 MHz, CDCl₃): δ = 8.25 (d, ³*J* = 7.7 Hz, 1H, C*H*Ar), 8.06 (d, ³*J* = 7.7 Hz, 1H, C*H*Ar), 7.85 (d, ³*J* = 7.9 Hz, 1H, C*H*Ar), 7.78 (td, ³*J* = 7.6 Hz, ⁴*J* = 1.3 Hz, 1H, C*H*Ar), 7.70 (td, ³*J* = 7.6 Hz, ⁴*J* = 1.3 Hz, 1H, C*H*Ar), 7.38 (dd, ³*J* = 7.9 Hz, ⁴*J* = 1.8 Hz, 1H, C*H*Ar), 7.32 (t, ³*J* = 7.6 Hz, 1H, C*H*Ar), 7.05 (td, ³*J* = 7.7 Hz, ⁴*J* = 1.7 Hz, 1H, C*H*Ar), 3.15–3.04 (m, 1H, C*H*), 2.94–2.72 (m, 3H, C*HH*), 2.64–2.53 (m, 1H, C*HH*), 1.72 (s, 3H, C*H₃*), 1.34 (s, 3H, C*H₃*) ppm. – ¹³C NMR (101 MHz, CDCl₃): δ = 199.4 (C_q, 1 × *C*=O), 194.6 (C_q, 1 × *C*=O), 194.5 (C_q, 1 × *C*=O), 138.6 (C_q, 1 × C_{qAr}), 136.5 (C_q, 1 × C_{qAr}), 135.1 (+, 1 × *C*HAr), 134.1 (+, 1 × *C*HAr), 133.7 (+, 1 × *C*HAr), 133.5 (C_q, 1 × C_{qAr}), 131.4 (+, 1 × *C*HAr), 127.9 (+, 1 × *C*HAr), 127.5 (+, 1 × *C*HAr), 127.4 (C_q, 1 × C_{qAr}), 126.4 (+, 1 × *C*HAr), 126.4 (+, 1 × *C*HAr), 123.1 (C_q, 1 × C_q), 120.0 (C_q, 1 × C_q), 70.6 (C_q, 1 × C_q), 50.7 (+, 1 × *C*H), 37.0 (–, 1 × *C*H₂), 30.5 (–, 1 × *C*H₂), 19.2 (+, 1 × *C*H₃), 19.1 (+, 1 × *C*H₃) ppm. – MS (EI, 70 eV), *m/z* (%): 423/425 (15/13) [M+H]⁺. – HRMS (EI, C₂₃H₂₀O₃⁷⁹Br): calc. 423.0596; found 423.0596.

The compound was isolated as a mixture of isomers. The signals of product **87bc-epi** were extracted from the NMR spectrum of the mixture.

Side product: **2,3-Dimethylanthracene-9,10-dione (117)**

Crystallized from the crude product. Pale yellow solid. – R_f (*c*Hex/EtOAc = 15:1) = 0.32. – ¹H NMR (300 MHz, CDCl₃): δ = 8.29 (dd, ³*J* = 5.8 Hz, ⁴*J* = 3.3 Hz, 2H, C*H*Ar), 8.05 (s, 2H, C*H*Ar), 7.77 (dd, ³*J* = 5.8 Hz, ⁴*J* = 3.3 Hz, 2H, C*H*Ar), 2.44 (s, 6H, C*H₃*) ppm. – Analytical data is in accordance with previously published literature.[198]

(5*R*,13a*R*)-15-((Trimethylsilyl)oxy)-6*H*-5,13a-methanobenzo[4,5]cycloocta[1,2-b]naphthalin-8,13,14(5*H*)-trione (113ad)

According to **GP4** the coupling was performed with (1*R*,4a*S*,9a*R*)-9a-(2-iodobenzoyl)-1-((trimethylsilyl)oxy)-1,4,4a,9a-tetrahydroanthracene-9,10-dione (**87ad**) (79.6 mg, 150 µmol, 1.00 equiv.), Pd(OAc)₂ (6.74 mg, 30.0 µmol, 20 mol%), PPh₃ (15.7 mg, 60.0 µmol, 40 mol%) and PMP (0.05 mL, 46.6 mg, 300 µmol, 2.00 equiv.) in dry DMA (1 mL). After 16 h the workup was done, and the crude product was purified *via* column chromatography on silica gel (*n*Hex/EtOAc = 9:1) to obtain isomer **113ad** as a colorless solid (35.0 mg, 87.0 µmol, 58%)

– R_f (*c*Hex/EtOAc = 9:1) = 0.24. – ^{1}H NMR (400 MHz, CDCl$_3$): δ = 8.30–8.19 (m, 2H, C*H*$_{Ar}$), 7.90 (dd, ^3J = 8.1 Hz, ^4J = 1.5 Hz, 1H, C*H*$_{Ar}$), 7.82–7.75 (m, 2H, C*H*$_{Ar}$), 7.59 (td, ^3J = 7.5 Hz, ^4J = 1.5 Hz, 1H, C*H*$_{Ar}$), 7.40–7.31 (m, 2H, C*H*$_{Ar}$), 7.19 (dd, ^3J = 4.9 Hz, ^4J = 2.8 Hz, 1H, =C*H*), 5.14 (d, ^3J = 4.3 Hz, 1H, C*H*), 3.37 (t, ^3J = 5.1 Hz, 1H, C*H*), 3.11 (ddd, ^2J = 20.2 Hz, ^3J = 6.2, ^4J = 2.8 Hz, 1H, C*H*H), 2.37 (ddd, ^2J = 20.2 Hz, ^3J = 4.8 Hz, ^4J = 1.4 Hz, 1H, CH*H*), 0.16 (s, 9H, OSi(C*H*$_3$)$_3$) ppm. – ^{13}C NMR (101 MHz, CDCl$_3$): δ = 194.5 (C$_q$, 1 × *C*=O), 193.2 (C$_q$, 1 × *C*=O), 183.5 (C$_q$, 1 × *C*=O), 145.0 (C$_q$, 1 × *C*$_{qAr}$), 139.5 (+, 1 × C=*C*H), 136.3 (C$_q$, 1 × *C*$_{qAr}$), 135.2 (C$_q$, 1 × *C*$_{qAr}$), 135.1 (+, 1 × *C*H$_{Ar}$), 134.5 (+, 1 × *C*H$_{Ar}$), 134.3 (+, 1 × *C*H$_{Ar}$), 130.2 (C$_q$, 1 × *C*$_{qAr}$), 129.1 (+, 1 × *C*H$_{Ar}$), 129.0 (C$_q$, 1 × *C*$_{qAr}$), 128.4 (+, 1 × *C*H$_{Ar}$), 128.1 (+, 1 × *C*H$_{Ar}$), 127.8 (+, 1 × *C*H$_{Ar}$), 126.8 (+, 1 × *C*H$_{Ar}$), 67.3 (+, 1 × *C*(OTMS)H), 65.9 (C$_q$, 1 × *C*$_q$), 40.1 (+, 1 × *C*H), 30.23 (–, 1 × *C*H$_2$), 0.52 (+, 3 × Si*C*H$_3$) ppm. – IR (ATR): $\tilde{v}$ = 2920 (w), 1698 (w), 1669 (m), 1591 (w), 1454 (w), 1407 (vw), 1264 (m), 1158 (w), 1107 (w), 949 (vw), 916 (w), 888 (w), 867 (w), 838 (m), 777 (w), 749 (w), 733 (w), 718 (w), 692 (w), 630 (w), 600 (w), 526 (w), 447 (w) cm^{-1}. – MS (FAB, 3-NBA), *m/z* (%): 403 (26) [M+H]$^+$, 387 (33), 313 (100) [M–OTMS]$^+$. – HRMS (FAB, C$_{24}$H$_{23}$O$_4$Si): calc. 403.1366; found 403.1368. – X-Ray: Crystallographic information on the product can be found in chapter 7.3.1.

(5*S*,13a*R*,15*S*)-5-Methyl-15-((trimethylsilyl)oxy)-6*H*-5,13a-methanobenzo[4,5]cycloocta[1,2-b]naphthalene-8,13,14(5*H*)-trione (113ae)

According to **GP4** the coupling was performed with (1*R*,4a*S*,9a*R*)-9a-(2-iodobenzoyl)-2-methyl-1-((trimethylsilyl)oxy)-1,4,4a,9a-tetrahydroanthracene-9,10-dione (**87ae**) (54.5 mg, 100 µmol, 1.00 equiv.), Pd(OAc)$_2$ (4.49 mg, 20.0 µmol, 20 mol%), PPh$_3$ (10.5 mg, 40.0 µmol, 40 mol%) and PMP (0.04 mL, 31.1 mg, 20.0 µmol, 2.00 equiv.) in dry DMA (1 mL). After 24 h the workup was done, and the crude product was purified *via* column chromatography on silica gel (*n*Hex/EtOAc = 9:1) to obtain **113ae** as a light yellow solid (29.0 mg, 69.6 µmol, 70%).

– R_f (*n*Hex/EtOAc = 9:1) = 0.32. – ^{1}H NMR (500 MHz, CDCl$_3$): δ = 8.28–8.22 (m, 2H, C*H*$_{Ar}$), 7.91 (dd, ^3J = 7.9 Hz, ^4J = 1.5 Hz, 1H, C*H*$_{Ar}$), 7.83–7.75 (m, 2H, C*H*$_{Ar}$), 7.63 (ddd, ^3J = 8.5, 7.1, ^4J = 1.5 Hz, 1H, C*H*$_{Ar}$), 7.56 (dd, ^3J = 8.1 Hz, ^4J = 1.2 Hz, 1H, C*H*$_{Ar}$), 7.32 (ddd, ^3J = 8.1, 7.1, ^4J = 1.2 Hz, 1H, C*H*$_{Ar}$), 7.17 (dd, ^3J = 4.9, 2.7 Hz, 1H, C=C*H*), 4.84 (s, 1H, C*H*), 2.76 (dd, ^2J =

20.1 Hz, ^3J = 2.8 Hz, 1H, CHH), 2.30 (dd, ^2J = 20.1 Hz, ^3J = 4.9 Hz, 1H, CHH), 1.59 (s, 3H, CH_3), 0.13 (s, 9H, OSi(CH_3)$_3$) ppm. – ^{13}C NMR (101 MHz, CDCl$_3$): δ = 194.9 (C$_q$, 1 × C=O), 194.1 (C$_q$, 1 × C=O), 183.7 (C$_q$, 1 × C=O), 148.4 (C$_q$, 1 × C_{qAr}), 141.0 (+, 1 × C=CH), 136.1 (C$_q$, 1 × C_{qAr}), 135.3 (+, 1 × CH$_{Ar}$), 135.3 (C$_q$, 1 × C_{qAr}), 134.6 (+, 1 × CH$_{Ar}$), 134.4 (+, 1 × CH$_{Ar}$), 130.0 (C$_q$, 1 × C_{qAr}), 129.0 (C$_q$, 1 × C_{qAr}), 128.0 (+, 1 × CH$_{Ar}$), 127.8 (+, 1 × CH$_{Ar}$), 127.5 (+, 1 × CH$_{Ar}$), 126.9 (+, 1 × CH$_{Ar}$), 126.3 (+, 1 × CH$_{Ar}$), 72.3 (+, 1 × CH), 67.4 (C$_q$, 1 × C_q), 39.5 (C$_q$, 1 × C_q), 38.0 (–, 1 × CH$_2$), 23.9 (+, 1 × CH$_3$), 0.7 (+, 3 × SiCH$_3$) ppm. – IR (ATR): $\tilde{v}$ = 2966 (w), 2914 (w), 1691 (s), 1672 (vs), 1619 (m), 1588 (s), 1456 (w), 1264 (vs), 1245 (vs), 1156 (w), 1111 (vs), 1086 (s), 1072 (s), 1033 (m), 894 (vs), 882 (vs), 841 (vs), 754 (vs), 720 (vs), 677 (vs), 622 (s), 535 (m), 438 (m), 402 (m), 377 (s) cm^{-1}. – MS (FAB, 3-NBA), m/z (%): 417 (44) [M+H]$^+$, 401 (28) , 327 (100) [M–OTMS]$^+$. – HRMS (FAB, C$_{25}$H$_{25}$O$_4$Si): calc. 417.1522; found 417.1523. – X-Ray: Crystallographic information on the product can be found in chapter 7.3.2.

(5R,13aS,15S)-15-((*tert*-Butyldimethylsilyl)oxy)-5-methyl-6H-5,13a-methanobenzo[4,5]cycloocta[1,2-b]naphthalin-8,13,14(5H)-trione (114bf)

According to **GP4** the coupling was performed with (4aR,9aS)-9a-(2-bromobenzoyl)-1-((*tert*-butyldimethylsilyl)oxy)-2-methyl-1,4,4a,9a-tetrahydroanthracene-9,10-dione (**87bf**) (52.8 mg, 90.0 μmol, 1.00 equiv.), Pd(OAc)$_2$ (4.04 mg, 18.0 μmol, 20 mol%), PPh$_3$ (9.44 mg, 36.0 μmol, 40 mol%) and PMP (0.03 mL, 28.0 mg, 180 μmol, 2.65 equiv.) in dry DMA (0.5 mL). After 16 h the workup was done, and the crude product was purified *via* column chromatography on silica gel (cHex/EtOAc = 9:1) to obtain **114bf** as a colorless solid (40.9 mg, 89.2 μmol, 99%).

– R_f (cHex/EtOAc = 9:1) = 0.31. – ^{1}H NMR (500 MHz, CDCl$_3$): δ = 8.29–8.23 (m, 2H, CH_{Ar}), 7.92 (dd, ^3J = 7.8 Hz, ^4J = 1.5 Hz, 1H, CH_{Ar}), 7.84–7.74 (m, 2H, CH_{Ar}), 7.64 (ddd, ^3J = 8.5, 7.1 Hz, ^4J = 1.5 Hz, 1H, CH_{Ar}), 7.57 (dd, ^3J = 8.2 Hz, ^4J = 1.2 Hz, 1H, CH_{Ar}), 7.33 (ddd, ^3J = 8.1, 7.0 Hz, ^4J = 1.2 Hz, 1H, CH_{Ar}), 7.19 (dd, ^3J = 4.7, 2.8 Hz, 1H, C=CH), 4.92 (s, 1H, CH), 2.81 (dd, ^2J = 20.2 Hz, ^3J = 2.9 Hz, 1H, CHH), 2.32 (dd, ^2J = 20.2 Hz, ^3J = 4.7 Hz, 1H, CHH), 1.65 (s, 3H, CH_3), 0.75 (s, 9H, SiC(CH_3)$_3$), 0.23 (s, 3H, SiCH_3), −0.05 (s, 3H, SiCH_3) ppm. – ^{13}C NMR (126 MHz, CDCl$_3$) δ = 195.1 (C$_q$, 1 × C=O), 194.1 (C$_q$, 1 × C=O), 183.6 (C$_q$, 1 × C=O), 148.7 (C$_q$, 1 × C_{qAr}), 140.9 (+, 1 × C=CH), 136.1 (C$_q$, 1 × C_{qAr}), 135.5 (+, 1 × CH$_{Ar}$), 135.2 (+, 1 × CH$_{Ar}$), 134.6 (C$_q$, 1 × C_{qAr}), 134.4 (+, 1 × CH$_{Ar}$), 130.1 (C$_q$, 1 × C_{qAr}), 128.8 (C$_q$, 1 × C_{qAr}), 128.0 (+, 1 × CH$_{Ar}$), 127.8 (+, 1 × CH$_{Ar}$), 127.5 (+, 1 × CH$_{Ar}$), 127.0 (+, 1 × CH$_{Ar}$), 126.3 (+, 1 × CH$_{Ar}$), 72.2 (+, 1 × CH), 67.4 (C$_q$, 1 × C_q), 39.7 (–, 1 × CH$_2$), 38.5 (C$_q$, 1 × C_q), 26.6 (+, 1 × CH$_3$), 24.0 (+,

$3 \times CH_3$), 18.9 (C_q, $1 \times C_q$), –3.18 (+, $1 \times SiCH_3$), –4.18 (+, $1 \times SiCH_3$) ppm. – IR (ATR): $\tilde{v}$ = 2926 (vw), 2854 (vw), 1702 (vw), 1678 (vw), 1623 (vw), 1593 (vw), 1472 (vw), 1264 (vw), 1221 (vw), 1158 (vw), 1095 (vw), 1002 (vw), 927 (vw), 884 (vw), 837 (vw), 778 (vw), 758 (vw), 716 (vw), 703 (vw), 668 (vw), 601 (vw) cm^{-1}. – MS (FAB, 3-NBA), m/z (%): 459 (41) [M+H]$^+$, 401 (100) [M–tBu]$^+$, 327 (63) [M–OTBDMS]$^+$. – HRMS (FAB, $C_{28}H_{31}O_4Si$): calc. 459.1992; found 459.1991. – X-Ray: Crystallographic information on the product can be found in chapter 7.3.

(5S,13aR,15R)-15-(($tert$-Butyldimethylsilyl)oxy)-5-methyl-6H-5,13a-methanobenzo[4,5]cycloocta[1,2-b]naphthalene-8,13,14(5H)-trione (113bf)

According to **GP4** the coupling was performed with (4aR,9aS)-9a-(2-bromobenzoyl)-1-(($tert$-butyldimethylsilyl)oxy)-2-methyl-1,4,4a,9a-tetrahydroanthracene-9,10-dione (**88bf**) (54.0 mg, 11.8 μmol, 1.00 equiv.), Pd(OAc)$_2$ (4.49 mg, 20.0 μmol, 20 mol%), PPh$_3$ (10.5 mg, 40.0 μmol, 40 mol%) and PMP (0.04 mL, 31.1 mg, 20.0 μmol, 2.00 equiv.) in dry DMA (0.5 mL). After 15 h the workup was done, and the crude product was purified *via* column chromatography on silica gel (nHex/EtOAc = 9:1) to obtain **113bf** as a colorless solid. The yield was not determined.

– R_f (cHex/EtOAc = 9:1) = 0.27. – ^{1}H NMR (500 MHz, CDCl$_3$): δ = 8.10 (dd, ^3J = 7.8 Hz, ^4J = 1.2 Hz, 1H, CH_{Ar}), 7.93 (dd, ^3J = 7.5 Hz, ^4J = 1.3 Hz, 1H, CH_{Ar}), 7.80 (td, ^3J = 7.5 Hz, ^4J = 1.3 Hz, 1H, CH_{Ar}), 7.78–7.70 (m, 2H, CH_{Ar}), 7.54 (td, ^3J = 7.6 Hz, ^4J = 1.5 Hz, 1H, CH_{Ar}), 7.32 (dd, ^3J = 7.9 Hz, ^4J = 1.2 Hz, 1H, CH_{Ar}), 7.24 (dd, ^3J = 7.7 Hz, ^4J = 1.1 Hz, 1H, CH_{Ar}), 7.12 (dd, ^3J = 5.7, 2.4 Hz, 1H, C=CH), 4.68 (s, 1H, CH), 2.75 (dd, ^2J = 20.3 Hz, ^3J = 2.4 Hz, 1H, CH-H), 2.62 (dd, ^2J = 20.4 Hz, ^3J = 5.7 Hz, 1H, CHH), 1.66 (s, 3H, CH_3), 0.63 (s, 9H, SiC(CH_3)$_3$), 0.21 (d, ^3J = 7.6 Hz, 6H, SiC(CH_3)$_2$) ppm. – ^{13}C NMR (126 MHz, CDCl$_3$) δ = 196.1 (C_q, $1 \times C$=O), 194.0 (C_q, $1 \times C$=O), 182.5 (C_q, $1 \times C$=O), 146.2 (C_q, $1 \times C_{qAr}$), 140.3 (+, $1 \times CH_{Ar}$), 140.0 (C_q, $1 \times C_{qAr}$), 134.8 (+, $1 \times CH_{Ar}$), 134.4 (+, $1 \times CH_{Ar}$), 134.0 (C_q, $1 \times C_{qAr}$), 133.5 (+, $1 \times CH_{Ar}$), 133.1 (C_q, $1 \times C_{qAr}$), 131.1 (C_q, $1 \times C_{qAr}$), 127.3 (+, $1 \times CH_{Ar}$), 127.0 (+, $1 \times CH_{Ar}$), 126.6 (+, $1 \times CH_{Ar}$), 126.4 (+, $1 \times CH_{Ar}$), 125.8 (+, $1 \times CH_{Ar}$), 76.9 (+, $1 \times CH$), 68.1 (C_q, $1 \times C_q$), 41.9 (–, $1 \times CH_2$), 40.6 (C_q, $1 \times C_q$), 27.1 (+, $1 \times CH_3$), 26.4 (+, $3 \times CH_3$), 19.1 (C_q, $1 \times C_q$), –3.4 (+, $1 \times SiCH_3$), –3.6 (+, $1 \times SiCH_3$) ppm. – IR (ATR): $\tilde{v}$ = 2928 (w), 1701 (s), 1677 (vs), 1622 (m), 1592 (s), 1264 (vs), 1221 (s), 1157 (w), 1106 (vs), 1092 (vs), 1072 (s), 1055 (s), 1001 (s), 926 (w), 884 (m), 836 (vs), 778 (vs), 758 (vs), 744 (vs), 715 (vs), 703 (vs), 669 (s), 643 (m), 602 (s), 578 (m), 460 (m), 385 (m) cm^{-1}. – MS (FAB, 3-NBA), m/z (%): 459 (62) [M+H]$^+$, 401 (100) [M–CMe$_3$]$^+$, 327 (22)

[M–OTBDMS]$^+$. – HRMS (FAB, $C_{28}H_{31}O_4Si$): calc. 459.1992; found 459.1992. – X-Ray: Crystallographic information on the product can be found in chapter 7.3.2.

(5S,13aS)-5-((Triisopropylsilyl)oxy)-6H-5,13a-methanobenzo[4,5]cycloocta[1,2-b]naphthalene-8,13,14(5H)-trione (119ag)

According to **GP4** the coupling was performed with (4aR,9aR)-4a-(2-iodobenzoyl)-2-((triisopropylsilyl)oxy)-1,4,4a,9a-tetrahydroanthracene-9,10-dione **(87ag)** (61.5 mg, 100 μmol, 1.00 equiv.), Pd(OAc)$_2$ (4.49 mg, 20.0 μmol, 20 mol%), PPh$_3$ (10.5 mg, 40.0 μmol, 40 mol%) and PMP (0.04 mL, 31.1 mg, 20.0 μmol, 2.00 equiv.) in dry DMA (1 mL). After 72 h the workup was done, and the crude product was subjected to column chromatography on silica gel (nHex/EtOAc = 9:1) to obtain a non-separable mixture. APCI mass spectrometry verified the existence of product in the mixture and [3.3.1]product **119ag** crystallized from the mixture with its molecular structure being verified by X-ray crystallography.

– R_f (nHex/EtOAc = 9:1) = 0.35. – X-Ray: Crystallographic information on the product can be found in chapter 7.3.

(5R,13aR)-5-((*tert*-Butyldiphenylsilyl)oxy)-6H-5,13a-methanobenzo[4,5]cycloocta[1,2-b]naphthalene-8,13,14(5H)-trione (119ah)

According to **GP4** the coupling was performed with (4aR,9aR)-2-((*tert*-butyldiphenylsilyl)oxy)-4a-(2-iodobenzoyl)-1,4,4a,9a-tetrahydroanthracene-9,10-dione **(87ah)** (209 mg, 300 μmol, 1.00 equiv.), Pd(OAc)$_2$ (13.5 mg, 60.0 μmol, 20 mol%), PPh$_3$ (31.5 mg, 120 μmol, 40 mol%) and PMP (0.11 mL, 93.2 mg, 60.0 μmol, 2.00 equiv.) in dry DMA (6 mL). After 18 h the workup was done, and the crude product was purified *via* column chromatography on silica gel (nHex/EtOAc = 9:1) to obtain **119ah** as a colorless solid. The yield was not determined.

– R_f (cHex/EtOAc = 9:1) = 0.22. – ^{1}H NMR (400 MHz, CDCl$_3$): δ = 8.15 (dd, ^3J = 7.4 Hz, ^4J = 1.7 Hz, 1H, CH_{Ar}), 8.05–7.97 (m, 2H, CH_{Ar}), 7.75 (m, 5H, CH_{Ar}), 7.66 (td, ^3J = 7.7 Hz, ^4J = 1.5 Hz, 1H, CH_{Ar}), 7.57–7.51 (m, 2H, CH_{Ar}), 7.44–7.32 (m, 4H, CH_{Ar}), 7.29 (d, ^3J = 7.5 Hz, 2H, CH_{Ar}), 7.05 (dd, ^3J = 5.2, 2.7 Hz, 1H, C=CH), 3.06–2.93 (m, 2H, CHH), 2.90–2.80 (m, 1H, CHH), 2.06 (d, ^2J = 12.8 Hz, 1H, CHH), 1.13 (s, 9H, C(CH_3)$_3$) ppm. – ^{13}C NMR (101 MHz, CDCl$_3$) δ = 194.2

(C_q, 1 × C=O), 194.0 (C_q, 1 × C=O), 182.5 (C_q, 1 × C=O), 149.4 (C_q, 1 × C_{qAr}), 140.4 (+, 1 × C=CH), 136.6 (C_q, 1 × C_{qAr}), 136.0 (+, 1 × CH_{Ar}), 135.9 (+, 2 × CH_{Ar}), 135.8 (C_q, 1 × C_{qAr}), 135.1 (+, 1 × CH_{Ar}), 134.9 (+, 1 × CH_{Ar}), 134.8 (C_q, 1 × C_{qAr}), 134.6 (+, 1 × CH_{Ar}), 134.4 (+, 1 × CH_{Ar}), 134.2 (C_q, 1 × C_{qAr}), 131.6 (C_q, 1 × C_{qAr}), 130.2 (+, 1 × CH_{Ar}), 130.0 (+, 1 × CH_{Ar}), 129.8 (C_q, 1 × C_q), 128.3 (+, 1 × CH_{Ar}), 128.0 (+, 1 × CH_{Ar}), 127.8 (+, 2 × CH_{Ar}), 127.8 (+, 1 × CH_{Ar}), 127.7 (+, 1 × CH_{Ar}), 127.6 (+, 1 × CH_{Ar}), 126.6 (+, 1 × CH_{Ar}), 126.0 (+, 1 × CH_{Ar}), 73.4 (C_q, 1 × C_q), 63.9 (C_q, 1 × C_q), 43.9 (–, 1 × CH_2), 39.3 (–, 1 × CH_2), 27.3 (+, 1 × $(CH_3)_3$), 19.6 (C_q, 1 × C_q) ppm. – IR (ATR): $\tilde{v}$ = 3473 (vw), 3070 (w), 2929 (w), 2893 (w), 2856 (w), 1701 (m), 1676 (m), 1616 (w), 1591 (m), 1470 (w), 1428 (w), 1361 (vw), 1300 (w), 1264 (vs), 1228 (w), 1140 (w), 1108 (vs), 1075 (m), 1028 (w), 1014 (w), 933 (w), 907 (w), 867 (w), 820 (m), 761 (w), 731 (m), 698 (vs), 637 (w), 608 (m), 548 (vw), 501 (vs), 489 (vs) cm⁻¹. – MS (FAB, 3-NBA), m/z (%): 569 (8) [M+H]⁺, 511 (10)) [M–C_4H_{10}]⁺, 136 (100). – HRMS (FAB, $C_{37}H_{33}O_4Si$): calc. 569.2148; found 569.2146. – X-Ray: Crystallographic information on the product can be found in chapter 7.3.

2,2,2-Trifluoro-N-((5S,13aR,15S)-8,13,14-trioxo-15-((trimethylsilyl)oxy)-5,8,13,14-tetrahydro-6H-5,13a-methanobenzo[4,5]cycloocta[1,2-b]naphthalen-2-yl)acetamide (119gd)

In a vial under argon atmosphere N-(4-bromo-3-((4S,4aS,9aR)-9,10-dioxo-4-((trimethylsilyl)oxy)-1,4,4a,9,9a,10-hexahydroanthracene-4a-carbonyl)phenyl)-2,2,2-trifluoroacetamide (**87gd**) (65.4 mg, 110 µmol, 1.00 equiv.) and Pd(PPh₃)₄ (25.4 mg, 22.0 µmol, 20 mol%) were dissolved in dry DMA (1 mL). PMP (0.04 mL, 34.2 mg, 220 µmol, 2.00 equiv.) was added and the mixture was stirred for 20 h. The reaction was quenched by the addition of water (2 mL) and the aqueous phase was extracted with EtOAc (3 × 5mL). The solvent was removed under reduced pressure and the crude product was purified via column chromatography on silica gel (cHex/EtOAc = 9:1 → 4:1) to obtain **119gd** as a light brown solid (13.6 mg, 26.5 µmol, 24%).

– ¹H NMR (500 MHz, CDCl₃): δ = 8.25–8.20 (m, 2H, CH_{Ar}), 8.18 (brs, 1H, NH), 8.11 (dd, ³J = 8.4 Hz, ⁴J = 2.4 Hz, 1H, CH_{Ar}), 7.81–7.76 (m, 2H, CH_{Ar}), 7.73 (d, ⁴J = 2.4 Hz, 1H, C=CH), 7.40 (d, ³J = 8.4 Hz, 1H, CH_{Ar}), 7.19–7.14 (m, 1H, CH_{Ar}), 5.10 (d, ³J = 4.2 Hz, 1H, CHC(OTMS)H), 3.37 (t, ³J = 5.1 Hz, 1H, CH₂CH), 3.10 (ddd, ²J = 20.3 Hz, J = 6.2, 2.8 Hz, 1H, CHH), 2.34 (ddd, ²J = 20.2 Hz, J = 4.8, 1.4 Hz, 1H, CHH), 0.15 (s, 9H, OSi(CH_3)₃) ppm. – ¹³C NMR (126 MHz, CDCl₃) δ = 193.7 (C_q, 1 × C=O), 192.9 (C_q, 1 × C=O), 183.5 (C_q, 1 × C=O), 155.3 (q, ²J = 38.0 Hz, 1 × COCF₃), 142.7 (C_q, 1 × C_{qAr}), 139.6 (+, 1 × CH_{Ar}), 136.1 (C_q, 1 × C_{qAr}), 135.2 (C_q, 1 × C_{qAr}),

135.1 (C_q, 1 × C_{qAr}), 134.7 (+, 1 × CH_{Ar}), 134.4 (+, 1 × CH_{Ar}), 130.4 (+, 1 × CH_{Ar}), 130.0 (C_q, 1 × C_{qAr}), 129.8 (C_q, 1 × C_{qAr}), 127.8 (+, 1 × CH_{Ar}), 127.4 (+, 1 × CH_{Ar}), 126.9 (+, 1 × CH_{Ar}), 119.4 (+, 1 × C=CH), 115.6 (q, ^1J = 290.3 Hz, 1 × CF_3), 67.2 (+, 1 × CH-C(OTMS)H), 65.6 (C_q, 1 × C_q), 39.6 (+, 1 × CH), 30.1 (−, 1 × CH_2), 0.5 (+, 3 × $SiCH_3$) ppm. – IR (ATR): $\tilde{v}$ = 3306 (w), 2918 (w), 1727 (m), 1698 (s), 1673 (vs), 1615 (m), 1592 (s), 1545 (m), 1497 (w), 1425 (w), 1408 (w), 1271 (vs), 1251 (vs), 1193 (s), 1154 (vs), 1109 (vs), 979 (w), 926 (m), 888 (s), 866 (vs), 840 (vs), 824 (vs), 795 (m), 756 (s), 737 (s), 724 (s), 693 (s), 680 (m), 630 (m), 602 (m), 562 (w), 543 (w), 518 (w) cm^{-1}. – MS (FAB, 3-NBA), m/z (%): 514 (62) [M+H]$^+$, 513 (15) [M]$^+$, 425 (39), 424 (98), 137 (100). – HRMS (FAB, $C_{26}H_{23}O_5NF_3Si$): calc. 514.1298; found 514.1297. – X-Ray: Crystallographic information on the product can be found in chapter 7.3.2.

7.2.7 Products of the Modifications

((4a*R*,9a*R*)-4-Hydroxy-9,10-dimethoxy-1,9,9a,10-tetrahydro-9,10-epoxyanthracen-4a(2*H*)-yl)(2-iodophenyl)methanone (120a)

In a round-bottomed flask equipped with a magnetic stirrer, a solution of trimethylsilyl ether **87ad** (53.0 mg, 100 μmol, 1.00 equiv.) in CH_2Cl_2 (0.30 mL) and MeOH (0.60 mL) was prepared at 0 °C. Then one drop of 1 M HCl (50 μL, 50 μmol, 0.50 equiv.) was added to the solution at 0 °C and the reaction mixture was stirred for 1 h at this temperature. After completion of the reaction (monitored by TLC), CH_2Cl_2 was added (4 mL), the reaction mixture was neutralized with 10% $NaHCO_3$ (0.5 mL) and washed with H_2O (5 mL).The organic layer was dried over Na_2SO_4 and concentrated *in vacuo* to give the crude product, which was purified *via* silica gel column chromatography on silica gel (*c*Hex/EtOAc = 3:1) to give product **120a** (6.30 mg, 14.0 μmol, 14%).

– R_f(*c*Hex/EtOAc = 4:1) = 0.25. – ^{1}H NMR (300 MHz, CDCl$_3$): δ = 13.87 (s, 1H, C=CO*H*), 8.50 (dt, ^3J = 8.3 Hz, ^4J = 1.0 Hz, 1H, C*H*$_{Ar}$), 8.07–7.82 (m, 2H, C*H*$_{Ar}$), 7.71–7.61 (m, 1H, C*H*$_{Ar}$), 7.58–7.44 (m, 2H, C*H*$_{Ar}$), 7.33–7.25 (m, 1H, C*H*$_{Ar}$), 7.20 (td, ^3J = 7.7 Hz, ^4J = 1.7 Hz, 1H, C*H*$_{Ar}$), 4.89–4.70 (m, 1H, C=C*H*), 4.59 (dd, ^3J = 7.5, 3.7 Hz, 1H, C*H*), 3.36 (s, 3H, OC*H*$_3$), 3.30 (s, 3H, OC*H*$_3$), 2.54 (brs, 1H, C*HH*), 2.05 (brs, 2H, C*HH*), 1.86–1.70 (m, 1H, C*HH*) ppm. – ^{13}C NMR (76 MHz, CDCl$_3$) δ = 201.5 (C$_q$, 1 × *C*=O), 160.4 (C$_q$, 1 × *C*$_{qAr}$), 147.4 (C$_q$, 1 × *C*$_{qAr}$), 146.1 (C$_q$, 1 × *C*$_{qAr}$), 139.4 (+, 1 × *C*H$_{Ar}$), 131.2 (+, 1 × *C*H$_{Ar}$), 130.7 (+, 1 × *C*H$_{Ar}$), 128.5 (+, 1 × *C*H$_{Ar}$), 127.1 (+, 1 × *C*H$_{Ar}$), 126.1 (+, 1 × *C*H$_{Ar}$), 125.8 (C$_q$, 1 × *C*$_q$), 125.4 (+, 1 × *C*H$_{Ar}$), 121.6 (+, 1 × *C*H$_{Ar}$), 114.9 (C$_q$, 2 × *C*$_q$), 110.8 (C$_q$, 1 × *C*$_q$), 102.2 (+, 1 × *C*H), 91.9 (C$_q$, 1 × *C*$_{qAr}$), 79.6 (+, 1 × *C*=CH), 54.0 (+, 1 × O*C*H$_3$), 53.0 (+, 1 × O*C*H$_3$), 39.4 (–, 1 × *C*H$_2$), 38.1 (–, 1 × *C*H$_2$) ppm. – IR (ATR): $\tilde{v}$ = 3051 (vw), 2952 (w), 2931 (w), 2832 (w), 1636 (s), 1595 (vs), 1564 (m), 1445 (s), 1392 (vs), 1327 (vs), 1262 (vs), 1245 (vs), 1123 (vs), 1052 (vs), 1026 (vs), 973 (w), 892 (w), 816 (w), 766 (vs), 730 (w), 645 (w) cm^{-1}. – MS (FAB, 3-NBA), *m/z* (%): 504 (97) [M]$^+$, 505 (26) [M+H]$^+$, 415 (74) [M–C$_2$HO$_4$]$^+$. – HRMS (FAB, C$_{23}$H$_{21}$O$_5$^{127}I): calc. 504.0434; found 504.0434.

((4a*R*,9a*R*)-4-Hydroxy-9,10-dimethoxy-1,9,9a,10-tetrahydro-9,10-epoxyanthracen-4a(2*H*)-yl)(2-bromophenyl)methanone (120b)

In a round-bottomed flask equipped with a magnetic stirrer, a solution of trimethylsilyl ether **87bd** (48.3 mg, 100 µmol, 1.00 equiv.) in CH_2Cl_2 (0.30 mL) and MeOH (0.60 mL) was prepared at 0 °C. Then one drop of 1 M HCl (50 µL, 50 µmol, 0.50 equiv.) was added to the solution at 0 °C and the reaction mixture was stirred for 1 h at this temperature. After completion of the reaction (monitored by TLC), CH_2Cl_2 was added (4 mL), the reaction mixture was neutralized with 10% $NaHCO_3$ (0.5 mL) and washed with H_2O (5 mL). The organic layer was dried over Na_2SO_4 and concentrated *in vacuo* to give the crude product, which was purified *via* silica gel column chromatography on silica gel (*c*Hex/EtOAc = 3:1) to give product **120b** (12.6 mg, 31.0 µmol, 31%).

– R_f(*c*Hex/EtOAc = 3:1) = 0.26. – ^{1}H NMR (300 MHz, CDCl$_3$): δ = 13.86 (s, 1H, C=CO*H*), 8.54–8.45 (m, 1H, C*H*$_{Ar}$), 7.91–7.82 (m, 1H, C*H*$_{Ar}$), 7.71–7.60 (m, 2H, C*H*$_{Ar}$), 7.58–7.49 (m, 1H, C*H*$_{Ar}$), 7.45 (td, ^3J = 7.4 Hz, ^4J = 1.2 Hz, 1H, C*H*$_{Ar}$), 7.38 (dd, ^3J = 7.7 Hz, ^4J = 1.9 Hz, 1H, C*H*$_{Ar}$), 7.35–7.28 (m, 1H, C*H*$_{Ar}$), 4.91–4.70 (m, 1H, C=C*H*), 4.59 (dd, *J* = 7.5, 3.7 Hz, 1H, C*H*), 3.36 (s, 3H, OC*H*$_3$), 3.29 (s, 3H, OC*H*$_3$), 2.57 (brs, 1H, C*H*H), 2.22 (brs, 1H, C*H*H), 2.05–1.95 (m, 1H, C*H*H), 1.89–1.69 (m, 1H, C*H*H) ppm. – ^{13}C NMR (76 MHz, CDCl$_3$) δ = 199.9 (C$_q$, 1 × *C*=O), 160.1 (C$_q$, 1 × *C*$_{qAr}$), 147.4 (C$_q$, 1 × *C*$_{qAr}$), 142.2 (C$_q$, 1 × *C*$_{qAr}$), 133.1 (+, 1 × *C*H$_{Ar}$), 131.3 (+, 1 × *C*H$_{Ar}$), 130.7 (+, 1 × *C*H$_{Ar}$), 127.9 (+, 1 × *C*H$_{Ar}$), 127.6 (+, 1 × *C*H$_{Ar}$), 126.1 (+, 1 × *C*H$_{Ar}$), 125.8 (C$_q$, 1 × *C*$_q$), 125.4 (+, 1 × *C*H$_{Ar}$), 125.3 (C$_q$, 1 × *C*$_q$), 121.6 (+, 1 × *C*H$_{Ar}$), 118.9 (C$_q$, 1 × *C*$_{qAr}$), 114.9 (C$_q$, 1 × *C*$_q$), 111.3 (C$_q$, 1 × *C*$_q$), 102.1 (+, 1 × *C*H), 79.6 (+, 1 × C=*C*H), 54.0 (+, 1 × O*C*H$_3$), 52.9 (+, 1 × O*C*H$_3$), 39.4 (–, 1 × *C*H$_2$), 37.7 (–, 1 × *C*H$_2$) ppm. – IR (ATR): $\tilde{v}$ = 3057 (vw), 2927 (w), 2830 (w), 1636 (s), 1594 (vs), 1562 (m), 1442 (s), 1431 (s), 1391 (vs), 1326 (vs), 1262 (s), 1242 (vs), 1193 (m), 1179 (w), 1159 (w), 1120 (vs), 1054 (vs), 1021 (vs), 973 (m), 952 (m), 912 (w), 892 (s), 866 (w), 816 (m), 789 (w), 762 (vs), 732 (s), 698 (m), 657 (w), 640 (m), 630 (s), 586 (w), 565 (w), 551 (w), 456 (w), 436 (w), 409 (vw), 388 (w) cm^{-1}. – MS (FAB, 3-NBA), *m/z* (%): 456/458 (98/100) [M]$^+$, 367/369 (93/92) [M–C$_2$HO$_4$]$^+$. – HRMS (FAB, C$_{23}$H$_{21}$O$_5$^{79}Br): calc. 456.0572; found 456.0572.

(4a*R*,9a*R*)-9a-(2-Iodobenzoyl)-3,3-dimethoxy-1,3,4,4a,9a,10-hexahydroanthracen-9(2*H*)-one (122)

To a solution of (4a*R*,9a*R*)-2-((*tert*-butyldiphenylsilyl)oxy)-4a-(2-iodobenzoyl)-1,4,4a,9a-tetrahydroanthracene-9,10-dione **(87ah)** (44.6 mg, 64.0 µmol, 1.00 equiv.) in a mixture of dry MeOH and dry CH$_2$Cl$_2$ (1:1) (0.4 mL) was added a drop of AcCl (50 µL, 55.3 mg, 700 µmol, 11 equiv.) at 0 °C and the reaction mixture was stirred for 3.5 h at this temperature. After completion of the reaction (monitored by TLC), CH$_2$Cl$_2$ was added (4 mL), the reaction mixture was neutralized with 10% NaHCO$_3$ (0.5 mL) and washed with H$_2$O (5 mL).The organic layer was dried over Na$_2$SO$_4$ and concentrated *in vacuo* to give the crude product, which was purified *via* silica gel column chromatography on silica gel (*c*Hex/EtOAc = 3:1) to give product **122** (31.0 mg, 61.5 µmol, 96%).

– R_f (*c*Hex/EtOAc = 3:1) = 0.37. – ^{1}H NMR (300 MHz, CDCl$_3$): δ = 8.21–8.15 (m, 1H, C*H*$_{Ar}$), 8.08–8.02 (m, 1H, C*H*$_{Ar}$), 7.85–7.68 (m, 3H, C*H*$_{Ar}$), 7.29 (td, ^3J = 7.6 Hz, ^4J = 1.2 Hz, 1H, C*H*$_{Ar}$), 7.07 (td, ^3J = 7.7 Hz, ^2J =1.6 Hz, 1H, C*H*$_{Ar}$), 6.91 (dd, ^3J = 7.7 Hz, ^4J = 1.6 Hz, 1H, C*H*$_{Ar}$), 3.63 (dd, *J* = 13.9, 4.1 Hz, 1H, C*H*), 3.21 (s, 3H, OC*H*$_3$), 3.14 (s, 3H, OC*H*$_3$), 2.69–2.58 (m, 1H, C*H*H), 2.27 (ddd, *J* = 13.6, 4.1, 3.0 Hz, 1H, CH*H*), 2.10–1.97 (m, 1H, C*H*H), 1.85 (td, *J* = 13.4, 4.1 Hz, 1H, CH*H*), 1.67–1.53 (m, 1H, C*H*H), 1.40 (t, ^2J = 13.8 Hz, 1H, CH*H*) ppm. – ^{13}C NMR (76 MHz, CDCl$_3$): δ = 201.0 (C$_q$, 1 × *C*=O), 195.8 (C$_q$, 1 × *C*=O), 193.1 (C$_q$, 1 × *C*=O), 143.6 (C$_q$, 1 × *C*$_{qAr}$), 140.2 (+, 1 × *C*H$_{Ar}$), 135.3 (C$_q$, 1 × *C*$_{qAr}$), 134.3 (+, 1 × *C*H$_{Ar}$), 134.0 (C$_q$, 1 × *C*$_{qAr}$), 133.4 (+, 1 × *C*H$_{Ar}$), 131.2 (+, 1 × *C*H$_{Ar}$), 127.3 (+, 3 × *C*H$_{Ar}$), 126.1 (+, 1 × *C*H$_{Ar}$), 98.2 (C$_q$, 1 × *C*$_q$OCH$_3$), 91.5 (C$_q$, 1 × *C*$_{qAr}$), 69.4 (C$_q$, 1 × *C*$_q$), 51.6 (+, 1 × *C*H), 47.9 (+, 1 × O*C*H$_3$), 47.7 (+, 1 × O*C*H$_3$), 34.9 (–, 1 × *C*H$_2$), 28.7 (–, 1 × *C*H$_2$), 26.3 (–, 1 × *C*H$_2$) ppm. IR (ATR, $\tilde{\nu}$) = 395 (w), 424 (w), 445 (m), 487 (w), 561 (w), 596 (w), 636 (m), 674 (w), 691 (s), 732 (vs), 744 (vs), 759 (s), 772 (s), 785 (m), 807 (w), 823 (m), 846 (m), 902 (m), 929 (vs), 982 (s), 1006 (s), 1045 (vs), 1081 (vs), 1112 (s), 1143 (vs), 1160 (w), 1218 (vs), 1255 (vs), 1358 (w), 1425 (w), 1459 (w), 1562 (w), 1591 (m), 1680 (vs), 2830 (w), 2936 (w), 2949 (w) cm^{-1}. – MS (EI, 70 eV), *m*/*z* (%): 504 (12) [M]$^+$, 473 (28) [M–OCH$_3$]$^+$, 231 (100). – HRMS (EI, C$_{23}$H$_{22}$O$_3$^{127}I): calc. 473.0614; found 473.0616.

7.2.8 Synthesis of a Xanthone Dienophile

2-Phenoxyterephthalic acid (129)

Bromoterephthalic acid (**127**) (1.47 g, 6.00 mmol, 1.00 equiv.) was added to 48 mL of DMF, followed by phenol (**128**) (1.06 mL, 1.13 g, 12.0 mmol, 2.00 equiv.), DBU (2.69 mL, 2.74 g, 18.0 mmol, 3.00 equiv.), pyridine (0.100 mL), copper(0) (50.0 mg, 0.787 mmol, 13 mol%) and copper(I) iodide (50.0 mg, 263 μmol, 4.4 mol%). The reaction was heated to reflux and monitored *via* TLC. After 2 h the reaction was cooled and diluted with 1 M HCl (200 mL) until no more precipitate had formed. The resulting precipitate was filtered, washed with water (50 mL) and dried under vacuum to give the desired product **129** as a light green solid (823 mg, 3.19 mmol, 53%).

– R_f (CH$_2$Cl$_2$/MeOH = 4:1) = 0.50. – ^{1}H NMR (300 MHz, DMSO-d_6): δ = 13.22 (brs, 2H, CO$_2$*H*), 7.92 (s, 1H, C*H*$_{Ar}$), 7.77 (d, J = 8.0 Hz, 1H, C*H*$_{Ar}$), 7.63 (dd, J = 8.1, 1.6 Hz, 1H, C*H*$_{Ar}$), 7.33–7.26 (m, 2H, C*H*$_{Ar}$), 7.09–7.00 (m, 1H, C*H*$_{Ar}$), 6.89 (dd, J = 7.6, 1.6 Hz, 2H, C*H*$_{Ar}$) ppm. – Analytical data is in accordance with previously published literature.[173]

9-Oxo-9*H*-xanthene-3-carboxylic acid (130)

2-Phenoxyterephthalic acid (**129**) (625 mg, 2.42 mmol, 1.00 equiv.) was added to 80% H$_2$SO$_4$ (19 mL) and heated to 80 °C while being stirred for 3 h. After completion of reaction, the mixture was cooled and poured over ice (100 mL) producing a gray-brown solid that was then filtered and washed with water. Column chromatography on silica gel (CH$_2$Cl$_2$/MeOH/formic acid = 93:6:1) yielded the desired product **130** as a yellow solid (135 mg, 563 μmol, 23%).

– R_f (CH$_2$Cl$_2$/MeOH/formic acid = 93:6:1) = 0.57. – ^{1}H NMR (300 MHz, DMSO-d_6): δ = 8.22 (dd, ^3J = 8.0 Hz, ^4J = 1.7 Hz, 1H, C*H*$_{Ar}$), 8.11 (d, ^4J = 1.7 Hz, 1H, C*H*$_{Ar}$), 8.04 (s, 1H, C*H*$_{Ar}$), 7.99–7.89 (m, 2H, C*H*$_{Ar}$), 7.72 (d, ^3J = 8.3 Hz, 1H, C*H*$_{Ar}$), 7.52 (t, ^3J = 7.5 Hz, 1H, C*H*$_{Ar}$) ppm. Signal missing (1H, COO*H*). – Analytical data is in accordance with previously published literature.[173]

3-(1,4-Dimethoxy-2-naphthoyl)-9*H*-xanthen-9-one (131)

According to **GP1** a mixture of trifluoroacetic anhydride (0.34 mL, 500 mg, 2.40 mmol, 10.0 equiv.), 1,4-dimethoxynaphthalene (**69**) (45.2 mg, 240 µmol, 1.00 equiv.) and 9-Oxo-9*H*-xanthene-3-carboxylic acid (**130**) (116 mg, 480 µmol, 2.00 equiv.) was used. The crude product was subjected to column chromatography on silica gel (*c*Hex/EtOAc = 9:1) which gave the product in a non-separable mixture.

– R_f (*c*Hex/EtOAc = 9:1) = 0.22. – MS (FAB, 3-NBA), *m/z* (%): 411 (43) [M+H]$^+$, 410 (33) [M]$^+$.
– HRMS (EI, $C_{26}19O_5$): calc. 411.1232; found 411.1232.

Dimethyl 2,5-dibromoterephthalate (135)

2,5-Dibromoterephthalic acid (**132**) (324 mg, 1.00 mmol, 1.00 equiv.) in $SOCl_2$ (1.10 mL, 1.80 g, 15.1 mmol, 15.1 equiv.) was slowly heated to 100 °C and stirred at this temperature for 5 h. After $SOCl_2$ was removed *in vacuo* and the flask was cooled to 0 °C, methanol (0.90 mL, 711 mg, 22.2 mmol, 22.2 equiv.) and triethylamine (0.45 mL, 329 mg, 3.25 mmol, 3.25 equiv.) were added slowly while stirring. The reaction mixture was stirred at room temperature for 2 h and then the solvent was removed by evaporation. The residue was dissolved in EtOAc (1 mL), washed with 1 M HCl (1 mL), sat. $NaHCO_3$ (1 mL), dried over Na_2SO_4 and the solvent was removed under reduced pressure. The product **135** was obtained as a colorless solid (307 mg, 872 µmol, 87%).

– R_f (*c*Hex/EtOAc = 9:1) = 0.42. – ^{1}H NMR (300 MHz, CDCl$_3$): δ = 8.06 (s, 2H, C*H*$_{Ar}$), 3.96 (s, 6H, OC*H*$_3$) ppm. – Analytical data is in accordance with previously published literature.[199]

Dimethyl 2-amino-3-bromoterephthalate (139) and dimethyl 2-amino-5-bromoterephthalate (140)

In a round-bottom flask dimethyl-2-aminoterephthalate (**138**) (3.14 g, 15.0 mmol, 1.00 equiv.) and NBS (2.94 g, 16.5 mmol, 1.10 equiv.) were dissolved in CHCl$_3$ (225 mL). The reaction mixture was stirred at room temperature for 24 h. After completion the crude product was washed with water (3 × 150 mL), dried over Na_2SO_4 and the solvent was removed under reduced pressure. After column chromatography on silica gel (*c*Hex/EtOAc = 9:1 → 6:1) the products were obtained: dimethyl-2-amino-3-bromoterephthalate **139** (1.12 g, 3.89 mmol, 26%) as a light yellow solid and dimethyl-2-amino-5-bromoterephthalate **140** (2.87 g, 9.96 mmol, 66%) as a yellow solid.

139: R_f (cHex/EtOAc = 5:1) = 0.43. – ^{1}H NMR (500 MHz, CDCl$_3$): δ = 7.89 (d, $^3J_{HH}$ = 8.3 Hz, 1H, CH$_{Ar}$), 6.87 (d, ^3J = 8.3 Hz, 1H, CH$_{Ar}$), 3.94 (s, 3H, CH$_3$), 3.90 (s, 3H, CH$_3$) ppm. Signals missing (2H, NH$_2$). – ^{13}C NMR (126 MHz, CDCl$_3$): δ = 167.7 (C$_q$, 1 × C=O), 167.4 (C$_q$, 1 × C=O), 148.3 (C$_q$, 1 × C$_{qAr}$), 138.3 (C$_q$, 1 × C$_{qAr}$), 130.5 (+, 1 × CH$_{Ar}$), 116.2 (+, 1 × CH$_{Ar}$), 113.1 (C$_q$, 1 × C$_{qAr}$), 109.0 (C$_q$, 1 × C$_{qAr}$), 52.9 (+, 1 × CH$_3$), 52.3 (+, 1 × CH$_3$) ppm. – Analytical data is in accordance with previously published literature.[174]

140: R_f (cHex/EtOAc = 5:1) = 0.24. – ^{1}H NMR (500 MHz, CDCl$_3$): δ = 8.09 (s, 1H, CH$_{Ar}$), 7.05 (s, 1H, CH$_{Ar}$), 5.82 (brs, 2H, NH$_2$), 3.92 (s, 3H, CH$_3$), 3.89 (s, 3H, CH$_3$) ppm. – ^{13}C NMR (126 MHz, CDCl$_3$): δ = 167.0 (C$_q$, 1 × C=O), 166.4 (C$_q$, 1 × C=O), 149.0 (C$_q$, 1 × C$_{qAr}$), 136.9 (C$_q$, 1 × C$_{qAr}$), 136.4 (+, 1 × CH$_{Ar}$), 119.3 (+, 1 × CH$_{Ar}$), 113.9 (C$_q$, 1 × C$_{qAr}$), 105.3 (C$_q$, 1 × C$_{qAr}$), 52.8 (+, 1 × CH$_3$), 52.2 (+, 1 × CH$_3$) ppm. – Analytical data is in accordance with previously published literature.[174]

Dimethyl-2-iodo-5-bromoterephthalate (141)

Dimethyl 2-amino-5-bromoterephthalate (**140**) (350 mg, 1.21 mmol, 1.00 equiv.) was dissolved in conc. H$_2$SO$_4$ (4 mL), and the solution was cooled to 0 °C. NaNO$_2$ (251 mg, 3.60 mmol, 3.00 equiv.) was added in portions to the cooled solution, and the mixture was stirred for 2 h at 0 °C. Then, ice (100 mL) and a solution of KI (1.01 g, 6.07 mmol, 5.00 equiv.) in H$_2$O (5 mL) was carefully added to the reaction mixture at 0 °C, and the mixture was heated to 80 °C for 30 min. The reaction mixture was cooled to 10 °C and the precipitate was filtered. The solid was dissolved in EtOAc (15 mL) and the organic layer was washed with sat. aqueous NaHCO$_3$ (10 mL), 10% aqueous Na$_2$S$_2$O$_3$ (10 mL), dried over Na$_2$SO$_4$ and the solvent was removed under reduced pressure. The product **141** was obtained as a light yellow solid (444 mg, 1.12 mmol, 92%).

– R_f (cHex/EtOAc = 5:1) = 0.58. – ^{1}H NMR (500 MHz, CDCl$_3$): δ = 8.34 (s, 1H, CH$_{Ar}$), 8.05 (s, 1H, CH$_{Ar}$), 3.95 (s, 6H, CH$_3$) ppm. – ^{13}C NMR (126 MHz, CDCl$_3$): δ = 165.1 (C$_q$, 1 × C=O), 164.6 (C$_q$, 1 × C=O), 143.5 (+, 1 × CH$_{Ar}$), 138.6 (C$_q$, 1 × C$_{qAr}$), 136.4 (+, 1 × CH$_{Ar}$), 135.6 (C$_q$, 1 × C$_{qAr}$), 121.5 (C$_q$, 1 × C$_{qAr}$), 91.8 (C$_q$, 1 × C$_{qAr}$), 53.2 (+, 1 × CH$_3$), 53.1 (+, 1 × CH$_3$) ppm. – Analytical data is in accordance with previously published literature.[175]

2-Iodo-5-bromoterephthalic acid (142)

Dimethyl-2-iodo-5-bromoterephthalate (141) (400 mg, 1.00 mmol, 1.00 equiv.) was suspended in a solution of NaOH (922 mg, 23.1 mmol, 23.0 equiv.) in EtOH (4.6 mL) and H_2O (1.2 mL), and the suspension was stirred for 12 h at room temperature. After H_2O (12 mL) was added, the solution was acidified with conc. HCl. The precipitate was collected by filtration and the mother liquor was thoroughly extracted with Et_2O. The combined organic layer was dried over Na_2SO_4 and the solvent was removed under reduced pressure to yield the product 142 as a colorless solid (367 mg, 990 µmol, 98%).

– R_f (CH_2Cl_2/MeOH/formic acid = 97:2:1) = 0.18. – ^1H NMR (500 MHz, MeOD): δ = 8.29 (s, 1H, CH_{Ar}), 7.99 (s, 1H, CH_{Ar}) ppm. Signals missing (2H, COOH). – ^{13}C NMR (126 MHz, MeOD): δ = 167.9 (C_q, 1 × C=O), 167.2 (C_q, 1 × C=O), 144.0 (+, 1 × CH_{Ar}), 141.2 (C_q, 1 × C_{qAr}), 137.7 (C_q, 1 × C_{qAr}), 136.5 (+, 1 × CH_{Ar}), 121.6 (C_q, 1 × C_{qAr}), 92.1 (C_q, 1 × C_{qAr}) ppm.– Analytical data is in accordance with previously published literature.[175]

Ullmann type reaction with 2-iodo-5-bromoterephthalic acid (142) and phenol (128)

133a, X = Br
133b, X = I

Under argon atmosphere DMF (13.5 mL) was added to 2-iodo-5-bromoterephthalic acid (142) (668 mg, 1.80 mmol, 1.00 equiv.), phenol (128) (220 mg, 2.30 mmol, 1.30 equiv.) and copper(0) (11.4 mg, 180 µmol, 10 mol%). Then 1,8-diazabicyclo[5.4.0]undec-7-ene (DBU) (0.81 mL, 822 mg, 5.40 mmol, 3.00 equiv.) was added. The reaction was stirred at rt and monitored *via* TLC and APCI. After all the starting material was consumed, the reaction was diluted with 1 M HCl (25 mL) and poured over ice (50 mL). The resulting precipitates containing the iodinated product 133b were filtered, and the mother liquid was extracted with ethyl acetate (3 × 50 mL) which enabled the isolation of the brominated Ullmann product 133a. All precipitates were dried under vacuum to give the desired product as a yellow solid (total yield: 392 mg, 1.16 mmol, 64%).

133a: – R_f (CH_2Cl_2/MeOH/formic acid = 98:1:1) = 0.39. – ^1H NMR (500 MHz, MeOD): δ = 8.13 (s, 1H, CH_{Ar}), 7.43–7.35 (m, 2H, CH_{Ar}), 7.28 (s, 1H, CH_{Ar}), 7.16 (tt, ^3J = 7.3 Hz, ^4J = 1.1 Hz, 1H, CH_{Ar}), 7.04–6.97 (m, 2H, CH_{Ar}) ppm. Signals missing (2H, COOH). – ^{13}C NMR (126 MHz, MeOD): δ = 168.0 (C_q, 1 × COOH), 166.9 (C_q, 1 × COOH), 157.8 (C_q, 1 × C_{qAr}), 156.8 (C_q, 1 × C_{qAr}), 138.8 (C_q, 1 × C_{qAr}), 137.9 (+, 1 × CH_{Ar}), 131.1 (+, 2 × CH_{Ar}), 128.2 (C_q, 1 × C_{qAr}), 125.3 (+, 1 × CH_{Ar}), 123.0 (+, 1 × CH_{Ar}), 119.9 (+, 2 × CH_{Ar}), 114.6 (C_q, 1 × C_{qAr}) ppm.

– IR (ATR): $\tilde{v}$ = 2798 (w), 2779 (w), 2502 (w), 1673 (vs), 1589 (s), 1545 (m), 1482 (s), 1434 (vs), 1404 (s), 1358 (s), 1303 (s), 1256 (s), 1222 (vs), 1187 (vs), 1162 (vs), 1099 (vs), 1065 (s), 1024 (s), 946 (vs), 905 (s), 887 (s), 839 (s), 813 (s), 779 (vs), 744 (vs), 694 (vs), 673 (vs), 657 (s), 629 (vs), 565 (s), 524 (vs), 455 (s), 416 (s) cm^{-1}. – MS (ESI–), m/z (%): 335/337 (100/98). – HRMS (ESI–, $C_{14}H_8{}^{79}BrO_5$): calc. 334.9555; found 334.9565.

133b: – R_f (CH$_2$Cl$_2$/MeOH/formic acid = 98:1:1) = 0.39. – ^{1}H NMR (500 MHz, MeOD): δ = 8.12 (s, 1H, CH_{Ar}), 7.43–7.34 (m, 2H, CH_{Ar}), 7.27 (s, 1H, CH_{Ar}), 7.20–7.10 (m, 1H, CH_{Ar}), 7.05–6.94 (m, 2H, CH_{Ar}) ppm. Signals missing (2H, COOH). – ^{13}C NMR (126 MHz, MeOD): δ = 168.0 (C$_q$, 1 × COOH), 166.9 (C$_q$, 1 × COOH), 157.8 (C$_q$, 1 × C_{qAr}), 157.7 (C$_q$, 1 × C_{qAr}), 156.8 (C$_q$, 1 × C_{qAr}), 137.8 (+, 1 × CH$_{Ar}$), 131.1 (+, 2 × CH$_{Ar}$), 125.3 (+, 1 × CH$_{Ar}$), 122.9 (+, 1 × CH$_{Ar}$), 122.4 (C$_q$, 1 × C_{qAr}), 120.0 (+, 2 × CH$_{Ar}$), 114.6 (C$_q$, 1 × C_{qAr}), 85.3 (C$_q$, 1 × C_{qAr}) ppm. – IR (ATR): $\tilde{v}$ = 2788 (w), 2520 (w), 1677 (vs), 1589 (m), 1545 (w), 1480 (s), 1434 (s), 1404 (s), 1357 (m), 1305 (s), 1256 (s), 1224 (vs), 1188 (vs), 1162 (vs), 1099 (s), 1065 (m), 1024 (m), 943 (s), 912 (s), 885 (s), 839 (m), 813 (s), 781 (vs), 744 (s), 694 (vs), 673 (s), 629 (m), 601 (w), 565 (s), 530 (s), 441 (m) cm^{-1}. – MS (ESI–), m/z (%): 383/384 (81/12), 127 (100). – HRMS (ESI–, $C_{14}H_8{}^{127}IO_5$): calc. 382.9416; found 382.9425.

Side product: **2,5-Diphenoxyterephthalic acid (134)**

– R_f (CH$_2$Cl$_2$/MeOH/formic acid = 98:1:1) = 0.39. – ^{1}H NMR (400 MHz, MeOD): δ = 7.43 (s, 2H, CH_{Ar}), 7.40–7.34 (m, 4H, CH_{Ar}), 7.16–7.09 (m, 2H, CH_{Ar}), 7.01–6.95 (m, 4H, CH_{Ar}) ppm. Signals missing (2H, OH). – ^{13}C NMR (101 MHz, MeOD): δ = 158.9 (C$_q$, 2 × COOH), 152.6 (C$_q$, 2 × C_{qAr}), 131.0 (+, 4 × CH$_{Ar}$), 124.6 (+, 4 × CH$_{Ar}$), 120.0 (C$_q$, 2 × C_{qAr}), 119.2 (+, 4 × CH$_{Ar}$) ppm. – MS (ESI–), m/z (%): 349 (100), 273 (90). – HRMS (ESI–, $C_{20}H_{13}O_6$): calc. 349.0712; found 349.0718.

2-Bromo-9-oxo-9H-xanthene-3-carboxylic acid (126a)

2-Bromo-5-phenoxyterephthalic acid **133a** (220 mg, 653 µmol, 1.00 equiv.) was added to 30 mL of 80 % H$_2$SO$_4$ and heated to 80 °C while being stirred for 3 h. After cooling to room temperature, the mixture was stirred for another 16 h. The reaction was cooled and poured over ice (50 mL), producing a solid that was filtered, washed with water (mL) and dried under vacuum. The crude was purified by HPLC (MeCN/H$_2$O 1:1) to give the desired product **126a** as a grey solid

(1 mg, 3.14 μmol, 0.5%). The amount of product was not sufficient for an appropriate ^{13}C NMR spectrum.

– ^{1}H NMR (500 MHz, MeOD): δ = 8.48 (s, 1H, CH_{Ar}), 8.28 (dd, ^3J = 8.0 Hz, ^4J = 1.7 Hz, 1H, CH_{Ar}), 7.93 (s, 1H, CH_{Ar}), 7.88 (ddd, ^3J = 8.7, 7.1 Hz, ^4J = 1.7 Hz, 1H, CH_{Ar}), 7.65 (dd, ^3J = 8.5 Hz, ^4J = 1.0 Hz, 1H, CH_{Ar}), 7.49 (ddd, ^3J = 8.1, 7.1 Hz, ^4J = 1.0 Hz, 1H, CH_{Ar}) ppm. – MS (EI, 70 eV), m/z (%): 318/320 (31/29) [M]$^+$, 241 (100), 240 (21) [M–Br]$^+$. – HRMS (EI, C$_{14}$H$_7$O$_4$^{79}Br): calc. 317.9528; found 317.9526.

7.3   Crystallographic Data

7.3.1   Crystallographic Data Solved by Dr. Martin Nieger

Crystal structures in this section were measured and solved by Dr. Martin Nieger at the University of Helsinki (Finland).

Table 9. Measured and solved crystal structures by Dr. Martin Nieger.

Numbering in this thesis	Sample code used by Dr. Nieger	CCDC
72b	SB1081	1992178
73a	SB1080	1992180
87aa	SB932	1992179
88ad	SB1077	1992181
87ae	SB1079	1992182
88af	SB898	1992183
87af	SB1164	1992184
87bh	SB1078	1992185
87aa-epi	SB1233	---
113ac	SB1239	---
113ad	SB1302	---
113af	SB960	---
119ag	SB1285	---
119ah	SB1288	---
87db-epi	SB1294	---

(2-Bromophenyl)(1,4-dimethoxy-1,4-dihydronaphthalen-2-yl)methanone (72b) SB1081_HY

Crystal data

$C_{19}H_{15}BrO_3$	$F(000) = 752$
$M_r = 371.22$	$D_x = 1.625$ Mg m^{-3}
Monoclinic, Pc (no.9)	Mo $K\alpha$ radiation, $\lambda = 0.71073$ Å
$a = 18.5274$ (9) Å	Cell parameters from 9776 reflections
$b = 7.6936$ (3) Å	$\theta = 2.8–27.5°$
$c = 10.7680$ (5) Å	$\mu = 2.72$ mm^{-1}
$\beta = 98.558$ (2)°	$T = 123$ K
$V = 1517.81$ (12) Å^3	Plates, colourless
$Z = 4$	$0.18 \times 0.09 \times 0.03$ mm

Data collection

Bruker D8 VENTURE diffractometer with PhotonII CPAD detector	6665 reflections with $I > 2\sigma(I)$
Radiation source: INCOATEC microfocus sealed tube	$R_{int} = 0.038$
rotation in ϕ and ω, 1°, shutterless scans	$\theta_{max} = 27.5°$, $\theta_{min} = 2.2°$
Absorption correction: multi-scan *SADABS* (Sheldrick, 2014)	$h = -24 \rightarrow 24$
$T_{min} = 0.764$, $T_{max} = 0.875$	$k = -9 \rightarrow 9$
54899 measured reflections	$l = -14 \rightarrow 14$
6945 independent reflections	

Refinement

Refinement on F^2	Secondary atom site location: difference Fourier map
Least-squares matrix: full	Hydrogen site location: difference Fourier map
$R[F^2 > 2\sigma(F^2)] = 0.020$	H-atom parameters constrained
$wR(F^2) = 0.043$	$w = 1/[\sigma^2(F_o^2) + (0.0178P)^2 + 0.6034P]$ where $P = (F_o^2 + 2F_c^2)/3$
$S = 1.04$	$(\Delta/\sigma)_{max} < 0.001$
6945 reflections	$\Delta)_{max} = 0.24$ e Å^{-3}
419 parameters	$\Delta)_{min} = -0.19$ e Å^{-3}
2 restraints	Absolute structure: Flack x determined using 3146 quotients [(I+)-(I-)]/[(I+)+(I-)] (Parsons, Flack and Wagner, Acta Cryst. B69 (2013) 249-259).
Primary atom site location: dual	Absolute structure parameter: -0.005 (2)

2-(2-Iodobenzoyl)naphthalene-1,4-dione (73a) SB1080_HY

disorder of the 2-iodobenzoyl moiety (approx. 9:1), two isomers/rotamers

Crystal data

C₁₇H₉IO₃	$F(000) = 752$
$M_r = 388.14$	$D_x = 1.904$ Mg m⁻³
Monoclinic, $P2_1/c$ *(no.14)*	Mo $K\alpha$ radiation, $\lambda = 0.71073$ Å
$a = 5.5188$ (3) Å	Cell parameters from 9884 reflections
$b = 10.5469$ (6) Å	$\theta = 2.6–27.5°$
$c = 23.2612$ (13) Å	$\mu = 2.37$ mm⁻¹
$\beta = 90.746$ (2)°	$T = 123$ K
$V = 1353.83$ (13) Å³	Blocks, yellow
$Z = 4$	$0.14 \times 0.08 \times 0.04$ mm

Data collection

Bruker D8 VENTURE diffractometer with PhotonII CPAD detector	2834 reflections with $I > 2\sigma(I)$
Radiation source: INCOATEC microfocus sealed tube	$R_{int} = 0.043$
rotation in ϕ and ω, 1°, shutterless scans	$\theta_{max} = 27.5°$, $\theta_{min} = 2.6°$
Absorption correction: multi-scan *SADABS* (Sheldrick, 2014)	$h = -7 \rightarrow 7$
$T_{min} = 0.793$, $T_{max} = 0.862$	$k = -13 \rightarrow 13$
34641 measured reflections	$l = -30 \rightarrow 30$
3104 independent reflections	

Refinement

Refinement on F^2	Primary atom site location: dual
Least-squares matrix: full	Secondary atom site location: difference Fourier map
$R[F^2 > 2\sigma(F^2)] = 0.045$	Hydrogen site location: inferred from neighbouring sites
$wR(F^2) = 0.096$	H-atom parameters constrained
$S = 1.16$	$w = 1/[\sigma^2(F_o^2) + 10.190P]$ where $P = (F_o^2 + 2F_c^2)/3$
3104 reflections	$(\Delta/\sigma)_{max} = 0.002$
151 parameters	$\Delta\rangle_{max} = 1.29$ e Å⁻³
21 restraints	$\Delta\rangle_{min} = -1.55$ e Å⁻³

4a-(2-Iodobenzoyl)-1,4,4a,9a-tetrahydroanthracene-9,10-dione (87aa) SB932_HY

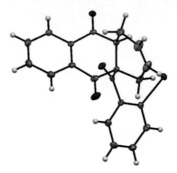

Crystal data

$C_{21}H_{15}IO_3$	$F(000) = 872$
$M_r = 442.23$	$D_x = 1.695$ Mg m^{-3}
Monoclinic, $P2_1/c$ *(no.14)*	Mo $K\alpha$ radiation, $\lambda = 0.71073$ Å
$a = 15.3153$ (7) Å	Cell parameters from 9480 reflections
$b = 8.2650$ (4) Å	$\theta = 2.6–27.5°$
$c = 15.5386$ (8) Å	$\mu = 1.87$ mm^{-1}
$\beta = 118.249$ (2)°	$T = 123$ K
$V = 1732.63$ (15) Å^3	Blocks, colourless
$Z = 4$	$0.24 \times 0.18 \times 0.06$ mm

Data collection

Bruker D8 VENTURE diffractometer with Photon100 detector	3998 independent reflections
Radiation source: INCOATEC microfocus sealed tube	3638 reflections with $I > 2\sigma(I)$
Detector resolution: 10.4167 pixels mm^{-1}	$R_{int} = 0.026$
rotation in ϕ and ω, 1°, shutterless scans	$\theta_{max} = 27.6°$, $\theta_{min} = 2.6°$
Absorption correction: multi-scan *SADABS* (Sheldrick, 2014)	$h = -19\rightarrow19$
$T_{min} = 0.757$, $T_{max} = 0.862$	$k = -10\rightarrow10$
41293 measured reflections	$l = -20\rightarrow19$

Refinement

Refinement on F^2	Secondary atom site location: difference Fourier map
Least-squares matrix: full	Hydrogen site location: difference Fourier map
$R[F^2 > 2\sigma(F^2)] = 0.015$	H-atom parameters constrained
$wR(F^2) = 0.036$	$w = 1/[\sigma^2(F_o^2) + (0.0133P)^2 + 1.3642P]$ where $P = (F_o^2 + 2F_c^2)/3$
$S = 1.06$	$(\Delta/\sigma)_{max} = 0.004$
3998 reflections	$\Delta\rangle_{max} = 0.43$ e Å^{-3}
227 parameters	$\Delta\rangle_{min} = -0.30$ e Å^{-3}
0 restraints	Extinction correction: *SHELXL2014/7* (Sheldrick 2014), $Fc^* = kFc[1+0.001xFc^2\lambda^3/\sin(2\theta)]^{-1/4}$
Primary atom site location: structure-invariant direct methods	Extinction coefficient: 0.00289 (17)

9a-(2-Iodobenzoyl)-1-((trimethylsilyl)-oxy)-1,4,4a,9a-tetrahydroanthracene-9,10-dione (88ad) SB1077_HY

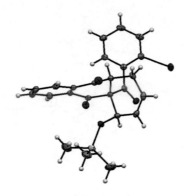

Crystal data

C₂₄H₂₃IO₄Si	$F(000) = 1064$
$M_r = 530.41$	$D_x = 1.558$ Mg m⁻³
Monoclinic, $P2_1/c$ *(no.14)*	Cu $K\alpha$ radiation, $\lambda = 1.54178$ Å
$a = 17.1926\ (4)$ Å	Cell parameters from 9910 reflections
$b = 9.0111\ (2)$ Å	$\theta = 2.7$–$72.0°$
$c = 15.3294\ (3)$ Å	$\mu = 11.86$ mm⁻¹
$\beta = 107.781\ (1)°$	$T = 123$ K
$V = 2261.45\ (9)$ Å³	Plates, colourless
$Z = 4$	$0.06 \times 0.04 \times 0.02$ mm

Data collection

Bruker D8 VENTURE diffractometer with PhotonII CPAD detector	3930 reflections with $I > 2\sigma(I)$
Radiation source: INCOATEC microfocus sealed tube	$R_{int} = 0.035$
rotation in ϕ and ω, 1°, shutterless scans	$\theta_{max} = 72.2°$, $\theta_{min} = 2.7°$
Absorption correction: multi-scan *SADABS* (Sheldrick, 2014)	$h = -21 \rightarrow 21$
$T_{min} = 0.613$, $T_{max} = 0.774$	$k = -10 \rightarrow 9$
23892 measured reflections	$l = -18 \rightarrow 18$
4439 independent reflections	

Refinement

Refinement on F^2	Primary atom site location: dual
Least-squares matrix: full	Secondary atom site location: difference Fourier map
$R[F^2 > 2\sigma(F^2)] = 0.031$	Hydrogen site location: difference Fourier map
$wR(F^2) = 0.069$	H-atom parameters constrained
$S = 1.03$	$w = 1/[\sigma^2(F_o^2) + (0.023P)^2 + 4.140P]$ where $P = (F_o^2 + 2F_c^2)/3$
4439 reflections	$(\Delta/\sigma)_{max} = 0.001$
271 parameters	$\Delta\rangle_{max} = 1.47$ e Å⁻³
0 restraints	$\Delta\rangle_{min} = -1.05$ e Å⁻³

9a-(2-Iodobenzoyl)-2-methyl-1-((trimethylsilyl)oxy)-1,4,4a,9a-tetrahydroanthracene-9,10-dione (87ae) SB1079_HY

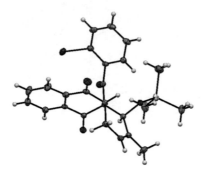

Crystal data

C$_{25}$H$_{25}$IO$_4$Si	F(000) = 1096
M_r = 544.44	D_x = 1.557 Mg m^{-3}
Monoclinic, $P2_1/c$ (no.14)	Mo $K\alpha$ radiation, λ = 0.71073 Å
a = 10.4516 (4) Å	Cell parameters from 9526 reflections
b = 12.0051 (4) Å	θ = 2.2–27.5°
c = 18.9173 (7) Å	μ = 1.46 mm^{-1}
β = 101.819 (1)°	T = 123 K
V = 2323.28 (15) Å^3	Blocks, colourless
Z = 4	0.20 × 0.16 × 0.10 mm

Data collection

Bruker D8 VENTURE diffractometer with PhotonII CPAD detector	4996 reflections with $I > 2\sigma(I)$
Radiation source: INCOATEC microfocus sealed tube	R_{int} = 0.026
rotation in ϕ and ω, 1°, shutterless scans	θ_{max} = 27.5°, θ_{min} = 2.2°
Absorption correction: multi-scan $SADABS$ (Sheldrick, 2014)	$h = -13 \rightarrow 13$
T_{min} = 0.806, T_{max} = 0.862	$k = -15 \rightarrow 15$
56591 measured reflections	$l = -24 \rightarrow 24$
5346 independent reflections	

Refinement

Refinement on F^2	Secondary atom site location: difference Fourier map
Least-squares matrix: full	Hydrogen site location: difference Fourier map
$R[F^2 > 2\sigma(F^2)]$ = 0.017	H-atom parameters constrained
$wR(F^2)$ = 0.043	$w = 1/[\sigma^2(F_o^2) + (0.0154P)^2 + 1.5947P]$ where $P = (F_o^2 + 2F_c^2)/3$
S = 1.08	$(\Delta/\sigma)_{max}$ = 0.003
5346 reflections	$\Delta\rangle_{max}$ = 0.43 e Å^{-3}
282 parameters	$\Delta\rangle_{min}$ = -0.33 e Å^{-3}
0 restraints	Extinction correction: $SHELXL2014/7$ (Sheldrick, 2014), Fc*=kFc[1+0.001xFc$^2\lambda^3$/sin(2θ)]$^{-1/4}$
Primary atom site location: dual	Extinction coefficient: 0.0019 (2)

1-((*tert*-Butyldimethylsilyl)oxy)-9a-(2-iodobenzoyl)-2-methyl-1,4,4a,9a-tetrahydroanthracene-9,10-dione (88af) SB898_HY

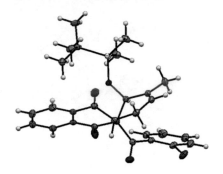

Crystal data

C$_{28}$H$_{31}$IO$_4$Si	$F(000) = 1192$
$M_r = 586.52$	$D_x = 1.503$ Mg m^{-3}
Monoclinic, $P2_1/n$ (no.14)	Cu $K\alpha$ radiation, $\lambda = 1.54178$ Å
$a = 12.4501$ (8) Å	Cell parameters from 9850 reflections
$b = 9.4569$ (6) Å	$\theta = 4.4–72.1°$
$c = 22.4286$ (14) Å	$\mu = 10.41$ mm^{-1}
$\beta = 101.113$ (2)°	$T = 123$ K
$V = 2591.2$ (3) Å^3	Blocks, yellow
$Z = 4$	0.16 × 0.12 × 0.08 mm

Data collection

Bruker D8 VENTURE diffractometer with Photon100 detector	5087 independent reflections
Radiation source: INCOATEC microfocus sealed tube	4969 reflections with $I > 2\sigma(I)$
Detector resolution: 10.4167 pixels mm^{-1}	$R_{int} = 0.024$
rotation in ϕ and ω, 1°, shutterless scans	$\theta_{max} = 72.2°$, $\theta_{min} = 3.8°$
Absorption correction: multi-scan *SADABS* (Sheldrick, 2014)	$h = -14 \rightarrow 15$
$T_{min} = 0.352$, $T_{max} = 0.470$	$k = -11 \rightarrow 11$
28040 measured reflections	$l = -27 \rightarrow 25$

Refinement

Refinement on F^2	Secondary atom site location: difference Fourier map
Least-squares matrix: full	Hydrogen site location: difference Fourier map
$R[F^2 > 2\sigma(F^2)] = 0.021$	H-atom parameters constrained
$wR(F^2) = 0.052$	$w = 1/[\sigma^2(F_o^2) + (0.0204P)^2 + 2.4086P]$ where $P = (F_o^2 + 2F_c^2)/3$
$S = 1.05$	$(\Delta/\sigma)_{max} = 0.001$
5087 reflections	$\Delta\rangle_{max} = 0.67$ e Å^{-3}
309 parameters	$\Delta\rangle_{min} = -0.54$ e Å^{-3}
0 restraints	Extinction correction: *SHELXL2014/7* (Sheldrick 2014, $Fc^* = kFc[1+0.001 \times Fc^2\lambda^3/\sin(2\theta)]^{-1/4}$
Primary atom site location: structure-invariant direct methods	Extinction coefficient: 0.00043 (3)

1-((*tert*-Butyldimethylsilyl)oxy)-9a-(2-iodobenzoyl)-2-methyl-1,4,4a,9a-

tetrahydroanthracene- 9,10-dione (87af) SB1164_HY

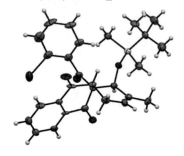

Crystal data

$C_{28}H_{31}IO_4Si$	$F(000) = 2384$
$M_r = 586.52$	$D_x = 1.469$ Mg m^{-3}
Monoclinic, $P2_1/c$ *(no.14)*	Cu $K\alpha$ radiation, $\lambda = 1.54178$ Å
$a = 19.1112$ (6) Å	Cell parameters from 9942 reflections
$b = 7.9277$ (3) Å	$\theta = 3.0-71.9°$
$c = 35.7482$ (11) Å	$\mu = 10.17$ mm^{-1}
$\beta = 101.728$ (2)°	$T = 123$ K
$V = 5303.1$ (3) Å^3	Needles, colourless
$Z = 8$	$0.12 \times 0.03 \times 0.01$ mm

Data collection

Bruker D8 VENTURE diffractometer with PhotonII CPAD detector	7491 reflections with $I > 2\sigma(I)$
Radiation source: INCOATEC microfocus sealed tube	$R_{int} = 0.131$
rotation in ϕ and ω, 1°, shutterless scans	$\theta_{max} = 72.3°$, $\theta_{min} = 2.4°$
Absorption correction: multi-scan *SADABS* (Sheldrick, 2014)	$h = -23 \rightarrow 23$
$T_{min} = 0.659$, $T_{max} = 0.864$	$k = -9 \rightarrow 9$
59783 measured reflections	$l = -44 \rightarrow 44$
10398 independent reflections	

Refinement

Refinement on F^2	Primary atom site location: dual
Least-squares matrix: full	Secondary atom site location: difference Fourier map
$R[F^2 > 2\sigma(F^2)] = 0.081$	Hydrogen site location: difference Fourier map
$wR(F^2) = 0.173$	H-atom parameters constrained
$S = 1.06$	$w = 1/[\sigma^2(F_o^2) + (0.037P)^2 + 52.7P]$ where $P = (F_o^2 + 2F_c^2)/3$
10398 reflections	$(\Delta/\sigma)_{max} = 0.001$
615 parameters	$\Delta\rangle_{max} = 1.84$ e Å^{-3}
0 restraints	$\Delta\rangle_{min} = -1.04$ e Å^{-3}

4a-(2-Bromobenzoyl)-2-((*tert*-butyldiphenylsilyl)oxy)-1,4,4a,9a-tetrahydroanthracene-9,10-dione (87bh) SB1078_HY

Crystal data

$C_{37}H_{33}BrO_4Si$	$F(000) = 1344$
$M_r = 649.63$	$D_x = 1.420$ Mg m^{-3}
Monoclinic, $P2_1/c$ *(no.14)*	Mo $K\alpha$ radiation, $\lambda = 0.71073$ Å
$a = 19.7258$ (5) Å	Cell parameters from 9416 reflections
$b = 9.8370$ (3) Å	$\theta = 2.3–27.5°$
$c = 16.3586$ (4) Å	$\mu = 1.43$ mm^{-1}
$\beta = 106.796$ (1)°	$T = 123$ K
$V = 3038.85$ (14) Å^3	Blocks, yellow
$Z = 4$	$0.40 \times 0.30 \times 0.20$ mm

Data collection

Bruker D8 VENTURE diffractometer with PhotonII CPAD detector	6448 reflections with $I > 2\sigma(I)$
Radiation source: INCOATEC microfocus sealed tube	$R_{int} = 0.032$
rotation in ϕ and ω, 1°, shutterless scans	$\theta_{max} = 27.5°$, $\theta_{min} = 2.3°$
Absorption correction: multi-scan *SADABS* (Sheldrick, 2014)	$h = -25 \rightarrow 25$
$T_{min} = 0.645$, $T_{max} = 0.746$	$k = -12 \rightarrow 12$
80591 measured reflections	$l = -21 \rightarrow 21$
6966 independent reflections	

Refinement

Refinement on F^2	Primary atom site location: dual
Least-squares matrix: full	Secondary atom site location: difference Fourier map
$R[F^2 > 2\sigma(F^2)] = 0.025$	Hydrogen site location: difference Fourier map
$wR(F^2) = 0.063$	H-atom parameters constrained
$S = 1.05$	$w = 1/[\sigma^2(F_o^2) + (0.0285P)^2 + 1.8188P]$ where $P = (F_o^2 + 2F_c^2)/3$
6966 reflections	$(\Delta/\sigma)_{max} = 0.001$
388 parameters	$\Delta\rangle_{max} = 0.34$ e Å^{-3}
0 restraints	$\Delta\rangle_{min} = -0.49$ e Å^{-3}

4a-(2-Iodobenzoyl)-1,4,4a,9a-tetrahydroanthracene-9,10-dione (87aa-epi) SB1233_HY

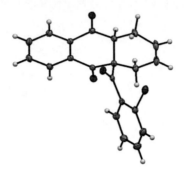

Crystal data

$C_{21}H_{15}IO_3$	$Z = 2$
$M_r = 442.23$	$F(000) = 436$
Triclinic, P-1 (no.2)	$D_x = 1.744$ Mg m^{-3}
$a = 8.0081$ (6) Å	Cu $K\alpha$ radiation, $\lambda = 1.54178$ Å
$b = 10.3030$ (8) Å	Cell parameters from 9947 reflections
$c = 10.6997$ (8) Å	$\theta = 4.2–72.2°$
$\alpha = 85.460$ (2)°	$\mu = 15.08$ mm^{-1}
$\beta = 78.146$ (2)°	$T = 123$ K
$\gamma = 77.273$ (2)°	Blocks, colourless
$V = 842.22$ (11) Å^3	$0.16 \times 0.08 \times 0.04$ mm

Data collection

Bruker D8 VENTURE diffractometer with PhotonII CPAD detector	3298 reflections with $I > 2\sigma(I)$
Radiation source: INCOATEC microfocus sealed tube	$R_{int} = 0.027$
rotation in ϕ and ω, 1°, shutterless scans	$\theta_{max} = 72.2°$, $\theta_{min} = 4.2°$
Absorption correction: multi-scan $SADABS$ (Sheldrick, 2014)	$h = -9\rightarrow9$
$T_{min} = 0.258$, $T_{max} = 0.523$	$k = -12\rightarrow12$
13410 measured reflections	$l = -13\rightarrow13$
3306 independent reflections	

Refinement

Refinement on F^2	Primary atom site location: dual
Least-squares matrix: full	Secondary atom site location: difference Fourier map
$R[F^2 > 2\sigma(F^2)] = 0.019$	Hydrogen site location: difference Fourier map
$wR(F^2) = 0.044$	H-atom parameters constrained
$S = 1.09$	$w = 1/[\sigma^2(F_o^2) + 0.8932P]$ where $P = (F_o^2 + 2F_c^2)/3$
3306 reflections	$(\Delta/\sigma)_{max} < 0.001$
226 parameters	$\Delta)_{max} = 0.49$ e Å^{-3}
0 restraints	$\Delta)_{min} = -0.33$ e Å^{-3}

5-Methyl-6-methylene-7,7a-dihydro-6*H*-5,13a-methanobenzo[4,5]cycloocta-[1,2-b]naphthalene-8,13,14(5*H*)-trione (113ac) SB1239_HY

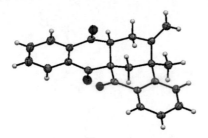

Crystal data

$C_{23}H_{18}O_3$	$D_x = 1.377$ Mg m^{-3}
$M_r = 342.37$	Cu $K\alpha$ radiation, $\lambda = 1.54178$ Å
Orthorhombic, *Pbca (no.61)*	Cell parameters from 9481 reflections
$a = 19.7103 (10)$ Å	$\theta = 4.3–72.2°$
$b = 8.2112 (4)$ Å	$\mu = 0.73$ mm^{-1}
$c = 20.4011 (10)$ Å	$T = 123$ K
$V = 3301.8 (3)$ Å^3	Plates, colourless
$Z = 8$	$0.24 \times 0.08 \times 0.04$ mm
$F(000) = 1440$	

Data collection

Bruker D8 VENTURE diffractometer with PhotonII CPAD detector	2990 reflections with $I > 2\sigma(I)$
Radiation source: INCOATEC microfocus sealed tube	$R_{int} = 0.048$
rotation in ϕ and ω, 1°, shutterless scans	$\theta_{max} = 72.4°$, $\theta_{min} = 4.3°$
Absorption correction: multi-scan *SADABS* (Sheldrick, 2014)	$h = -24 \rightarrow 24$
$T_{min} = 0.803$, $T_{max} = 0.971$	$k = -9 \rightarrow 10$
34231 measured reflections	$l = -23 \rightarrow 25$
3264 independent reflections	

Refinement

Refinement on F^2	Primary atom site location: dual
Least-squares matrix: full	Secondary atom site location: difference Fourier map
$R[F^2 > 2\sigma(F^2)] = 0.036$	Hydrogen site location: difference Fourier map
$wR(F^2) = 0.097$	H-atom parameters constrained
$S = 1.03$	$w = 1/[\sigma^2(F_o^2) + (0.0477P)^2 + 1.3004P]$ where $P = (F_o^2 + 2F_c^2)/3$
3264 reflections	$(\Delta/\sigma)_{max} = 0.001$
235 parameters	$\Delta\rangle_{max} = 0.31$ e Å^{-3}
0 restraints	$\Delta\rangle_{min} = -0.15$ e Å^{-3}

15-((Trimethylsilyl)oxy)-6*H*-5,13a-methanobenzo[4,5]cycloocta-[1,2-b]naphthalene-8,13,14(5*H*)-trione (113ad) SB1302_HY

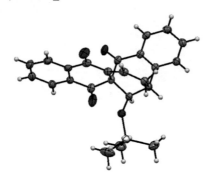

Crystal data

$C_{24}H_{22}O_4Si$	$F(000) = 1696$
$M_r = 402.50$	$D_x = 1.302$ Mg m^{-3}
Monoclinic, $P2_1/c$ (no.14)	Cu $K\alpha$ radiation, $\lambda = 1.54178$ Å
$a = 16.2129$ (7) Å	Cell parameters from 9633 reflections
$b = 12.8085$ (5) Å	$\theta = 4.3–72.1°$
$c = 19.7840$ (8) Å	$\mu = 1.24$ mm^{-1}
$\beta = 91.499$ (2)°	$T = 123$ K
$V = 4107.0$ (3) Å^3	Blocks, yellow
$Z = 8$	$0.25 \times 0.20 \times 0.10$ mm

Data collection

Bruker D8 VENTURE diffractometer with PhotonII CPAD detector	7792 reflections with $I > 2\sigma(I)$
Radiation source: INCOATEC microfocus sealed tube	$R_{int} = 0.025$
rotation in ϕ and ω, 1°, shutterless scans	$\theta_{max} = 72.2°$, $\theta_{min} = 2.7°$
Absorption correction: multi-scan *SADABS* (Sheldrick, 20149	$h = -20 \rightarrow 20$
$T_{min} = 0.801$, $T_{max} = 0.864$	$k = -15 \rightarrow 15$
77592 measured reflections	$l = -24 \rightarrow 24$
8071 independent reflections	

Refinement

Refinement on F^2	Primary atom site location: dual
Least-squares matrix: full	Secondary atom site location: difference Fourier map
$R[F^2 > 2\sigma(F^2)] = 0.034$	Hydrogen site location: difference Fourier map
$wR(F^2) = 0.092$	H-atom parameters constrained
$S = 1.03$	$w = 1/[\sigma^2(F_o^2) + (0.0454P)^2 + 1.8867P]$ where $P = (F_o^2 + 2F_c^2)/3$
8071 reflections	$(\Delta/\sigma)_{max} = 0.002$
523 parameters	$\Delta)_{max} = 0.34$ e Å^{-3}
0 restraints	$\Delta)_{min} = -0.37$ e Å^{-3}

15-((*tert*-Butyldimethylsilyl)-oxy)-5-methyl-6*H*-5,13a-methanobenzo[4,5]cycloocta-[1,2-b]naphthalene-8,13,14(5*H*)-trione (113af) SB960_HY

Crystal data

$C_{28}H_{30}O_4Si$	$F(000) = 976$
$M_r = 458.61$	$D_x = 1.276$ Mg m^{-3}
Monoclinic, $P2_1/c$ *(no.14)*	Cu $K\alpha$ radiation, $\lambda = 1.54178$ Å
$a = 16.9227$ (12) Å	Cell parameters from 4177 reflections
$b = 17.7462$ (11) Å	$\theta = 3.6–71.3°$
$c = 7.9931$ (5) Å	$\mu = 1.13$ mm^{-1}
$\beta = 96.137$ (5)°	$T = 123$ K
$V = 2386.7$ (3) Å^3	Plates, yellow
$Z = 4$	$0.12 \times 0.06 \times 0.02$ mm

Data collection

Bruker D8 VENTURE diffractometer with Photon100 detector	4643 independent reflections
Radiation source: INCOATEC microfocus sealed tube	3118 reflections with $I > 2\sigma(I)$
Detector resolution: 10.4167 pixels mm^{-1}	$R_{int} = 0.111$
rotation in ϕ and ω, 1°, shutterless scans	$\theta_{max} = 71.9°$, $\theta_{min} = 2.6°$
Absorption correction: multi-scan *SADABS* (Sheldrick, 2014)	$h = -19 \rightarrow 20$
$T_{min} = 0.814$, $T_{max} = 0.971$	$k = -21 \rightarrow 20$
18207 measured reflections	$l = -9 \rightarrow 9$

Refinement

Refinement on F^2	Primary atom site location: structure-invariant direct methods
Least-squares matrix: full	Secondary atom site location: difference Fourier map
$R[F^2 > 2\sigma(F^2)] = 0.074$	Hydrogen site location: difference Fourier map
$wR(F^2) = 0.176$	H-atom parameters constrained
$S = 1.02$	$w = 1/[\sigma^2(F_o^2) + (0.061P)^2 + 4.4177P]$ where $P = (F_o^2 + 2F_c^2)/3$
4643 reflections	$(\Delta/\sigma)_{max} < 0.001$
298 parameters	$\Delta\rangle_{max} = 0.99$ e Å^{-3} (close to SI15)
0 restraints	$\Delta\rangle_{min} = -0.50$ e Å^{-3}

5-((Triisopropylsilyl)oxy)-6*H*-5,13a-methanobenzo[4,5]cycloocta-[1,2-b]naphthalene-8,13,14(5*H*)-trione (119ag) SB1285_HY

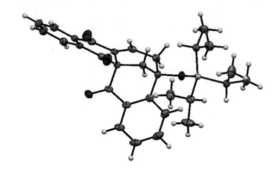

Crystal data

$C_{30}H_{34}O_4Si$	$F(000) = 1040$
$M_r = 486.66$	$D_x = 1.277$ Mg m^{-3}
Monoclinic, $P2_1/c$ *(no.14)*	Cu $K\alpha$ radiation, $\lambda = 1.54178$ Å
$a = 15.2706$ (7) Å	Cell parameters from 9757 reflections
$b = 11.9203$ (6) Å	$\theta = 4.9–72.2°$
$c = 15.5395$ (8) Å	$\mu = 1.09$ mm^{-1}
$\beta = 116.515$ (2)°	$T = 123$ K
$V = 2531.1$ (2) Å^3	Plates, colourless
$Z = 4$	$0.20 \times 0.18 \times 0.06$ mm

Data collection

Bruker D8 VENTURE diffractometer with PhotonII CPAD detector	4764 reflections with $I > 2\sigma(I)$
Radiation source: INCOATEC microfocus sealed tube	$R_{int} = 0.027$
rotation in ϕ and ω, 1°, shutterless scans	$\theta_{max} = 72.2°$, $\theta_{min} = 4.9°$
Absorption correction: multi-scan *SADABS* (Sheldrick, 2014)	$h = -18 \rightarrow 18$
$T_{min} = 0.807$, $T_{max} = 0.942$	$k = -14 \rightarrow 14$
46675 measured reflections	$l = -19 \rightarrow 19$
4981 independent reflections	

Refinement

Refinement on F^2	Primary atom site location: dual
Least-squares matrix: full	Secondary atom site location: difference Fourier map
$R[F^2 > 2\sigma(F^2)] = 0.033$	Hydrogen site location: difference Fourier map
$wR(F^2) = 0.092$	H-atom parameters constrained
$S = 1.03$	$w = 1/[\sigma^2(F_o^2) + (0.0503P)^2 + 1.0289P]$ where $P = (F_o^2 + 2F_c^2)/3$
4981 reflections	$(\Delta/\sigma)_{max} = 0.001$
316 parameters	$\Delta\rangle_{max} = 0.36$ e Å^{-3}
0 restraints	$\Delta\rangle_{min} = -0.30$ e Å^{-3}

5-((*tert*-Butyldiphenylsilyl)oxy)-6*H*-5,13a-methanobenzo[4,5]cycloocta-[1,2-b]naphthalene-8,13,14(5*H*)-trione (119ah) SB1288_HY

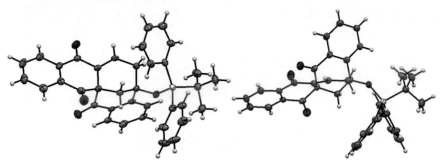

Crystal data

$C_{37}H_{32}O_4Si$	$F(000) = 1200$
$M_r = 568.71$	$D_x = 1.290$ Mg m^{-3}
Monoclinic, $P2_1/c$ *(no.14)*	Cu $K\alpha$ radiation, $\lambda = 1.54178$ Å
$a = 21.6364$ (12) Å	Cell parameters from 9796 reflections
$b = 8.9177$ (5) Å	$\theta = 4.2–72.1°$
$c = 15.6705$ (8) Å	$\mu = 1.03$ mm^{-1}
$\beta = 104.390$ (3)°	$T = 123$ K
$V = 2928.7$ (3) Å^3	Blocks, yellow
$Z = 4$	$0.16 \times 0.10 \times 0.06$ mm

Data collection

Bruker D8 VENTURE diffractometer with PhotonII CPAD detector	5294 reflections with $I > 2\sigma(I)$
Radiation source: INCOATEC microfocus sealed tube	$R_{int} = 0.057$
rotation in ϕ and ω, 1°, shutterless scans	$\theta_{max} = 72.6°$, $\theta_{min} = 4.2°$
Absorption correction: multi-scan *SADABS* (Sheldrick, 2014)	$h = -24 \rightarrow 26$
$T_{min} = 0.757$, $T_{max} = 0.942$	$k = -10 \rightarrow 10$
28460 measured reflections	$l = -19 \rightarrow 19$
5683 independent reflections	

Refinement

Refinement on F^2	Primary atom site location: dual
Least-squares matrix: full	Secondary atom site location: difference Fourier map
$R[F^2 > 2\sigma(F^2)] = 0.096$	Hydrogen site location: difference Fourier map
$wR(F^2) = 0.219$	H-atom parameters constrained
$S = 1.09$	$w = 1/[\sigma^2(F_o^2) + (0.0187P)^2 + 18.906P]$ where $P = (F_o^2 + 2F_c^2)/3$
5683 reflections	$(\Delta/\sigma)_{max} < 0.001$
379 parameters	$\Delta\rangle_{max} = 0.66$ e Å^{-3}
0 restraints	$\Delta\rangle_{min} = -1.03$ e Å^{-3}

4a-(2-Bromo-4,5-dimethoxybenzoyl)-2-methyl-1,4,4a,9a-tetrahydroanthracene-9,10-dione and 9a-(2-bromo-4,5-dimethoxybenzoyl)-2-methyl-1,4,4a,9a-tetrahydroanthracene-9,10-dione (87db-epi) SB1294_HY

disorder Me-moieties at C2/C3, approx. 2:1

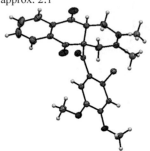

Crystal data

$C_{24}H_{21}BrO_5$	$D_x = 1.519$ Mg m^{-3}
$M_r = 469.32$	Cu $K\alpha$ radiation, $\lambda = 1.54178$ Å
Orthorhombic, *Pbca (no.61)*	Cell parameters from 9941 reflections
$a = 14.0870$ (3) Å	$\theta = 4.8–72.0°$
$b = 16.6071$ (3) Å	$\mu = 3.02$ mm^{-1}
$c = 17.5448$ (3) Å	$T = 123$ K
$V = 4104.50$ (13) Å^3	Plates, colourless
$Z = 8$	$0.36 \times 0.24 \times 0.06$ mm
$F(000) = 1920$	

Data collection

Bruker D8 VENTURE diffractometer with PhotonII CPAD detector	4019 reflections with $I > 2\sigma(I)$
Radiation source: INCOATEC microfocus sealed tube	$R_{int} = 0.061$
rotation in ϕ and ω, 1°, shutterless scans	$\theta_{max} = 72.4°, \theta_{min} = 4.8°$
Absorption correction: multi-scan *SADABS* (Sheldrick, 2014)	$h = -17\rightarrow17$
$T_{min} = 0.580, T_{max} = 0.841$	$k = -17\rightarrow20$
93968 measured reflections	$l = -21\rightarrow21$
4045 independent reflections	

Refinement

Refinement on F^2	Primary atom site location: dual
Least-squares matrix: full	Secondary atom site location: difference Fourier map
$R[F^2 > 2\sigma(F^2)] = 0.042$	Hydrogen site location: mixed
$wR(F^2) = 0.091$	H-atom parameters constrained
$S = 1.28$	$w = 1/[\sigma^2(F_o^2) + 9.9045P]$ where $P = (F_o^2 + 2F_c^2)/3$
4045 reflections	$(\Delta/\sigma)_{max} < 0.001$
274 parameters	$\Delta\rangle_{max} = 0.39$ e Å^{-3}
3 restraints	$\Delta\rangle_{min} = -0.49$ e Å^{-3}

7.3.2   Crystallographic Data Solved by Dr. Olaf Fuhr

The following crystal structures were measured and solved by Dr. Olaf Fuhr at the Institute of Nanotechnology (KIT).

Table 10. Measured and solved crystal structures by Dr. Olaf Fuhr.

Numbering in this thesis	Sample code used by Dr. Fuhr	CCDC
72a	QP093	1992878
72c	JB234	1992874
72d	JB265	1992879
72f	JB263 F3	1992880
72g	JB244	1992881
73b	JB205.1	1992875
73d	JB254	1992882
87ba	JB249	1992876
87ac	JB246	1992877
87bc	JB366	1992883
87ad	JB252 F2.2.2	1992884
88bd	JB211 F1	1992885
87bd	JB243 F2.3	1992886
88ae	JB228 F1	1992887
88bf	JB369 F2	1992888
87ag	JB201 F1	1992889
87fc	JB274	1992890
87gc	JB264	1992891
88gd	JB280 F1	1992892
161a	JB136	---
111a	JB250 F1	---
113ab	JB288 F1	---
115ab	JB288 F3	---
87ab-epi	JB289 F1_2	---
113ae	JB380 B4	---
113bf	JB384 B10 = B9	---
114bf	JB098 F2	---
119gd	JB378 C6/C12	---

(1,4-Dimethoxynaphthalen-2-yl)(2-iodophenyl)methanone (72a)

Identification code	QB093_tw
Empirical formula	$C_{19}H_{15}IO_3$
Formula weight	418.21
Temperature/K	180.15
Crystal system	monoclinic
Space group	Pc
a/Å	18.4714(9)
b/Å	8.0153(2)
c/Å	10.8747(4)
α/°	90
β/°	98.540(4)
γ/°	90
Volume/Å³	1592.19(11)
Z	4
ρ_{calc}g/cm³	1.745
μ/mm⁻¹	2.024
F(000)	824.0
Crystal size/mm³	0.29 × 0.25 × 0.04
Radiation	MoKα ($\lambda = 0.71073$)
2Θ range for data collection/°	4.46 to 53.996
Index ranges	$-23 \leq h \leq 23, -10 \leq k \leq 10, -13 \leq l \leq 13$
Reflections collected	5516
Independent reflections	5516 [R_{int} = merged, R_{sigma} = 0.0160]
Indep. refl. with I>=2σ (I)	5347
Data/restraints/parameters	5516/2/420
Goodness-of-fit on F^2	1.032
Final R indexes [I>=2σ (I)]	$R_1 = 0.0275$, $wR_2 = 0.0732$
Final R indexes [all data]	$R_1 = 0.0298$, $wR_2 = 0.0768$
Largest diff. peak/hole / e Å⁻³	0.58/-0.70
Flack parameter	0.05(2)

(2-Bromo-5-methoxyphenyl)(1,4-dimethoxynaphthalen-2-yl)methanone (72c)

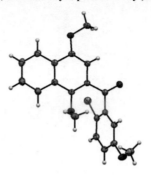

Identification code	JB234
Empirical formula	$C_{20}H_{17}BrO_4$
Formula weight	401.24
Temperature/K	150
Crystal system	monoclinic
Space group	$P2_1/c$
a/Å	8.0208(3)
b/Å	10.7247(4)
c/Å	20.3313(8)
α/°	90
β/°	100.599(3)
γ/°	90
Volume/Å^3	1719.07(11)
Z	4
ρ_{calc}g/cm^3	1.550
μ/mm^{-1}	2.358
F(000)	816.0
Crystal size/mm^3	0.23 × 0.2 × 0.04
Radiation	GaKα (λ = 1.34143)
2Θ range for data collection/°	9.76 to 123.996
Index ranges	-3 ≤ h ≤ 10, -14 ≤ k ≤ 13, -26 ≤ l ≤ 25
Reflections collected	10370
Independent reflections	4015 [R_{int} = 0.0170, R_{sigma} = 0.0139]
Data/restraints/parameters	4015/0/295
Goodness-of-fit on F^2	1.098
Final R indexes [I>=2σ (I)]	R_1 = 0.0272, wR_2 = 0.0757
Final R indexes [all data]	R_1 = 0.0296, wR_2 = 0.0773
Largest diff. peak/hole / e Å^{-3}	0.39/-0.47

(2-Bromo-4,5-dimethoxyphenyl)(1,4-dimethoxynaphthalen-2-yl)methanone (72d)

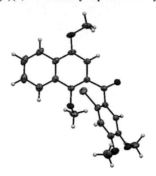

Identification code	JB265
Empirical formula	$C_{21}H_{19}BrO_5$
Formula weight	431.27
Temperature/K	150
Crystal system	monoclinic
Space group	$P2_1/c$
a/Å	12.4751(4)
b/Å	13.4745(3)
c/Å	12.7582(4)
α/°	90
β/°	117.121(2)
γ/°	90
Volume/Å^3	1908.79(10)
Z	4
ρ_{calc}g/cm^3	1.501
μ/mm^{-1}	2.176
F(000)	880.0
Crystal size/mm^3	0.26 × 0.24 × 0.22
Radiation	GaKα (λ = 1.34143)
2Θ range for data collection/°	8.86 to 124.99
Index ranges	-15 ≤ h ≤ 16, -17 ≤ k ≤ 6, -16 ≤ l ≤ 16
Reflections collected	13459
Independent reflections	4551 [R_{int} = 0.0136, R_{sigma} = 0.0113]
Data/restraints/parameters	4551/0/248
Goodness-of-fit on F^2	1.074
Final R indexes [I>=2σ (I)]	R_1 = 0.0293, wR_2 = 0.0768
Final R indexes [all data]	R_1 = 0.0318, wR_2 = 0.0794
Largest diff. peak/hole / e Å^{-3}	0.44/-0.55

N-(4-Bromo-3-(1,4-dimethoxy-2-naphthoyl)phenyl)acetamide (72f)

Identification code	JB263-F3
Empirical formula	$C_{21}H_{20}BrNO_5$
Formula weight	446.29
Temperature/K	150.15
Crystal system	monoclinic
Space group	$P2_1/c$
a/Å	18.5103(5)
b/Å	7.30020(10)
c/Å	14.5567(3)
$\alpha/°$	90
$\beta/°$	91.803(2)
$\gamma/°$	90
Volume/Å^3	1966.06(7)
Z	4
ρ_{calc}g/cm^3	1.508
μ/mm^{-1}	2.136
F(000)	912.0
Crystal size/mm^3	0.22 × 0.2 × 0.12
Radiation	GaKα (λ = 1.34143)
2Θ range for data collection/°	4.156 to 115
Index ranges	-23 ≤ h ≤ 23, -7 ≤ k ≤ 9, -9 ≤ l ≤ 18
Reflections collected	15260
Independent reflections	3978 [R_{int} = 0.0171, R_{sigma} = 0.0096]
Indep. refl. with I>=2σ (I)	3896
Data/restraints/parameters	3978/0/333
Goodness-of-fit on F^2	1.066
Final R indexes [I>=2σ (I)]	R_1 = 0.0256, wR_2 = 0.0719
Final R indexes [all data]	R_1 = 0.0261, wR_2 = 0.0722
Largest diff. peak/hole / e Å^{-3}	0.36/-0.62

N-(4-Bromo-3-(1,4-dimethoxy-2-naphthoyl)phenyl)-2,2,2-trifluoroacetamide (72g)

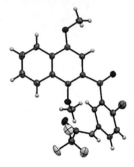

Identification code	JB244
Empirical formula	$C_{21}H_{15}BrF_3NO_4$
Formula weight	482.25
Temperature/K	150.15
Crystal system	triclinic
Space group	P-1
a/Å	8.0894(2)
b/Å	10.0616(3)
c/Å	12.6050(3)
α/°	97.484(2)
β/°	101.001(2)
γ/°	103.175(2)
Volume/Å^3	964.25(5)
Z	2
ρ_{calc}g/cm^3	1.661
μ/mm^{-1}	2.334
F(000)	484.0
Crystal size/mm^3	0.26 × 0.16 × 0.13
Radiation	GaKα (λ = 1.34143)
2Θ range for data collection/°	6.32 to 114.98
Index ranges	-3 ≤ h ≤ 10, -12 ≤ k ≤ 12, -15 ≤ l ≤ 15
Reflections collected	13918
Independent reflections	3925 [R_{int} = 0.0147, R_{sigma} = 0.0084]
Indep. refl. with I>=2σ (I)	3852
Data/restraints/parameters	3925/0/331
Goodness-of-fit on F^2	1.051
Final R indexes [I>=2σ (I)]	R$_1$ = 0.0234, wR$_2$ = 0.0594
Final R indexes [all data]	R$_1$ = 0.0238, wR$_2$ = 0.0597
Largest diff. peak/hole / e Å^{-3}	0.36/-0.43

2-(2-Bromobenzoyl)naphthalene-1,4-dione (73b)

Identification code	JB205-1
Empirical formula	$C_{17}H_9O_3Br$
Formula weight	341.15
Temperature/K	180.15
Crystal system	monoclinic
Space group	$P2_1/c$
a/Å	5.42720(10)
b/Å	10.5371(4)
c/Å	23.1383(6)
$\alpha/°$	90
$\beta/°$	91.114(2)
$\gamma/°$	90
Volume/Å^3	1322.96(7)
Z	4
$\rho_{calc}g/cm^3$	1.713
μ/mm^{-1}	2.951
F(000)	680.0
Crystal size/mm^3	0.23 × 0.05 × 0.04
Radiation	GaKα (λ = 1.34143)
2Θ range for data collection/°	12.374 to 118.834
Index ranges	$-3 \leq h \leq 6, -13 \leq k \leq 13, -25 \leq l \leq 29$
Reflections collected	6867
Independent reflections	2723 [R_{int} = 0.0214, R_{sigma} = 0.0161]
Indep. refl. with I>=2σ (I)	2574
Data/restraints/parameters	2723/0/190
Goodness-of-fit on F^2	1.106
Final R indexes [I>=2σ (I)]	R_1 = 0.0454, wR_2 = 0.1207
Final R indexes [all data]	R_1 = 0.0471, wR_2 = 0.1220
Largest diff. peak/hole / e Å^{-3}	1.43/-0.74

2-(2-Bromo-4,5-dimethoxybenzoyl)naphthalene-1,4-dione (73d)

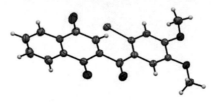

Identification code	JB254
Empirical formula	$C_{19}H_{13}BrO_5$
Formula weight	431.27
Temperature/K	180.15
Crystal system	monoclinic
Space group	$P2_1/c$
a/Å	12.2153(4)
b/Å	10.1360(2)
c/Å	14.2168(5)
α/°	90
β/°	112.292(2)
γ/°	90
Volume/Å^3	1628.69(9)
Z	4
ρ_{calc}g/cm^3	1.759
μ/mm^{-1}	2.551
F(000)	880.0
Crystal size/mm^3	0.22 × 0.18 × 0.04
Radiation	GaKα (λ = 1.34143)
2Θ range for data collection/°	6.804 to 118.402
Index ranges	-15 ≤ h ≤ 14, -12 ≤ k ≤ 12, -9 ≤ l ≤ 17
Reflections collected	7926
Independent reflections	3418 [R_{int} = 0.0215, R_{sigma} = 0.0154]
Indep. refl. with I>=2σ (I)	3279
Data/restraints/parameters	3418/0/228
Goodness-of-fit on F^2	1.085
Final R indexes [I>=2σ (I)]	R_1 = 0.0406, wR_2 = 0.1096
Final R indexes [all data]	R_1 = 0.0419, wR_2 = 0.1108
Largest diff. peak/hole / e Å^{-3}	0.44/-0.68

(4aR,9aR)-4a-(2-Bromobenzoyl)-1,4,4a,9a-tetrahydroanthracene-9,10-dione (87ba)

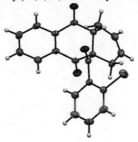

Identification code	JB249
Empirical formula	$C_{21}H_{15}BrO_3$
Formula weight	395.24
Temperature/K	180.15
Crystal system	monoclinic
Space group	P2$_1$/c
a/Å	14.4170(5)
b/Å	7.8476(2)
c/Å	15.3699(5)
α/°	90
β/°	104.275(3)
γ/°	90
Volume/Å^3	1685.24(9)
Z	4
ρ_{calc}g/cm^3	1.558
μ/mm^{-1}	2.373
F(000)	800.0
Crystal size/mm^3	0.26 × 0.24 × 0.04
Radiation	GaKα (λ = 1.34143)
2Θ range for data collection/°	10.334 to 107.988
Index ranges	-17 ≤ h ≤ 17, -9 ≤ k ≤ 7, -18 ≤ l ≤ 10
Reflections collected	6226
Independent reflections	2951 [R_{int} = 0.0171, R_{sigma} = 0.0131]
Indep. refl. with I>=2σ (I)	2778
Data/restraints/parameters	2951/0/226
Goodness-of-fit on F^2	1.034
Final R indexes [I>=2σ (I)]	R_1 = 0.0328, wR_2 = 0.0886
Final R indexes [all data]	R_1 = 0.0344, wR_2 = 0.0897
Largest diff. peak/hole / e Å^{-3}	0.84/-1.09

(4a*R*,9a*R*)-4a-(2-Iodobenzoyl)-2,3-dimethyl-1,4,4a,9a-tetrahydroanthracene-9,10-dione (87ac)

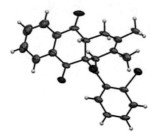

Identification code	JB246
Empirical formula	$C_{23}H_{19}IO_3$
Formula weight	470.28
Temperature/K	180.15
Crystal system	monoclinic
Space group	$P2_1/n$
a/Å	7.7683(5)
b/Å	20.5514(14)
c/Å	11.8395(9)
α/°	90
β/°	90.887(6)
γ/°	90
Volume/Å^3	1889.9(2)
Z	4
ρ_{calc}g/cm^3	1.653
μ/mm^{-1}	13.480
F(000)	936.0
Crystal size/mm^3	0.38 × 0.32 × 0.24
Radiation	CuKα (λ = 1.54186)
2Θ range for data collection/°	13.538 to 141.084
Index ranges	-9 ≤ h ≤ 6, -24 ≤ k ≤ 19, -13 ≤ l ≤ 14
Reflections collected	7433
Independent reflections	3486 [R_{int} = 0.0704, R_{sigma} = 0.0571]
Indep. refl. with I>=2σ (I)	3058
Data/restraints/parameters	3486/0/246
Goodness-of-fit on F^2	1.466
Final R indexes [I>=2σ (I)]	R_1 = 0.1294, wR_2 = 0.3134
Final R indexes [all data]	R_1 = 0.1341, wR_2 = 0.3224
Largest diff. peak/hole / e Å^{-3}	6.39/-4.22

(4aR,9aR)-4a-(2-Bromobenzoyl)-2,3-dimethyl-1,4,4a,9a-tetrahydroanthracene-9,10-dione (87bc)

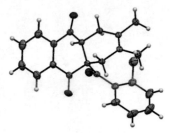

Identification code	JB366
Empirical formula	$C_{23}H_{19}O_3Br$
Formula weight	423.29
Temperature/K	150
Crystal system	orthorhombic
Space group	Pca2$_1$
a/Å	15.1575(5)
b/Å	16.7865(7)
c/Å	7.2918(2)
α/°	90
β/°	90
γ/°	90
Volume/Å^3	1855.34(11)
Z	4
ρ_{calc}g/cm^3	1.515
μ/mm^{-1}	2.180
F(000)	864.0
Crystal size/mm^3	0.24 × 0.22 × 0.04
Radiation	GaKα (λ = 1.34143)
2Θ range for data collection/°	6.836 to 124.96
Index ranges	-20 ≤ h ≤ 8, -22 ≤ k ≤ 21, -8 ≤ l ≤ 9
Reflections collected	11374
Independent reflections	3836 [R_{int} = 0.0128, R_{sigma} = 0.0097]
Data/restraints/parameters	3836/1/247
Goodness-of-fit on F^2	1.054
Final R indexes [I>=2σ (I)]	R_1 = 0.0260, wR_2 = 0.0699
Final R indexes [all data]	R_1 = 0.0264, wR_2 = 0.0702
Largest diff. peak/hole / e Å^{-3}	0.21/-0.58
Flack parameter	0.41(2)

(1*S*,4a*S*,9a*R*)-9a-(2-Iodobenzoyl)-1-((trimethylsilyl)oxy)-1,4,4a,9a-tetrahydroanthracene-9,10-dione (87ad)

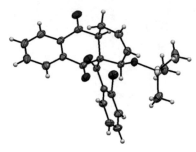

Identification code	JB252
Empirical formula	$C_{24}H_{23}O_4SiI$
Formula weight	530.41
Temperature/K	180.15
Crystal system	triclinic
Space group	P-1
a/Å	10.0925(3)
b/Å	10.8789(3)
c/Å	11.9319(3)
α/°	116.360(2)
β/°	102.313(2)
γ/°	93.973(2)
Volume/Å^3	1126.55(6)
Z	2
ρ_{calc}g/cm^3	1.564
μ/mm^{-1}	7.958
F(000)	532.0
Crystal size/mm^3	0.21 × 0.18 × 0.17
Radiation	GaKα (λ = 1.34143)
2Θ range for data collection/°	7.478 to 119.99
Index ranges	-11 ≤ h ≤ 13, -11 ≤ k ≤ 14, -15 ≤ l ≤ 14
Reflections collected	12471
Independent reflections	4912 [R_{int} = 0.0113, R_{sigma} = 0.0086]
Indep. refl. with I>=2σ (I)	4833
Data/restraints/parameters	4912/0/274
Goodness-of-fit on F^2	1.039
Final R indexes [I>=2σ (I)]	R_1 = 0.0193, wR_2 = 0.0505
Final R indexes [all data]	R_1 = 0.0196, wR_2 = 0.0506
Largest diff. peak/hole / e Å^{-3}	0.82/-0.70

(1*R*,4a*S*,9a*R*)-9a-(2-Bromobenzoyl)-1-((trimethylsilyl)oxy)-1,4,4a,9a-tetrahydroanthracene-9,10-dione (88bd)

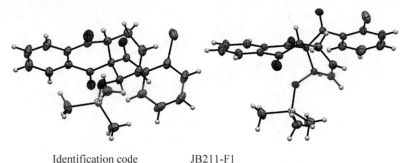

Identification code	JB211-F1
Empirical formula	$C_{24}H_{23}BrO_4Si$
Formula weight	483.42
Temperature/K	180.15
Crystal system	monoclinic
Space group	$P2_1/n$
a/Å	13.2366(3)
b/Å	8.60730(10)
c/Å	19.3654(4)
α/°	90
β/°	91.976(2)
γ/°	90
Volume/Å^3	2205.01(7)
Z	4
ρ_{calc}g/cm^3	1.456
μ/mm^{-1}	2.230
F(000)	992.0
Crystal size/mm^3	0.39 × 0.36 × 0.34
Radiation	GaKα (λ = 1.34143)
2Θ range for data collection/°	6.928 to 125.986
Index ranges	-14 ≤ h ≤ 17, -11 ≤ k ≤ 8, -25 ≤ l ≤ 25
Reflections collected	14639
Independent reflections	5292 [R_{int} = 0.0156, R_{sigma} = 0.0116]
Data/restraints/parameters	5292/0/274
Goodness-of-fit on F^2	1.067
Final R indexes [I>=2σ (I)]	R_1 = 0.0419, wR_2 = 0.1077
Final R indexes [all data]	R_1 = 0.0426, wR_2 = 0.1082
Largest diff. peak/hole / e Å^{-3}	0.39/-1.27

(1*R*,4a*R*,9a*S*)-9a-(2-Bromobenzoyl)-1-((trimethylsilyl)oxy)-1,4,4a,9a-tetrahydroanthracene-9,10-dione (87bd)

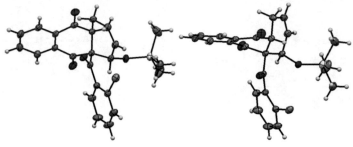

Identification code	JB243
Empirical formula	$C_{24}H_{23}BrO_4Si$
Formula weight	483.42
Temperature/K	150
Crystal system	triclinic
Space group	P-1
a/Å	9.3741(5)
b/Å	11.0638(6)
c/Å	11.8758(6)
α/°	69.145(4)
β/°	74.144(4)
γ/°	85.795(4)
Volume/Å^3	1106.80(11)
Z	2
ρ_{calc}g/cm^3	1.451
μ/mm^{-1}	2.221
F(000)	496.0
Crystal size/mm^3	0.2 × 0.19 × 0.18
Radiation	GaKα (λ = 1.34143)
2Θ range for data collection/°	7.188 to 125.348
Index ranges	-12 ≤ h ≤ 12, -10 ≤ k ≤ 14, -15 ≤ l ≤ 14
Reflections collected	13753
Independent reflections	5240 [R_{int} = 0.0147, R_{sigma} = 0.0138]
Data/restraints/parameters	5240/0/364
Goodness-of-fit on F^2	1.073
Final R indexes [I>=2σ (I)]	R_1 = 0.0267, wR_2 = 0.0728
Final R indexes [all data]	R_1 = 0.0298, wR_2 = 0.0767
Largest diff. peak/hole / e Å^{-3}	0.40/-0.46

(1*R*,4a*S*,9a*R*)-9a-(2-Iodobenzoyl)-2-methyl-1-((trimethylsilyl)oxy)-1,4,4a,9a-tetrahydroanthracene-9,10-dione (88ae)

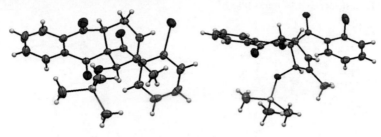

Identification code	JB228-F1
Empirical formula	$C_{25}H_{25}IO_4Si$
Formula weight	544.44
Temperature/K	180.15
Crystal system	monoclinic
Space group	$P2_1/n$
a/Å	13.5865(6)
b/Å	8.6619(3)
c/Å	19.7422(9)
α/°	90
β/°	95.241(4)
γ/°	90
Volume/Å^3	2313.65(17)
Z	4
ρ_{calc}g/cm^3	1.563
μ/mm^{-1}	1.465
F(000)	1096.0
Crystal size/mm^3	0.28 × 0.26 × 0.24
Radiation	MoKα (λ = 0.71073)
2Θ range for data collection/°	3.808 to 63.39
Index ranges	-18 ≤ h ≤ 19, -12 ≤ k ≤ 12, -28 ≤ l ≤ 28
Reflections collected	32349
Independent reflections	7193 [R_{int} = 0.0275, R_{sigma} = 0.0237]
Data/restraints/parameters	7193/0/380
Goodness-of-fit on F^2	1.022
Final R indexes [I>=2σ (I)]	R_1 = 0.0291, wR_2 = 0.0653
Final R indexes [all data]	R_1 = 0.0450, wR_2 = 0.0712
Largest diff. peak/hole / e Å^{-3}	0.96/-1.00

(1R,4aS,9aR)-9a-(2-Bromobenzoyl)-1-((*tert*-butyldimethylsilyl)oxy)-2-methyl-1,4,4a,9a-tetrahydroanthracene-9,10-dione (88bf)

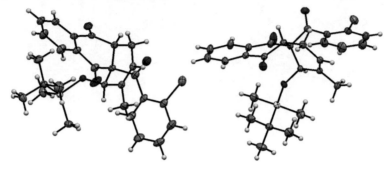

Identification code	JB369-F2
Empirical formula	$C_{28}H_{31}BrO_4Si$
Formula weight	539.53
Temperature/K	150
Crystal system	monoclinic
Space group	P2$_1$/n
a/Å	11.9246(2)
b/Å	14.6618(2)
c/Å	15.2528(3)
α/°	90
β/°	104.8900(10)
γ/°	90
Volume/Å^3	2577.19(8)
Z	4
ρ_{calc}g/cm^3	1.391
μ/mm^{-1}	1.945
F(000)	1120.0
Crystal size/mm^3	0.34 × 0.33 × 0.32
Radiation	GaKα (λ = 1.34143)
2Θ range for data collection/°	7.34 to 124.992
Index ranges	-8 ≤ h ≤ 15, -18 ≤ k ≤ 19, -19 ≤ l ≤ 20
Reflections collected	37678
Independent reflections	6219 [R$_{int}$ = 0.0136, R$_{sigma}$ = 0.0082]
Ind. refl. with [I>=2σ (I)]	5899
Data/restraints/parameters	6219/0/313
Goodness-of-fit on F^2	1.100
Final R indexes [I>=2σ (I)]	R$_1$ = 0.0279, wR$_2$ = 0.0741
Final R indexes [all data]	R$_1$ = 0.0290, wR$_2$ = 0.0748
Largest diff. peak/hole / e Å^{-3}	0.31/-0.52

(4aR,9aR)-4a-(2-Iodobenzoyl)-2-((triisopropylsilyl)oxy)-1,4,4a,9a-tetrahydroanthracene-9,10-dione (87ag)

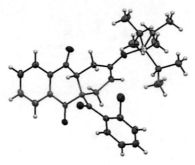

Identification code	JB201-F1
Empirical formula	$C_{30}H_{35}IO_4Si$
Formula weight	614.57
Temperature/K	150.15
Crystal system	monoclinic
Space group	$P2_1/n$
a/Å	9.0381(3)
b/Å	11.1267(3)
c/Å	27.9237(11)
$\alpha/°$	90
$\beta/°$	97.084(3)
$\gamma/°$	90
Volume/Å^3	2786.69(16)
Z	4
$\rho_{calc}g/cm^3$	1.465
μ/mm^{-1}	6.485
F(000)	1256.0
Crystal size/mm^3	0.14 × 0.12 × 0.05
Radiation	GaKα (λ = 1.34143)
2Θ range for data collection/°	7.448 to 114.994
Index ranges	-11 ≤ h ≤ 11, -8 ≤ k ≤ 13, -35 ≤ l ≤ 31
Reflections collected	20060
Independent reflections	5657 [R_{int} = 0.0275, R_{sigma} = 0.0174]
Indep. refl. with I>=2σ (I)	5379
Data/restraints/parameters	5657/0/331
Goodness-of-fit on F^2	1.100
Final R indexes [I>=2σ (I)]	R_1 = 0.0283, wR_2 = 0.0812
Final R indexes [all data]	R_1 = 0.0301, wR_2 = 0.0818
Largest diff. peak/hole / e Å^{-3}	1.46/-0.90

N-(4-Bromo-3-((4a*R*,9a*R*)-2,3-dimethyl-9,10-dioxo-1,4,4a,9,9a,10-hexahydroanthracene-4a-carbonyl)phenyl)acetamide (87fc)

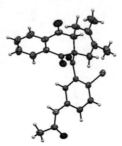

Identification code	JB274
Empirical formula	$C_{25}H_{22}BrNO_4$
Formula weight	480.34
Temperature/K	150
Crystal system	hexagonal
Space group	$P6_5$
a/Å	23.8314(3)
b/Å	23.8314(3)
c/Å	7.72040(10)
$\alpha/°$	90
$\beta/°$	90
$\gamma/°$	120
Volume/Å^3	3797.25(11)
Z	6
ρ_{calc}g/cm^3	1.260
μ/mm^{-1}	1.665
F(000)	1476.0
Crystal size/mm^3	0.28 × 0.27 × 0.26
Radiation	GaKα (λ = 1.34143)
2Θ range for data collection/°	3.724 to 124.896
Index ranges	-31 $\leq$ h $\leq$ 31, -30 $\leq$ k $\leq$ 31, -10 $\leq$ l $\leq$ 3
Reflections collected	43479
Independent reflections	4150 [R_{int} = 0.0203, R_{sigma} = 0.0114]
Ind. refl. with [I>=2σ (I)]	3941
Data/restraints/parameters	4150/1/283
Goodness-of-fit on F^2	1.098
Final R indexes [I>=2σ (I)]	R_1 = 0.0237, wR_2 = 0.0677
Final R indexes [all data]	R_1 = 0.0251, wR_2 = 0.0685
Largest diff. peak/hole / e Å^{-3}	0.21/-0.29
Flack parameter	0.015(15)

N-(4-Bromo-3-((4a*S*,9a*S*)-2,3-dimethyl-9,10-dioxo-1,4,4a,9,9a,10-hexahydroanthracene-4a-carbonyl)phenyl)-2,2,2-trifluoroacetamide (87gc)

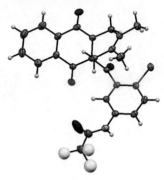

Identification code	JB264
Empirical formula	$C_{25}H_{19}NO_4F_3Br$
Formula weight	534.32
Temperature/K	180.15
Crystal system	monoclinic
Space group	$P2_1/c$
a/Å	19.8014(7)
b/Å	13.6268(4)
c/Å	9.0030(3)
α/°	90
β/°	96.975(3)
γ/°	90
Volume/Å^3	2411.30(14)
Z	4
ρ_{calc}g/cm^3	1.472
μ/mm^{-1}	1.759
F(000)	1080.0
Crystal size/mm^3	0.25 × 0.18 × 0.1
Radiation	MoKα (λ = 0.71073)
2Θ range for data collection/°	3.636 to 59.998
Index ranges	-27 ≤ h ≤ 27, -19 ≤ k ≤ 19, -12 ≤ l ≤ 12
Reflections collected	31134
Independent reflections	7014 [R_{int} = 0.0204, R_{sigma} = 0.0190]
Data/restraints/parameters	7014/0/306
Goodness-of-fit on F^2	1.041
Final R indexes [I>=2σ (I)]	R_1 = 0.0570, wR_2 = 0.1523
Final R indexes [all data]	R_1 = 0.0673, wR_2 = 0.1595
Largest diff. peak/hole / e Å^{-3}	1.81/-0.81

N-(4-Bromo-3-((4*S*,4a*S*,9a*R*)-9,10-dioxo-4-((trimethylsilyl)oxy)-1,4,4a,9,9a,10-hexahydroanthracene-4a-carbonyl)phenyl)-2,2,2-trifluoroacetamide (88gd)

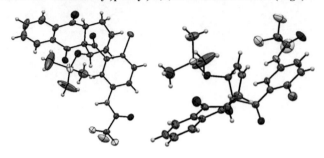

Identification code	JB280-F1
Empirical formula	$C_{26.33}H_{23.33}BrClF_3NO_5Si$
Formula weight	634.24
Temperature/K	180.15
Crystal system	trigonal
Space group	R3c
a/Å	34.6858(6)
b/Å	34.6858(6)
c/Å	11.9046(2)
α/°	90
β/°	90
γ/°	120
Volume/Å³	12403.6(5)
Z	18
ρ_{calc}g/cm³	1.528
μ/mm⁻¹	2.590
F(000)	5784.0
Crystal size/mm³	0.17 × 0.04 × 0.03
Radiation	GaKα (λ = 1.34143)
2Θ range for data collection/°	4.432 to 119.972
Index ranges	-44 ≤ h ≤ 44, -44 ≤ k ≤ 44, -15 ≤ l ≤ 5
Reflections collected	40108
Independent reflections	4281 [R_{int} = 0.0306, R_{sigma} = 0.0176]
Indep. refl. with I>=2σ (I)	3975
Data/restraints/parameters	4281/2/341
Goodness-of-fit on F²	1.078
Final R indexes [I>=2σ (I)]	R_1 = 0.0550, wR_2 = 0.1540
Final R indexes [all data]	R_1 = 0.0585, wR_2 = 0.1581
Largest diff. peak/hole / e Å⁻³	2.88/-1.61
Flack parameter	-0.02(3)

(4aS,9aS)-2-Chloro-4a-(2-iodobenzoyl)-3-methyl-1,4,4a,9a-tetrahydroanthracene-9,10-dione (161a)

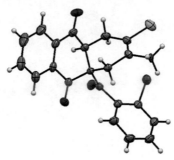

Identification code	JB136-F1-193-176
Empirical formula	$C_{22}H_{16}ClIO_3$
Formula weight	490.70
Temperature/K	180.15
Crystal system	monoclinic
Space group	$P2_1/n$
a/Å	7.7822(17)
b/Å	20.617(2)
c/Å	11.6941(15)
α/°	90
β/°	91.392(14)
γ/°	90
Volume/Å^3	1875.7(5)
Z	4
ρ_{calc}g/cm^3	1.738
μ/mm^{-1}	1.870
F(000)	968.0
Crystal size/mm^3	0.18 × 0.16 × 0.15
Radiation	MoKα (λ = 0.71073)
2Θ range for data collection/°	5.268 to 66.446
Index ranges	-11 ≤ h ≤ 11, -31 ≤ k ≤ 27, -17 ≤ l ≤ 13
Reflections collected	27950
Independent reflections	6716 [R_{int} = 0.0233, R_{sigma} = 0.0275]
Data/restraints/parameters	6716/0/245
Goodness-of-fit on F^2	1.040
Final R indexes [I>=2σ (I)]	R_1 = 0.0365, wR_2 = 0.1008
Final R indexes [all data]	R_1 = 0.0549, wR_2 = 0.1087
Largest diff. peak/hole / e Å^{-3}	1.30/-1.24

(5S,13aR)-6H-5,13a-Methanobenzo[4,5]cycloocta[1,2-b]naphthalene-8,13,14(5H)-trione (111a)

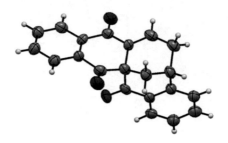

Identification code	JB375_C10
Empirical formula	$C_{21}H_{14}O_3$
Formula weight	314.32
Temperature/K	180.15
Crystal system	monoclinic
Space group	Pc
a/Å	15.8412(8)
b/Å	7.9531(3)
c/Å	11.9997(5)
α/°	90
β/°	100.414(3)
γ/°	90
Volume/Å^3	1486.90(11)
Z	4
ρ_{calc}g/cm^3	1.404
μ/mm^{-1}	0.485
F(000)	656.0
Crystal size/mm^3	0.36 × 0.18 × 0.01
Radiation	GaKα (λ = 1.34143)
2Θ range for data collection/°	4.934 to 123.304
Index ranges	-20 ≤ h ≤ 20, -10 ≤ k ≤ 8, -15 ≤ l ≤ 11
Reflections collected	17148
Independent reflections	5806 [R_{int} = 0.0663, R_{sigma} = 0.0539]
Data/restraints/parameters	5806/2/434
Goodness-of-fit on F^2	1.324
Final R indexes [I>=2σ (I)]	R_1 = 0.1271, wR_2 = 0.2919
Final R indexes [all data]	R_1 = 0.1642, wR_2 = 0.3182
Largest diff. peak/hole / e Å^{-3}	1.07/-0.50
Flack parameter	-0.9(10)

(5S,7aS,13aS)-6-Methylene-7,7a-dihydro-6H-5,13a-methanobenzo[4,5]cycloocta[1,2-b]naphthalene-8,13,14(5H)-trione (113ab)

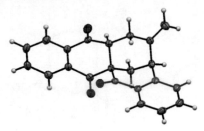

Identification code	JB288-F1
Empirical formula	$C_{22}H_{16}O_3$
Formula weight	328.35
Temperature/K	150.15
Crystal system	orthorhombic
Space group	$P2_12_12_1$
a/Å	6.6655(2)
b/Å	14.8931(5)
c/Å	15.7134(6)
α/°	90
β/°	90
γ/°	90
Volume/Å³	1559.87(9)
Z	4
$\rho_{calc}g/cm^3$	1.398
μ/mm⁻¹	0.478
F(000)	688.0
Crystal size/mm³	0.25 × 0.08 × 0.06
Radiation	GaKα (λ = 1.34143)
2Θ range for data collection/°	9.794 to 114.998
Index ranges	-8 ≤ h ≤ 3, -16 ≤ k ≤ 18, -18 ≤ l ≤ 19
Reflections collected	7830
Independent reflections	3142 [R_{int} = 0.0149, R_{sigma} = 0.0120]
Indep. refl. with I>=2σ (I)	3069
Data/restraints/parameters	3142/0/291
Goodness-of-fit on F^2	1.024
Final R indexes [I>=2σ (I)]	R_1 = 0.0265, wR_2 = 0.0693
Final R indexes [all data]	R_1 = 0.0272, wR_2 = 0.0697
Largest diff. peak/hole / e Å⁻³	0.18/-0.13
Flack parameter	0.5(3)

(5*R*,13a*S*)-5-Methyl-6*H*-5,13a-methanobenzo[4,5]cycloocta[1,2-b]naphthalene-8,13,14(5*H*)-trione (115ab)

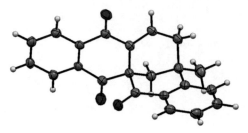

Identification code	Jb288-F3
Empirical formula	$C_{22}H_{16}O_3$
Formula weight	328.35
Temperature/K	150.15
Crystal system	monoclinic
Space group	$P2_1/c$
a/Å	5.7532(2)
b/Å	12.2528(5)
c/Å	21.7939(8)
α/°	90
β/°	92.733(3)
γ/°	90
Volume/Å^3	1534.57(10)
Z	4
ρ_{calc}g/cm^3	1.421
μ/mm^{-1}	0.486
F(000)	688.0
Crystal size/mm^3	0.23 × 0.21 × 0.17
Radiation	GaKα (λ = 1.34143)
2Θ range for data collection/°	31.342 to 111.994
Index ranges	$-5 \leq h \leq 7, -15 \leq k \leq 12, -26 \leq l \leq 26$
Reflections collected	9473
Independent reflections	2875 [R_{int} = 0.0174, R_{sigma} = 0.0225]
Indep. refl. with I>=2σ (I)	2388
Data/restraints/parameters	2875/0/290
Goodness-of-fit on F^2	1.061
Final R indexes [I>=2σ (I)]	R_1 = 0.0440, wR_2 = 0.1154
Final R indexes [all data]	R_1 = 0.0532, wR_2 = 0.1190
Largest diff. peak/hole / e Å^{-3}	0.35/-0.19

(4aS,9aR)-9a-(2-Iodobenzoyl)-2-methyl-1,4,4a,9a-tetrahydroanthracene-9,10-dione (87ab-epi)

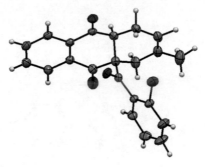

Identification code	JB289-F1
Empirical formula	$C_{22}H_{17}IO_3$
Formula weight	456.25
Temperature/K	180.15
Crystal system	monoclinic
Space group	P2$_1$/n
a/Å	7.8456(2)
b/Å	9.6015(2)
c/Å	24.0162(6)
α/°	90
β/°	95.335(2)
γ/°	90
Volume/Å^3	1801.29(7)
Z	4
ρ_{calc}g/cm^3	1.682
μ/mm^{-1}	1.797
F(000)	904.0
Crystal size/mm^3	0.22 × 0.2 × 0.18
Radiation	MoKα (λ = 0.71073)
2Θ range for data collection/°	4.572 to 65
Index ranges	-11 ≤ h ≤ 9, -14 ≤ k ≤ 13, -36 ≤ l ≤ 36
Reflections collected	28401
Independent reflections	6516 [R$_{int}$ = 0.0174, R$_{sigma}$ = 0.0114]
Indep. refl. with I>=2σ (I)	5789
Data/restraints/parameters	6516/0/303
Goodness-of-fit on F^2	1.092
Final R indexes [I>=2σ (I)]	R$_1$ = 0.0287, wR$_2$ = 0.0725
Final R indexes [all data]	R$_1$ = 0.0341, wR$_2$ = 0.0776
Largest diff. peak/hole / e Å^{-3}	1.82/-0.86

(5S,13aR,15S)-5-Methyl-15-((trimethylsilyl)oxy)-6H-5,13a-methanobenzo[4,5]cycloocta[1,2-b]naphthalene-8,13,14(5H)-trione (113ae)

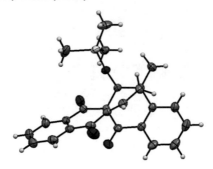

Identification code	JB380_B4
Empirical formula	$C_{25}H_{24}O_4Si$
Formula weight	416.53
Temperature/K	180.15
Crystal system	monoclinic
Space group	$P2_1/n$
a/Å	8.0664(2)
b/Å	16.4629(4)
c/Å	16.1099(3)
α/°	90
β/°	94.371(2)
γ/°	90
Volume/Å^3	2133.11(8)
Z	4
ρ_{calc}g/cm^3	1.297
μ/mm^{-1}	0.780
F(000)	880.0
Crystal size/mm^3	0.36 × 0.35 × 0.34
Radiation	GaKα (λ = 1.34143)
2Θ range for data collection/°	10.37 to 124
Index ranges	-5 ≤ h ≤ 10, -18 ≤ k ≤ 21, -21 ≤ l ≤ 20
Reflections collected	17609
Independent reflections	4998 [R_{int} = 0.0220, R_{sigma} = 0.0122]
Data/restraints/parameters	4998/0/275
Goodness-of-fit on F^2	1.077
Final R indexes [I>=2σ (I)]	R_1 = 0.0398, wR_2 = 0.1082
Final R indexes [all data]	R_1 = 0.0409, wR_2 = 0.1091
Largest diff. peak/hole / e Å^{-3}	0.28/-0.37

(5*S*,13a*R*,15*R*)-15-((*tert*-Butyldimethylsilyl)oxy)-5-methyl-6*H*-5,13a-
methanobenzo[4,5]cycloocta[1,2-b]naphthalene-8,13,14(5*H*)-trione (113bf)

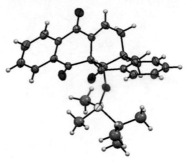

Identification code	JB384-B10_P21c
Empirical formula	$C_{28}H_{30}O_4Si$
Formula weight	458.61
Temperature/K	150
Crystal system	monoclinic
Space group	$P2_1/c$
a/Å	17.4328(5)
b/Å	10.4520(2)
c/Å	13.5501(3)
α/°	90
β/°	101.267(2)
γ/°	90
Volume/Å^3	2421.35(10)
Z	4
ρ_{calc}g/cm^3	1.258
μ/mm^{-1}	0.717
F(000)	976.0
Crystal size/mm^3	0.24 × 0.22 × 0.03
Radiation	GaKα (λ = 1.34143)
2Θ range for data collection/°	4.496 to 121.994
Index ranges	-22 ≤ h ≤ 22, -13 ≤ k ≤ 13, -8 ≤ l ≤ 17
Reflections collected	39760
Independent reflections	5622 [R_{int} = 0.1220, R_{sigma} = 0.0586]
Ind. refl. with [I>=2σ (I)]	3922
Data/restraints/parameters	5622/6/303
Goodness-of-fit on F^2	1.050
Final R indexes [I>=2σ (I)]	R_1 = 0.0730, wR_2 = 0.1904
Final R indexes [all data]	R_1 = 0.0975, wR_2 = 0.2005
Largest diff. peak/hole / e Å^{-3}	0.65/-0.64

(5*R*,13a*S*,15*S*)-15-(((*tert*-Butyldimethylsilyl)oxy)-5-methyl-6*H*-5,13a-
methanobenzo[4,5]cycloocta[1,2-b]naphthalin-8,13,14(5*H*)-trione (114bf)

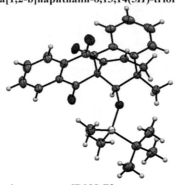

Identification code	JB098-F2
Empirical formula	$C_{28}H_{30}O_4Si$
Formula weight	458.61
Temperature/K	180.15
Crystal system	monoclinic
Space group	$P2_1/c$
a/Å	16.9616(13)
b/Å	17.7649(9)
c/Å	8.0390(6)
α/°	90
β/°	96.301(6)
γ/°	90
Volume/Å^3	2407.7(3)
Z	4
ρ_{calc}g/cm^3	1.265
μ/mm^{-1}	0.130
F(000)	976.0
Crystal size/mm^3	0.21 × 0.17 × 0.04
Radiation	MoKα (λ = 0.71073)
2Θ range for data collection/°	4.832 to 53.986
Index ranges	-21 ≤ h ≤ 21, -19 ≤ k ≤ 22, -10 ≤ l ≤ 10
Reflections collected	11321
Independent reflections	5130 [R_{int} = 0.0737, R_{sigma} = 0.0951]
Data/restraints/parameters	5130/0/304
Goodness-of-fit on F^2	0.922
Final R indexes [I>=2σ (I)]	R_1 = 0.0718, wR_2 = 0.1750
Final R indexes [all data]	R_1 = 0.1203, wR_2 = 0.1956
Largest diff. peak/hole / e Å^{-3}	0.61/-0.65

2,2,2-Trifluoro-*N*-((5*S*,13a*R*,15*S*)-8,13,14-trioxo-15-((trimethylsilyl)oxy)-5,8,13,14-tetrahydro-6*H*-5,13a-methanobenzo[4,5]cycloocta[1,2-b]naphthalen-2-yl)acetamide (119gd)

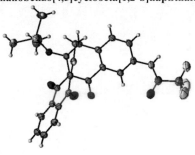

Identification code	JB378_C6
Empirical formula	$C_{26}H_{22}F_3NO_5Si$
Formula weight	513.53
Temperature/K	180.15
Crystal system	tetragonal
Space group	I4$_1$/a
a/Å	37.5781(5)
b/Å	37.5781(5)
c/Å	7.68370(10)
α/°	90
β/°	90
γ/°	90
Volume/Å^3	10850.3(3)
Z	16
ρ_{calc}g/cm^3	1.257
μ/mm^{-1}	0.807
F(000)	4256.0
Crystal size/mm^3	0.38 × 0.16 × 0.15
Radiation	GaKα (λ = 1.34143)
2Θ range for data collection/°	4.092 to 131.772
Index ranges	-49 ≤ h ≤ 45, -51 ≤ k ≤ 35, -10 ≤ l ≤ 3
Reflections collected	44439
Independent reflections	6980 [R$_{int}$ = 0.0188, R$_{sigma}$ = 0.0138]
Indep. refl. with I>=2σ (I)	5770
Data/restraints/parameters	6980/0/329
Goodness-of-fit on F^2	1.096
Final R indexes [I>=2σ (I)]	R$_1$ = 0.0469, wR$_2$ = 0.1308
Final R indexes [all data]	R$_1$ = 0.0548, wR$_2$ = 0.1359
Largest diff. peak/hole / e Å^{-3}	0.39/-0.41

8   List of Abbreviations

A

AIBN .... azobisisobutyronitrile

APCI .... atmospheric pressure chemical ionization

aq .... aqueous

ATP .... adenosine triphosphate

ATR .... attenuated total reflection

B

B_2pin_2 .... bis(pinacolato)diboron

BHT .... butylated hydroxytoluene

BINAP .... 2,2'-bis(diphenylphosphino)-1,1'-binaphthyl

bp .... boiling point

Bu_3SnH .... tributyltin hydride

C

CAAC .... cyclic (alkyl)(amino)carbene

calc .... calculated

CAN .... ceric ammonium nitrate

*c*Hex .... cyclohexane

Cy_2NMe .... *N,N*-dicyclohexylmethylamine

D

DABCO .... 1,4-diazabicyclo[2.2.2]octane

DBU .... 1,8-diazabicyclo[5.4.0]undec-7-ene

DDQ 2,3-dichloro-5,6-dicyano-1,4-benzoquinone

DEPT .... distortionless enhancement by polarization transfer

$(DHQ)_2AQN$ ...... hydroquinine anthraquinone-1,4-diyl diether

DME .... dimethoxyethane

DMF .... *N,N*-dimethylformamide

DMP .... Dess-Martin periodinane

dppe .... 1,2-bis(diphenylphosphino)ethane

dppf ...... 1,1'-bis(diphenylphosphino)ferrocene

dr .... diastereomeric ratio

E

EDG .... electron-donating group

EI .... electron ionization

ESI .... electrospray ionization

Et .... ethyl

EWG .... electron-withdrawing group

F

FAB .... fast atom bombardment

G

GP .... general procedure

H

HMBC .... heteronuclear multiple quantum correlation

HOMO ...... highest occupied molecular orbital

HPLC .... high pressure liquid chromatography

HRMS .... high resolution mass spectrometry

HSQC .... heteronuclear single quantum coherence

I

IBX .... 2-iodoxybenzoic acid

i-Pr_2NEt .... *N,N*-diisopropylethylamine

IR .... infrared

L

LDA .... lithium diisopropylamide

LiHMDS .... lithium bis(trimethylsilyl)amide

LUMO ... lowest unoccupied molecular orbital

M

m-CPBA .... *meta*-chloroperoxybenzoic acid

Me .... methyl

MeCN .... acetonitrile

MeI .... methyliodide

MEMCl.... (2-methoxyethoxy)methyl chloride

MeOH .... methanol

N

NBS .... *N*-bromosuccinimide

n-BuLi .... *n*-butyllithium

*n*Hex .... *n*-hexane

NMO .... 4-methylmorpholine *N*-oxide

NMR .... nuclear magnetic resonance

P

P(Cy)3 .... tricyclohexylphosphine

Pd2(dba)3 .... tris(dibenzylideneacetone)
dipalladium(0)

Pd(OTFA)2 .... Palladium(II) trifluoroacetate

PIFA .... phenyliodine bis(trifluoroacetate)

PMP .... pentamethylpiperidine

POT .... tris(*o*-tolyl)phosphine

prepTLC .... preparative thin layer
chromatography

p-TsCl .... 4-toluenesulfonyl chloride

PXRD .... powder X-ray diffraction

R

rt .... room temperature

T

TBAF .... tetrabutylammonium fluoride

TBDMS .... *tert*-butyldimethylsilyl

TBDPS .... *tert*-butyldiphenylsilyl

TFAA .... trifluoroacetic anhydride

THF .... tetrahydrofurane

TIPS .... triisopropylsilyl

TIPSOTf .... triisopropylsilyl
trifluoromethanesulfonate

TMEDA .... tetramethylethylenediamine

TMS .... trimethylsilyl

TPAP .... tetrapropylammonium perruthenate

U

UV/Vis .... ultraviolet-visible

X

XRD .... X-ray Diffraction

9 References

[1]   J. L. Richard, *Int. J. Food Microbiol.* **2007**, *119*, 3–10.
[2]   N. W. Turner, S. Subrahmanyam, S. A. Piletsky, *Anal. Chim. Acta* **2009**, *632*, 168–180.
[3]   J. L. Richard, *Toxin Rev.* **2008**, *27*, 171–201.
[4]   J. W. Bennett, M. Klich, *Clin. Mirobiol. Rev.* **2003**, *16*, 497–516.
[5]   H. S. Hussein, J. M. Brasel, *Toxicology* **2001**, *167*, 101–134.
[6]   J. Weiland, G. Koch, *Mol. Plant Pathol.* **2004**, *5*, 157–166.
[7]   R. W. Hoffmann, *Angew. Chem. Int. Ed.* **2013**, *52*, 123–130.
[8]   F. Wöhler, *Ann. Phys.* **1828**, *88*, 253–256.
[9]   M. Iinuma, K. Moriyama, H. Togo, *Synlett* **2013**, *24*, 1707–1711.
[10]   H. Kolbe, *Justus Liebigs Ann. Chem.* **1845**, *54*, 145–188.
[11]   A. Baeyer, V. Drewsen, *Ber. Dtsch. Chem. Ges.* **1882**, *15*, 2856–2864.
[12]   A. Kekulé, *Justus Liebigs Ann. Chem.* **1857**, *104*, 129–150.
[13]   A. Kekulé, *Justus Liebigs Ann. Chem.* **1858**, *106*, 129–159.
[14]   A. M. Butlerov, *Z. Chem.* **1861**, *4*, 549–560.
[15]   E. Fischer, in *Untersuchungen Über Kohlenhydrate und Fermente (1884–1908)*, Springer, **1909**, pp. 355–361.
[16]   R. B. Woodward, W. v. E. Doering, *J. Am. Chem. Soc.* **1944**, *66*, 849–849.
[17]   E. J. Corey, N. M. Weinshenker, T. K. Schaaf, W. Huber, *J. Am. Chem. Soc.* **1969**, *91*, 5675–5677.
[18]   E. J. Corey, *Pure Appl. Chem.* **1967**, *14*, 19–38.
[19]   E. J. Corey, X. M. Cheng, *The Logic of Chemical Synthesis*, Wiley, **1989**.
[20]   X.-T. Chen, B. Zhou, S. K. Bhattacharya, C. E. Gutteridge, T. R. R. Pettus, S. J. Danishefsky, *Angew. Chem. Int. Ed.* **1998**, *37*, 789–792.
[21]   K. C. Nicolaou, W.-M. Dai, *Angew. Chem. Int. Ed. Engl.* **1991**, *30*, 1387–1416.
[22]   K. C. Nicolaou, F. van Delft, T. Ohshima, D. Vourloumis, J. Xu, S. Hosokawa, J. Pfefferkorn, S. Kim, T. Li, *Angew. Chem. Int. Ed. Engl.* **1997**, *36*, 2520–2524.
[23]   X.-T. Chen, C. E. Gutteridge, S. K. Bhattacharya, B. Zhou, T. R. R. Pettus, T. Hascall, S. J. Danishefsky, *Angew. Chem. Int. Ed.* **1998**, *37*, 185–186.
[24]   S. Hanessian, N. G. Cooke, B. DeHoff, Y. Sakito, *J. Am. Chem. Soc.* **1990**, *112*, 5276–5290.
[25]   D. A. Evans, R. L. Dow, T. L. Shih, J. M. Takacs, R. Zahler, *J. Am. Chem. Soc.* **1990**, *112*, 5290–5313.
[26]   R. A. Sheldon, *Chem. Commun.* **2008**, 3352–3365.
[27]   B. Trost, *Science* **1991**, *254*, 1471–1477.
[28]   P. A. Wender, V. A. Verma, T. J. Paxton, T. H. Pillow, *Acc. Chem. Res.* **2008**, *41*, 40–49.
[29]   P. A. Wender, M. P. Croatt, B. Witulski, *Tetrahedron* **2006**, *62*, 7505–7511.
[30]   M. B. Plutschack, B. Pieber, K. Gilmore, P. H. Seeberger, *Chem. Rev.* **2017**, *117*, 11796–11893.
[31]   M. Elkin, T. R. Newhouse, *Chem. Soc. Rev.* **2018**, *47*, 7830–7844.
[32]   F. Peiretti, J. M. Brunel, *ACS Omega* **2018**, *3*, 13263–13266.
[33]   D. Merk, F. Grisoni, L. Friedrich, G. Schneider, *Commun. Chem.* **2018**, *1*, 68.
[34]   A. F. de Almeida, R. Moreira, T. Rodrigues, *Nat. Rev. Chem.* **2019**, *3*, 589–604.
[35]   R. Verpoorte, R. van der Heijden, J. Memelink, *Transgenic Res.* **2000**, *9*, 323–343.
[36]   J. Hamill, M. Rhodes, in *Biosynthesis and Manipulation of Plant Products*, Springer, **1993**, pp. 178–209.
[37]   A. L. Demain, A. Fang, in *History of modern biotechnology I*, Springer, **2000**, pp. 1–39.
[38]   S. Bräse, A. Encinas, J. Keck, C. F. Nising, *Chem. Rev.* **2009**, *109*, 3903–3990.

[39]   P. S. Steyn, M. A. Stander, *J. Toxicol. Toxin Rev.* **1999**, *18*, 229–243.

[40]   B. R. Rushing, M. I. Selim, *Food Chem. Toxicol.* **2019**, *124*, 81–100.

[41]   L. Geiger, M. Nieger, S. Bräse, *Chem. Sel.* **2017**, *2*, 3268–3275.

[42]   L. Geiger, M. Nieger, S. Bräse, *Adv. Synth. Catal.* **2017**, *359*, 3421–3427.

[43]   T. Qin, J. A. Porco Jr, *Angew. Chem. Int. Ed.* **2014**, *53*, 3107–3110.

[44]   C. F. Nising, U. K. Ohnemüller, S. Bräse, *Angew. Chem. Int. Ed.* **2006**, *45*, 307–309.

[45]   D. Ganapathy, J. R. Reiner, L. E. Loeffler, L. Ma, B. Gnanaprakasam, B. Niepoetter, I. Koehne, L. F. Tietze, *Chem. Eur. J.* **2015**, *21*, 16807–16810.

[46]   B. Jiang, D. Xu, J. Allocco, C. Parish, J. Davison, K. Veillette, S. Sillaots, W. Hu, R. Rodriguez-Suarez, S. Trosok, *Chem. Biol.* **2008**, *15*, 363–374.

[47]   C. A. Parish, S. K. Smith, K. Calati, D. Zink, K. Wilson, T. Roemer, B. Jiang, D. Xu, G. Bills, G. Platas, *J. Am. Chem. Soc.* **2008**, *130*, 7060–7066.

[48]   F. Kraft, *Arch. Pharm.* **1906**, *244*, 336–359.

[49]   C. C. Howard, R. A. Johnstone, T. J. King, L. Lessinger, *J. Chem. Soc., Perkin Trans. 1* **1976**, 1820–1822.

[50]   J. Hooper, W. Marlow, W. Whalley, A. Borthwick, R. Bowden, *J. Chem. Soc. C* **1971**, 3580–3590.

[51]   I. Kurobane, S. Iwahashi, A. Fukuda, *Drugs Exp. Clin. Res.* **1987**, *13*, 339–344.

[52]   C. Reddy, A. Hayes, W. Williams, A. Ciegler, *J. Toxicol. Environ. Health* **1979**, *5*, 1159–1169.

[53]   B. H. Wang, G. M. Polya, *Planta Med.* **1996**, *62*, 111–114.

[54]   M. Isaka, S. Palasarn, P. Auncharoen, S. Komwijit, E. G. Jones, *Tetrahedron Lett.* **2009**, *50*, 284–287.

[55]   S. Ayers, T. N. Graf, A. F. Adcock, D. J. Kroll, Q. Shen, S. M. Swanson, S. Matthew, E. J. C. De Blanco, M. C. Wani, B. A. Darveaux, *J. Antibiot.* **2012**, *65*, 3.

[56]   H. He, R. Bigelis, E. H. Solum, M. Greenstein, G. T. Carter, *J. Antibiot.* **2003**, *56*, 923–930.

[57]   N. Tabata, Y. Suzumura, H. Tomoda, R. Masuma, K. Haneda, M. KISNI, Y. Iwai, S. Omura, *J. Antibiot.* **1993**, *46*, 749–755.

[58]   N. Tabata, H. Tomoda, Y. Iwai, S. Omura, *J. Antibiot.* **1996**, *49*, 267–271.

[59]   B. Wu, J. Wiese, A. Wenzel-Storjohann, S. Malien, R. Schmaljohann, J. F. Imhoff, *Chem. Eur. J.* **2016**, *22*, 7452–7462.

[60]   C. Intaraudom, N. Bunbamrung, A. Dramae, N. Boonyuen, S. Komwijit, P. Rachtawee, P. Pittayakhajonwut, *Tetrahedron* **2016**, *72*, 1415–1421.

[61]   K. Matsuzaki, N. Tabata, H. Tomoda, Y. Iwai, H. Tanaka, S. Ōmura, *Tetrahedron Lett.* **1993**, *34*, 8251–8254.

[62]   S. D. Holmbo, S. V. Pronin, *J. Am. Chem. Soc.* **2018**, *140*, 5065–5068.

[63]   C. S. Kramer, M. Nieger, S. Bräse, *Eur. J. Org. Chem.* **2014**, *2014*, 2150–2159.

[64]   Y. Hirano, K. Tokudome, H. Takikawa, K. Suzuki, *Synlett* **2017**, *28*, 214–220.

[65]   E. Schlösser, *Phytopathol. Mediterr.* **1971**, *10*, 154–158.

[66]   J.-P. Blein, I. Bourdil, M. Rossignol, R. Scalla, *Plant Physiol.* **1988**, *88*, 429–434.

[67]   M. A. F. Jalal, M. B. Hossain, D. J. Robeson, D. Vanderhelm, *J. Am. Chem. Soc.* **1992**, *114*, 5967–5971.

[68]   E. Schlösser, *J. Phytopathol.* **1962**, *44*, 295–312.

[69]   M. Milat, T. Prangé, P. Ducrot, J. Tabet, J. Einhorn, J. Blein, J. Lallemand, *J. Am. Chem. Soc.* **1992**, *114*, 1478–1479.

[70]   M. A. Jalal, M. B. Hossain, D. J. Robeson, D. Van der Helm, *J. Am. Chem. Soc.* **1992**, *114*, 5967–5971.

[71]   P.-H. Ducrot, J.-Y. Lallemand, M.-L. Milat, J.-P. Blein, *Tetrahedron Lett.* **1994**, *35*, 8797–8800.

[72] P.-H. Ducrot, J. Einhorn, L. Kerhoas, J.-Y. Lallemand, M.-L. Milat, J.-P. Blein, A. Neuman, T. Prange, *Tetrahedron Lett.* **1996**, *37*, 3121–3124.

[73] G. A. Secor, V. V. Rivera, M. F. R. Khan, N. C. Gudmestad, *Plant Dis.* **2010**, *94*, 1272–1282.

[74] S. Haley, N. Suarez, *Sugar and sweetener situation and outlook yearbook, Vol. SSS-2004*, Economic Research Service/USDA, **2004**.

[75] Rasbak, *Suikerbiet planten Cercospora beticola*, https://commons.wikimedia.org/wiki/File:Suikerbiet_planten_Cercospora_beticola.jpg https://creativecommons.org/licenses/by-sa/3.0/legalcode

[76] Rasbak, *Suikerbiet Cercospora beticola*, https://commons.wikimedia.org/wiki/File:Suikerbiet_Cercospora_beticola.jpg https://creativecommons.org/licenses/by-sa/3.0/legalcode

[77] F. Macri, A. Vianello, *Physiol. Plant Pathol.* **1979**, *15*, 161–170.

[78] F. Simon-Plas, E. Gomès, M.-L. Milat, A. Pugin, J.-P. Blein, *Plant Physiol.* **1996**, *111*, 773–779.

[79] E. Gomès, F. Simon-Plas, M. L. Milat, I. Gapillout, V. Mikès, A. Pugin, J. P. Blein, *Phys. Plant.* **1996**, *98*, 133–139.

[80] I. Gapillout, V. Mikes, M.-L. Milat, F. Simon-Plas, A. Pugin, J.-P. Blein, *Phytochemistry* **1996**, *43*, 387–392.

[81] A. Arnone, G. Nasini, L. Merlini, E. Ragg, G. Assante, *J. Chem. Soc., Perkin Trans. 1* **1993**, 145–151.

[82] E. Gomès, R. Gordon-Weeks, F. Simon-Plas, A. Pugin, M.-L. Milat, R. A. Leigh, J.-P. Blein, *Biochim. Biophys. Acta, Biomembr.* **1996**, *1285*, 38–46.

[83] V. Mikes, M.-L. Milat, A. Pugin, J.-P. Blein, *Biochim. Biophys. Acta, Biomembr.* **1994**, *1195*, 124–130.

[84] C. Goudet, J.-P. Benitah, M.-L. Milat, H. Sentenac, J.-B. Thibaud, *Biophys. J.* **1999**, *77*, 3052–3059.

[85] T. Prangé, A. Neuman, M.-L. Milat, J.-P. Blein, *J. Chem. Soc., Perkin Trans. 2* **1997**, 1819–1826.

[86] P.-H. Ducrot, *C. R. Acad. Sci.* **2001**, *4*, 273–283.

[87] C. Goudet, A. A. Véry, M. L. Milat, M. Ildefonse, J. B. Thibaud, H. Sentenac, J. P. Blein, *Plant J.* **1998**, *14*, 359–364.

[88] C. Goudet, M.-L. Milat, H. Sentenac, J.-B. Thibaud, *Mol. Plant-Microbe Interact.* **2000**, *13*, 203–209.

[89] C. Rusterucci, M. L. Milat, J. P. Blein, *Phytochemistry* **1996**, *42*, 979–983.

[90] G. Ding, G. Maume, M. L. Milat, C. Humbert, J. P. Blein, B. F. Maume, *Cell Biol. Int.* **1996**, *20*, 523–530.

[91] J. S. Holker, E. O'Brien, T. J. Simpson, *J. Chem. Soc., Perkin Trans. 1* **1983**, 1365–1368.

[92] W. Zhang, K. Krohn, U. Flörke, G. Pescitelli, L. Di Bari, S. Antus, T. Kurtán, J. Rheinheimer, S. Draeger, B. Schulz, *Chem. Eur. J.* **2008**, *14*, 4913–4923.

[93] J. Antonovics, *Proc. R. Soc. London, Ser. B* **2009**, *276*, 1443–1448.

[94] W. B. Turner, *J. Chem. Soc., Perkin Trans. 1* **1978**, 1621–1621.

[95] E. Gérard, S. Bräse, *Chem. Eur. J.* **2008**, *14*, 8086–8089.

[96] L. F. Tietze, S. Jackenkroll, J. Hierold, L. Ma, B. Waldecker, *Chem. Eur. J.* **2014**, *20*, 8628–8635.

[97] T. J. Monks, R. P. Hanzlik, G. M. Cohen, D. Ross, D. G. Graham, *Toxicol. Appl. Pharmacol.* **1992**, *112*, 2–16.

[98] G. Bonadonna, S. Monfardini, M. de Lena, F. Fossati-Bellani, *Br. Med. J.* **1969**, *3*, 503–506.

[99] N. Gessler, A. Egorova, T. Belozerskaya, *Appl. Biochem. Microbiol.* **2013**, *49*, 85–99.

216   References

[100] F. Pankewitz, A. Zöllmer, Y. Gräser, M. Hilker, *Arch. Insect Biochem. Physiol.* **2007**, *66*, 98–108.
[101] D. Hsieh, R. Singh, R. C. Yao, J. Bennett, *Appl. Environ. Microbiol.* **1978**, *35*, 980–982.
[102] K. Kawai, Y. Nozawa, Y. Maebayashi, M. Yamazaki, T. Hamasaki, *Appl. Environ. Microbiol.* **1984**, *47*, 481–483.
[103] A. Laurent, *Justus Liebigs Ann. Chem.* **1840**, *34*.
[104] J. W. Lown, *Pharmacol. Ther.* **1993**, *60*, 185–214.
[105] F. Arcamone, G. Cassinelli, G. Fantini, A. Grein, P. Orezzi, C. Pol, C. Spalla, *Biotechnol. Bioeng.* **2000**, *67*, 704–713.
[106] K. Kiyoshi, N. Yoshinori, M. Hideki, O. Yukio, *Toxicol. Lett.* **1986**, *30*, 105–111.
[107] G. Powis, *Pharmacol. Ther.* **1987**, *35*, 57–162.
[108] E. M. Malik, C. E. Müller, *Med. Chem. Res.* **2016**, *36*, 705–748.
[109] C. C. Nawrat, C. J. Moody, *Angew. Chem. Int. Ed.* **2014**, *53*, 2056–2077.
[110] M. A. Brimble, R. J. Elliott, *Tetrahedron* **1997**, *53*, 7715–7730.
[111] K.-S. Masters, S. Bräse, *Chem. Rev.* **2012**, *112*, 3717–3776.
[112] P. A. Turner, E. M. Griffin, J. L. Whatmore, M. Shipman, *Org. Lett.* **2011**, *13*, 1056–1059.
[113] K. Tatsuta, S. Yoshihara, N. Hattori, S. Yoshida, S. Hosokawa, *J. Antibiot.* **2009**, *62*, 469–470.
[114] T. Qin, R. P. Johnson, J. A. Porco Jr, *J. Am. Chem. Soc.* **2011**, *133*, 1714–1717.
[115] K. Sakamoto, S. Kiriki, K. Okuyama, K. Taira, WO2014010325A1, Zeon Corporation Japan, **2014**, p. 235pp.
[116] M. Buccini, M. J. Piggott, *Org. Lett.* **2014**, *16*, 2490–2493.
[117] M. T. Gieseler, M. Kalesse, *Org. Lett.* **2011**, *13*, 2430–2432.
[118] P. Duhamel, D. Cahard, J. M. Poirier, *J. Chem. Soc., Perkin Trans. 1* **1993**, 2509–2511.
[119] A. J. Penwell, *Sequential Cycloadditions for the Synthesis of Taxoid Mimics and Stereoselective Synthesis of Tamoxifen Analogues via a New Carbomegnesiation-Palladium Coupling Reaction,* **2002**.
[120] H. Xu, J. L. Hu, L. Wang, S. Liao, Y. Tang, *J. Am. Chem. Soc.* **2015**, *137*, 8006–8009.
[121] O. Diels, K. Alder, *Justus Liebigs Ann. Chem.* **1928**, *460*, 98–122.
[122] K. C. Nicolaou, S. A. Snyder, T. Montagnon, G. Vassilikogiannakis, *Angew. Chem. Int. Ed.* **2002**, *41*, 1668–1698.
[123] M. J. S. Dewar, S. Olivella, J. J. P. Stewart, *J. Am. Chem. Soc.* **1986**, *108*, 5771–5779.
[124] K. Houk, Y. T. Lin, F. K. Brown, *J. Am. Chem. Soc.* **1986**, *108*, 554–556.
[125] F. A. Carey, R. J. Sundberg, *Advanced Organic Chemistry: Part A: Structure and Mechanisms*, Springer Science & Business Media, **2007**.
[126] T. E. Sample Jr, L. F. Hatch, *J Chem Educ* **1968**, *45*, 55.
[127] C. O. Kappe, S. S. Murphree, A. Padwa, *Tetrahedron* **1997**, *53*, 14179–14233.
[128] Y. Garcia-Rodeja, I. Fernandez, *J. Org. Chem.* **2016**, *81*, 6554–6562.
[129] R. Hoffmann, R. Woodward, *J. Am. Chem. Soc.* **1965**, *87*, 4388–4389.
[130] A. Wassermann, *J. Chem. Soc.* **1935**, 1511–1514.
[131] R. Woodward, H. Baer, *J. Am. Chem. Soc.* **1944**, *66*, 645–649.
[132] T. Inukai, T. Kojima, *J. Org. Chem.* **1966**, *31*, 1121–1123.
[133] P. Vermeeren, T. A. Hamlin, I. Fernández, F. M. Bickelhaupt, *Angew. Chem. Int. Ed.,* *59*, 6201–6206.
[134] U. Pindur, G. Lutz, C. Otto, *Chem. Rev.* **1993**, *93*, 741–761.
[135] D.-S. Hsu, J.-Y. Huang, *J. Org. Chem.* **2012**, *77*, 2659–2666.
[136] I. Fleming, *Frontier Orbitals and Organic Chemical Reactions*, Wiley, **1977**.
[137] W. T. Borden, *Modern Molecular Orbital Theory for Organic Chemists*, Prentice Hall, **1975**.
[138] J. Beck, *Master Thesis,* Karlsruhe Institute of Technology (KIT), **2016**.
[139] P. Merino, E. Marqués-López, T. Tejero, R. P. Herrera, *Synthesis* **2010**, *2010*, 1–26.

[140] F. J. Duarte, A. G. Santos, *J. Org. Chem.* **2012**, *77*, 3252–3261.
[141] H. Jiang, C. Rodríguez-Escrich, T. K. Johansen, R. L. Davis, K. A. Jørgensen, *Angew. Chem. Int. Ed.* **2012**, *51*, 10271–10274.
[142] A. Dieckmann, M. Breugst, K. Houk, *J. Am. Chem. Soc.* **2013**, *135*, 3237–3242.
[143] I. D. Jurberg, I. Chatterjee, R. Tannert, P. Melchiorre, *Chem. Commun.* **2013**, *49*, 4869–4883.
[144] C. S. Kramer, S. Bräse, *Beilstein J. Org. Chem.* **2013**, *9*, 1414–1418.
[145] J.-M. Duffault, F. Tellier, *Synth. Commun.* **1998**, *28*, 2467–2481.
[146] N. K. Garg, D. D. Caspi, B. M. Stoltz, *Synlett* **2006**, *2006*, 3081–3087.
[147] B. Giese, B. Kopping, T. Göbel, J. Dickhaut, G. Thoma, K. Kulicke, F. Trach, *Org. React.* **1996**.
[148] T. Mizoroki, K. Mori, A. Ozaki, *Bull. Chem. Soc. Jpn.* **1971**, *44*, 581–581.
[149] R. F. Heck, J. P. Nolley, *J. Org. Chem.* **1972**, *37*, 2320–2322.
[150] R. F. Heck, *Acc. Chem. Res.* **1979**, *12*, 146–151.
[151] R. F. Heck, J. Nolley Jr, *J. Org. Chem.* **1972**, *37*, 2320–2322.
[152] H. A. Dieck, R. F. Heck, *J. Org. Chem.* **1975**, *40*, 1083–1090.
[153] M. Mori, K. Chiba, Y. Ban, *Tetrahedron Lett.* **1977**, *18*, 1037–1040.
[154] J. T. Link, L. E. Overman, in *Metal-Catalyzed Cross-Coupling Reactions*, **1998**, pp. 230–269.
[155] B. M. Trost, S. A. Godleski, J. P. Genet, *J. Am. Chem. Soc.* **1978**, *100*, 3930–3931.
[156] C. Y. Hong, N. Kado, L. E. Overman, *J. Am. Chem. Soc.* **1993**, *115*, 11028–11029.
[157] S. A. Kelly, Y. Foricher, J. Mann, J. M. Bentley, *Org Biomol Chem* **2003**, *1*, 2865–2876.
[158] W. A. Herrmann, C. Brossmer, K. Öfele, C. P. Reisinger, T. Priermeier, M. Beller, H. Fischer, *Angew. Chem. Int. Ed. Engl.* **1995**, *34*, 1844–1848.
[159] W. A. Herrmann, C. Broßmer, K. Öfele, M. Beller, H. Fischer, *J. Mol. Catal. A: Chem.* **1995**, *103*, 133–146.
[160] J. Potier, S. Menuel, J. Rousseau, V. Tumkevicius, F. Hapiot, E. Monflier, *Appl. Catal., A* **2014**, *479*, 1–8.
[161] N. T. S. Phan, M. Van Der Sluys, C. W. Jones, *Adv. Synth. Catal.* **2006**, *348*, 609–679.
[162] T. Kawano, T. Shinomaru, I. Ueda, *Org. Lett.* **2002**, *4*, 2545–2547.
[163] C. Liu, D. J. Burnell, *J. Org. Chem.* **1997**, *62*, 3683–3687.
[164] K. Hayakawa, K. Ueyama, K. Kanematsu, *J. Org. Chem.* **1985**, *50*, 1963–1969.
[165] T. Jeffery, M. David, *Tetrahedron Lett.* **1998**, *39*, 5751–5754.
[166] T. Jeffery, *Tetrahedron* **1996**, *52*, 10113–10130.
[167] C. Gürtler, S. L. Buchwald, *Chem. Eur. J.* **1999**, *5*, 3107–3112.
[168] T. Norin, L. Westfelt, *Acta Chem. Scand.* **1963**, *17*, 1828–1830.
[169] R. Heath, J. Tumlinson, R. Doolittle, A. T. Proveaux, *J. Chromatogr. Sci.* **1975**, *13*, 380–382.
[170] M. Özcimder, W. Hammers, *J. Chromatogr. A* **1980**, *187*, 307–317.
[171] T.-S. Li, J.-T. Li, H.-Z. Li, *J. Chromatogr. A* **1995**, *715*, 372–375.
[172] A. T. Khan, E. Mondal, *Synlett* **2003**, *2003*, 0694–0698.
[173] R. Giri, J. R. Goodell, C. Xing, A. Benoit, H. Kaur, H. Hiasa, D. M. Ferguson, *Bioorg. Med. Chem.* **2010**, *18*, 1456–1463.
[174] M. Kim, J. A. Boissonnault, P. V. Dau, S. M. Cohen, *Angew. Chem. Int. Ed.* **2011**, *50*, 12193–12196.
[175] Y. Yamaguchi, T. Ochi, Y. Matsubara, Z.-i. Yoshida, *J. Phys. Chem. A* **2015**, *119*, 8630–8642.
[176] N. Miyaura, K. Yamada, A. Suzuki, *Tetrahedron Lett.* **1979**, *20*, 3437–3440.
[177] N. Miyaura, A. Suzuki, *J. Chem. Soc., Chem. Commun.* **1979**, 866–867.
[178] N. Miyaura, A. Suzuki, *Chem. Rev.* **1995**, *95*, 2457–2483.
[179] H. Leicht, I. Göttker-Schnetmann, S. Mecking, *ACS Macro Lett.* **2016**, *5*, 777–780.

[180] V. Sridharan, L. Fan, S. Takizawa, T. Suzuki, H. Sasai, *Org. Biomol. Chem.* **2013**, *11*, 5936–5943.
[181] T. Nishiyama, T. Esumi, Y. Iwabuchi, H. Irie, S. Hatakeyama, *Tetrahedron Lett.* **1998**, *39*, 43–46.
[182] B. Tao, S. C. Goel, J. Singh, D. W. Boykin, *Synthesis* **2002**, *2002*, 1043–1046.
[183] B. Akgun, C. Li, Y. Hao, G. Lambkin, R. Derda, D. G. Hall, *J. Am. Chem. Soc.* **2017**, *139*, 14285–14291.
[184] S. S. Moleele, J. P. Michael, C. B. de Koning, *Tetrahedron* **2006**, *62*, 2831–2844.
[185] G. Köbrich, *Angew. Chem. Int. Ed. Engl.* **1973**, *12*, 464–473.
[186] J. Bredt, *Justus Liebigs Ann. Chem.* **1924**, *437*, 1–13.
[187] L. Ling, Y. He, X. Zhang, M. Luo, X. Zeng, *Angew. Chem. Int. Ed.* **2019**, *58*, 6554–6558.
[188] M. C. Redondo, M. Veguillas, M. Ribagorda, M. C. Carreño, *Angew. Chem. Int. Ed.* **2009**, *48*, 370–374.
[189] H. E. Gottlieb, V. Kotlyar, A. Nudelman, *J. Org. Chem.* **1997**, *62*, 7512–7515.
[190] G. M. Sheldrick, *Acta Crystallogr., Sect. A* **2015**, *A71*, 3–8.
[191] G. M. Sheldrick, *Acta Crystallogr., Sect. C* **2015**, *C71*, 3–8.
[192] O. V. Dolomanov, L. J. Bourhis, R. J. Gildea, J. A. K. Howard, H. Puschmann, *J. Appl. Crystallogr.* **2009**, *42*, 339–341.
[193] G. M. Sheldrick, *Acta Crystallogr., Sect. A* **2008**, *A64*, 112–122.
[194] M. K. Miraki, E. Yazdani, L. Ghandi, K. Azizi, A. Heydari, *Appl. Organomet. Chem.* **2017**, *31*, e3744.
[195] Y. Cui, H. Jiang, Z. Li, N. Wu, Z. Yang, J. Quan, *Org. Lett.* **2009**, *11*, 4628–4631.
[196] H. J. Liu, G. Ulibarri, E. N. C. Browne, *Can. J. Chem.* **1992**, *70*, 1545–1554.
[197] S. Basak, D. Mal, *J. Org. Chem.* **2017**, *82*, 11035–11051.
[198] S. Akine, D. Kusama, Y. Takatsuki, T. Nabeshima, *Tetrahedron Lett.* **2015**, *56*, 4880–4884.
[199] M. Hempe, M. Reggelin, *RSC Adv.* **2017**, *7*, 47183–47189.

Synthesis of Functionalized
β-Peptoid Foldamers

Research stay at the

Department of Drug Design and Pharmacology

University of Copenhagen

Denmark

August – December 2019

Under the supervision of

Prof. Dr. Christian Adam Olsen

Project in cooperation with Prof. Dr. Stefan Bräse, KIT

Table of Contents

1   Introduction 223

    1.1   Foldamers 223

    1.2   Peptoids 223

    1.3   *β*-Peptoid Foldamers 224

2   Objective 227

3   Results and Discussion 229

    3.1   Functionalization of *β*-Peptoid Helices 229

        3.1.1   Incorporation of a Hydroxy Group 229

        3.1.2   Incorporation of an Azide Residue 231

    3.2   Metal Complexes with *β*-Peptoids 238

4   Summary and Outlook 243

5   Experimental Part 245

    5.1   General Remarks 245

        5.1.1   Preparative Work 245

        5.1.2   Solvents and Reagents 245

        5.1.3   Analytics and Equipment 245

    5.2   Syntheses and Characterizations 248

        5.2.1   General Procedures (GPs) 248

        5.2.2   Synthesis of Building Blocks 249

        5.2.3   Synthesis of Trimers and Hexamers 253

6   List of Abbreviations 259

7   References 261

1 Introduction

Natural polymers mostly represented by proteins are entrusted with crucial tasks, for instance catalysis, specific binding or directed flow of electrons. These molecules adopt unique thermodynamically and kinetically stable conformations that contain "active sites" with a precise arrangement of functional groups. Due to the remarkable chemical capabilities of proteins, there is a high interest in designing non-natural molecules having the ability to adopt stabilized three-dimensional structures as well as inheriting specific advantageous properties. The proteolytic stability of peptidomimetics such as peptoids and their capability to fold into secondary structures makes them attractive as drug candidates.[1-2] An application in catalysis or as functional materials could be realized with structures which exhibit folding in non-aqueous environments.[3-5] Crucial for the applicability of folded peptoids however is the incorporation of functionalities which are displayed to well-defined three-dimensional space.

Therefore, this project focused on the synthesis and functionalization of rarely studied β-peptoid foldamers in order to make them applicable for various purposes.

1.1 Foldamers

The term "foldamer" was created by Samuel H. Gellman in 1998 in order "*to describe any polymer with a strong tendency to adopt a specific compact conformation*".[6] With the design of artificial tertiary structures it is envisioned to mimick the performance of natural polymers in terms of complex chemical tasks. Foldamers represent a diverse group of molecules with a wide variety of structures derived from for example peptides, aromatics or oligourea.[7-9] This great amount of structural diversity opens up possibilities for applications in various fields. Synthetic peptidomimetics for example have advantages, namely better proteolytic resistance and larger structural diversity. Well-known examples for this compound class are β-peptides as well as α-peptoids (oligo *N*-alkylglycines).[10-12]

1.2 Peptoids

Peptoids are important mimics of polypeptides, especially with their ability to form well-defined folded architectures.[13] Remarkable propensity for folding has been proven for various α-peptoid sequences at different lengths.[14-18] α-Peptoids are a class of non-natural peptide analogs with a tertiary amide bond giving rise to high backbone flexibility due to the hampered formation of intramolecular hydrogen-bonds. Moreover, isomerization between *transoid* and *cisoid* conformations is facilitated.

The combination of the properties of β-peptides and α-peptoids can be found in β-peptoids which are oligomers of *N*-alkyl-β-alanine residues (Figure 1).[19] In comparison to peptides, in β-peptoids an additional methylene group is inserted and the side chain is connected to the nitrogen of the backbone. Thus, the oligomers inherit a stability towards proteolytic degradation and with the absence of backbone hydrogen bonding, the molecules have increased flexibility.

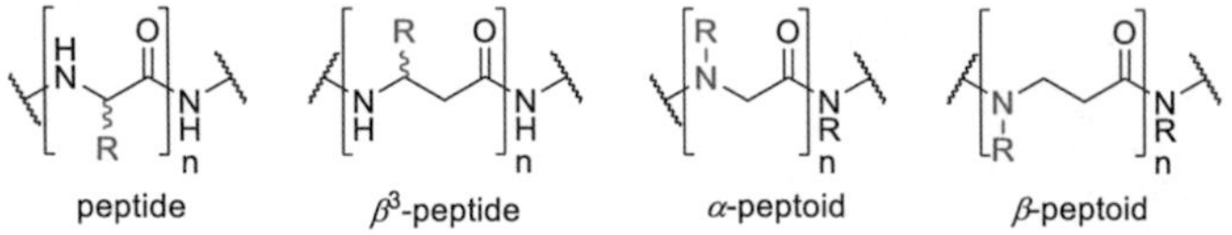

Figure 1. Comparison of the generic structures of peptides, β³-peptides, α- and β-peptoids.

Despite this backbone flexibility of β-peptoids in comparison to peptides, computational simulations predicted the ability of β-peptoid backbones to form helical secondary structures.[20] Moreover, both parental architectures, meaning β-peptides as well as α-peptoids, have shown high folding propensity.[14, 21-22]

1.3   β-Peptoid Foldamers

Encouraged by the results of the computational simulations, the ability of β-peptoids to show helical folding propensity was investigated. Inspired by studies on the equilibria of *trans-* and *cis-* amide bonds of proline residues, a nuclear magnetic resonance (NMR) study with monomeric β-peptoid model systems was performed (Figure 2).[23-26] With the incorporation of specific side chains on the nitrogen atom, highly *cis*-amide bond-inducing side chain *N*-(*S*)-1-(1-naphthyl)ethyl (*N*s1npe) was identified.

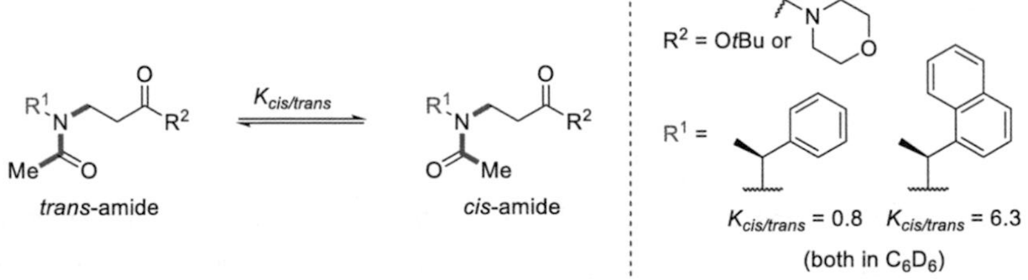

Figure 2. Rotamer equilibrium constants ($K_{cis/trans}$) for β-peptoid monomers in deuterated benzene investigated by Laursen *et al.* The figure was adapted from the given reference.[26]

When homo-oligomers bearing *N*s1npe residues were synthesized, the ability of these β-peptoids to fold into unique triangular prism-shaped helices was proven with a molecular structure verified by X-ray crystallography (Figure 3, A/B).[27-28] The highly *cis*-amide bond-inducing residues gave rise to stabilized secondary structure formation with a predictable display of side chains. A right-handed helical conformation with exactly three residues per turn and a helical pitch of 9.6 – 9.8 Å between turns was observed. It was not possible to use NMR spectroscopy for secondary structure determination in solution due to pronounced signal overlap. Analysis of the three-dimensional structure of the hexamer *via* circular dichroism (CD) spectroscopy gave a spectrum with a trace bearing a maximum at around 216 nm, a minimum at around 225 nm and another positive signal at around 232 nm (Figure 3, D). Therefore, these signals represent the characteristic signals of such helical β-peptoid foldamers.

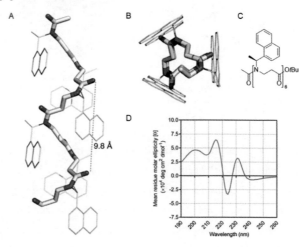

Figure 3. A) Crystal structure of a β-peptoid helix comprising exclusively Ns1npe residues.[27] B) End view of the β-peptoid helix shown in A.[27] C) The chemical structure of the β-peptoid shown in A. D) The corresponding CD spectrum. The figure was adapted from the given reference and reprinted with permission. Further permissions related to the material excerpted should be directed to the ACS.

With the incorporation of longer hydrocarbons as well as amino groups into the β-peptoid foldamers, solubility was significantly increased along with retained folding.[29] Moreover, with the crystal structure of a hexamer containing a butyl side chain, it was shown that the functional side chains are presented towards the outside of the helix (Figure 4). The successful incorporation of functional side chains into β-peptoid foldamers opens up the potential for the development of new functional materials as well as applications in medical and materials sciences.

A   B

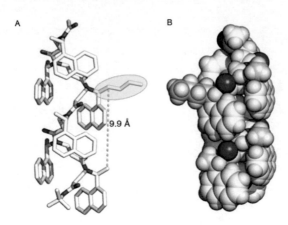

9.9 Å

Figure 4. Crystal structure of a β-peptoid helix containing a protruding butyl side chain highlighted in light green. A) View perpendicular to the helical axis.[29] B) Side view in space filling representation: the naphthyl groups are arranged along the faces of the helix, the methyl groups are displayed in light orange.[29] The figure was adapted from the given reference and reprinted with permission. Copyright 2019 American Chemical Society.

For the synthesis of the β-peptoid oligomers, a solution-phase approach using the iterative submonomer method was applied.[28, 30] First, an acryloyl group **1** was subjected to an *aza*-Michael addition with the side chain-bearing amine **2** (Scheme 1, A). The resulting secondary amine **3** at the N-terminus of the growing peptoid chain, was then reacted with acryloyl chloride **4**. With another *aza*-Michael addition between acrylamide **5** and a primary amine bearing a side chain **2** and repetition of the remaining previously described steps, the backbone of the β-peptoid was elongated. After completion of the synthesis of β-peptoid trimer **7** bearing a C-terminal *tert*-butyl ester, the N-terminal trimer was capped at the N-terminus **8** (e.g. with an acetyl) and the C-terminal ester hydrolysed (Scheme 1, B). Peptide coupling of the trimers using HATU (hexafluorophosphate azabenzotriazole tetramethyl uronium) yielded β-peptoid hexamer **9**.

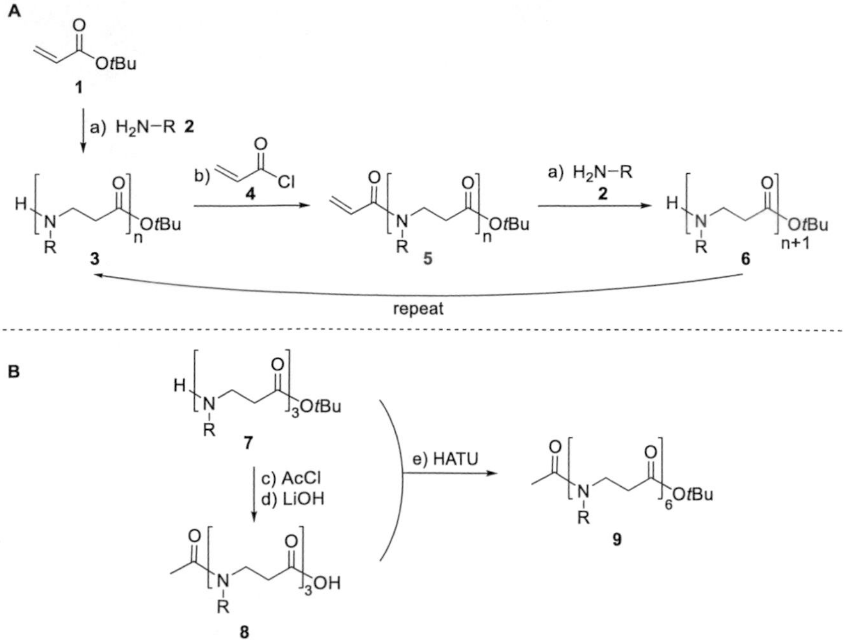

A

B

Scheme 1. General synthetic procedure of β-peptoid hexamers **9** in solution, following previously reported conditions.[28, 30] Reagents and conditions: a) H₂N-R (2 equiv.), MeOH, 50 °C, 18 h; b) acryloyl chloride (1.4 equiv.), Et₃N (1.2 equiv.), CH₂Cl₂, N₂, 0 °C, 1 h; c) AcCl (1.2 equiv.), Et₃N (1.4 equiv.), CH₂Cl₂, N₂, 0 °C, 1 h; d) 1 M LiOH_aq–MeOH 1:1, rt, 48 h; e) HATU (1.2 equiv.), DIPEA (3 equiv.), CH₂Cl₂, rt, 18 h. HATU = hexafluorophosphate azabenzotriazole tetramethyl uranium.

2   Objective

Peptidomimetics with folding propensity and stability towards proteolysis are of high interest for biomedical research, chemical biology, and materials science. For the applicability of such foldamers, well-defined positioning of side chain functionalities in three dimensions is important. Thus, in this project the aim was the design and synthesis of β-peptoid backbones containing functional groups, with retained folding of the oligomers. In order to increase the solubility of the oligomers, polar side chains were supposed to be incorporated as well as azide handles for late-stage functionalization *via* click chemistry. In addition, the formation of helical secondary structures should be investigated by CD spectroscopy.

3   Results and Discussion[1]

3.1   Functionalization of β-Peptoid Helices

In previous studies conducted in the laboratory of C. A. Olsen, it was found that β-peptoids containing N-(S)-1-(1-naphthyl)ethyl side chains provide a stable folding.[28] The incorporation of many aromatic residues however can lead to low solubility of the peptoids in polar solvents. A solution was provided by the integration of polar side chains, in particular amino groups which yielded β-peptoid oligomers with retained folding and increased solubility.[29]

3.1.1   Incorporation of a Hydroxy Group

In order to expand the selection of functional groups and further increase the polarity of the oligomers, a hydroxy group was supposed to be incorporated into a β-peptoid hexamer. For this purpose, it was necessary to synthesize a monomeric peptoid building block containing a hydroxy residue. Following a known procedure, chiral amine monomer building block **12** was prepared enantioselectively *via* an Ellman-type Mannich reaction. In the presence of a lewis acid, 1-naphthaldehyde (**10**) was subjected to a reaction with Ellman's auxiliary (R)-*tert*-butanesulfinamide (**11**) to give *tert*-butanesulfinimine **12** (Scheme 2, A).

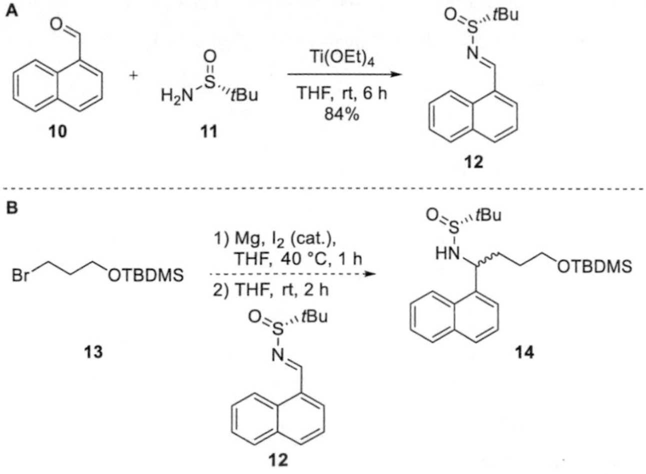

Scheme 2. A) Synthesis of (R)-configured butanesulfinimine **12**. B) Attempted Grignard reaction towards sulfinamide **14** containing a TBDMS protected OH group.

---

[1] Excerpts of this chapter are published in
I. Wellhöfer, J. Beck, K. Frydenvang, S. Bräse, C. A. Olsen, *J. Org. Chem.* **2020**, *85*, 16, 10466–10478 as well as in the PhD thesis of Isabelle Wellhöfer (University of Copenhagen, DK) "Functionalization of Foldamers with Peptidomimetic Backbone Architectures".

Bromide **13** containing a *tert*-butyldimethylsilyl (TBDMS) ether group, synthesized from 3-bromopropanol, should be attached to an (*S*)-configured sulfinimide with *n*-BuLi in previous experiments performed in the Olsen laboratory. As that method was not successful, it was envisioned to incorporate the residue into sulfinimide **12** by nucleophilic attack of a Grignard reagent. In this case the opposite stereoselectivity occurs as compared to a Lithium-organyl. A first test reaction of (*R*)-configured butanesulfinimine **12** with bromide **13** under standard Grignard conditions gave 52% yield, with a diastereomeric ratio of 1 : 1.2, as estimated by ^{1}H NMR (Scheme 2, B). The obtained result however could not be reproduced in further attempts. Thereupon, reaction conditions such as temperature or equivalents of reagents, as well as the procedure for the formation of the Grignard reagent and the subsequent enantioselective addition to sulfinimide **12** were systematically varied. Presumably the Grignard reagent was not formed due to high air- and moisture sensitivity, as none of the investigated conditions gave the desired product **14**.

Therefore, a synthetic strategy involving the integration of a vinyl group on the sulfinamide and subsequent hydroboration was pursued. Following a procedure developed by Isabelle Wellhöfer (group of C. A. Olsen), a Grignard reaction between vinylmagnesium bromide **15** and the (*R*)-enantiomer of sulfinimide **12** was performed to obtain sulfinamide **16** in 77% yield with high diastereoselectivity (Scheme 3).

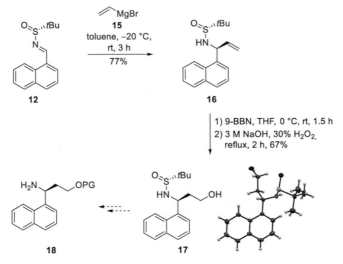

Scheme 3. Synthesis of vinyl-sulfinamide **16** and subsequent hydroboration. 9-BBN = 9-borabicyclo[3.3.1]nonane.

Hydroboration using 9-borabicyclo[3.3.1]nonane (9-BBN) gave hydroxy-containing sulfinamide **17** in 67% yield. The absolute stereochemistry of compound **17** was confirmed by an X-ray crystal structure solved by Karla Frydenvang (University of Copenhagen). The successful implementation of a hydroxy group in sulfinamide **17** enables the synthesis of hexamers containing hydroxy side chains as well as further derivatization of the compounds, for example by oxidation to carboxylic acids. For this purpose, the free alcohol functionality needs to be protected and subsequent removal of the auxiliary gives rise to chiral β-peptoid monomer **18**. Due to the limited amount of time, this strategy was not further pursued.

3.1.2 Incorporation of an Azide Residue

The incorporation of an azide residue into the helices was intended, as azides represent useful chemical handles enabling ready functionalization of the foldamer surface *via* Cu(I)-catalyzed azide–alkyne cycloaddition (CuAAC, "click" chemistry) with alkynes.[31-32]

Initially, a synthetic route towards a chiral amine building block was investigated. Therefore, (*S*)-configured sulfinimide **20** was prepared with (*S*)-*tert*-butanesulfinamide as the auxiliary, following the same procedure as for the (*R*)-configured sulfinimide **12**. For the incorporation of an amine side chain into the building block, 2,5-dimethylpyrrole protected amine **19**, which is stable in the presence of *n*-BuLi, was synthesized from 3-bromopropylamine in one step.[29] Through lithium-halogen exchange of 2,5-dimethylpyrrole protected amine **19** with *n*-BuLi and subsequent reaction with *tert*-butanesulfinimine **20**, sulfinamide **21** was obtained in 48% yield (Scheme 4). Deprotection of the latter provided amine **22** in the desired (*S*)-configuration with good yield.

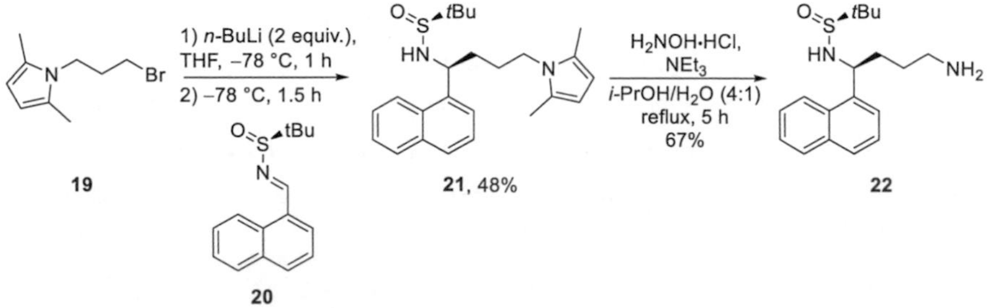

Scheme 4. Incorporation of an amine side chain to obtain sulfinamide **22**.

Literature known diazotransfer reagent imidazole-1-sulfonyl azide (**25**) was prepared following a safe and facile route by Ye *et al.* (Scheme 5).[33] Starting from bench-stable *N*,*N*'-sulfuryldiimidazole **23**, triflate salt **24** was generated which was subsequently reacted with sodium azide in cold water to give product **25**. By extraction with ethyl acetate, imidazole-1-sulfonyl azide (**25**) was obtained and diretly used in solution for diazotransfer reactions.

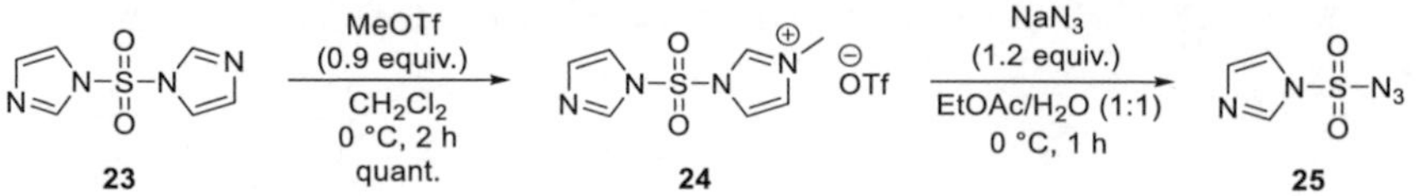

Scheme 5. Preparation of diazotransfer reagent imidazole-1-sulfonyl azide **25** from *N,N'*-sulfuryldiimidazole **23**.[33]

In the presence of catalytic amounts of copper sulfate and base, the diazo transfer between imidazole-1-sulfonyl azide (**25**) and sulfinamide **22** took place (Scheme 6). The thereby generated azide-containing sulfinamide **26** was subjected to auxiliary cleavage under acidic conditions which furnished chiral azide building block **27** in 86% yield.

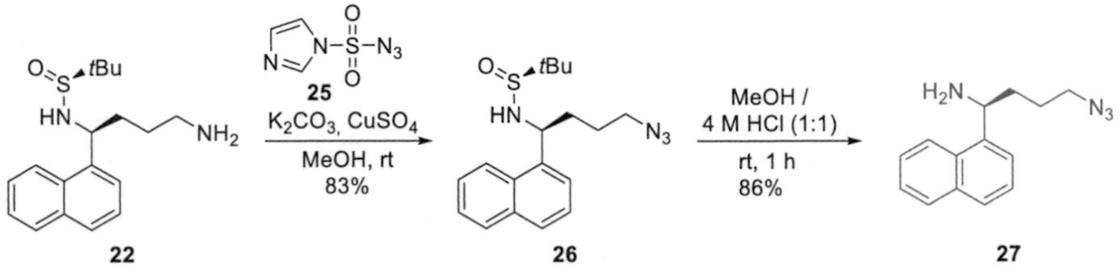

Scheme 6. Diazo transfer to form chiral azide building block **27**.

With this monomer building block in hand, the synthesis of β-peptoid trimers was pursued. Acrylamide **28**, which was synthesized from *N*-(*S*)-1-(1-naphthyl)ethyl amine in an *aza*-Michael addition with *tert*-butyl acrylate and subsequent reaction with acryloyl chloride, was provided by Isabelle Wellhöfer. Following a procedure developed in the group of C. A. Olsen, acrylamide **28** was subjected to an *aza*-Michael addition with azide building block **27** to give dimer **29** (Scheme 7). The resulting secondary amine **29** was reacted with acryloyl chloride **4** in the presence of triethylamine at 0 °C, yielding acrylamide **30**. Another *aza*-Michael addition finally provided azide-containing trimer **32**, which will represent the C-terminus in the hexamer, in good yield.

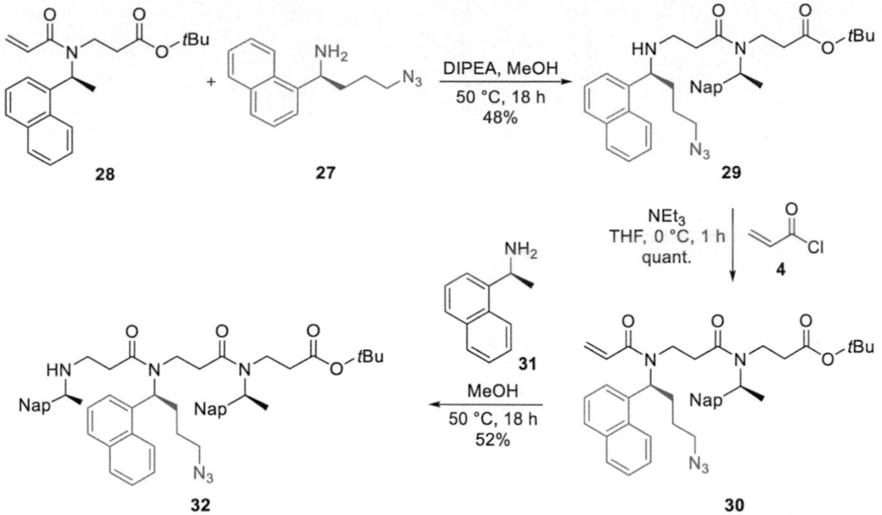

Scheme 7. Synthesis of azide-containing C-terminal trimer **32**. DIPEA = *N,N*-diisopropylethylamine; Nap = naphthyl.

To increase the solubility of the oligomeric compounds as well as to avoid a restriction to organic solvents for NMR and CD spectra, the group of C. A. Olsen investigated the incorporation of polar capping groups. By taking inspiration from Maayan and co-workers, the attachment of a piperazine moiety was implemented.[29, 34] As this strategy successfully increased the polarity of the hexamer without affecting the folding propensity, it was equally applied for the synthesis of the azide-containing hexamers. Allyloxycarbonyl protected piperazine salt **34** was provided by Isabelle Wellhöfer and incorporated at the N-terminal part of trimer **33** (Scheme 8). Hydrolysis of the C-terminal *tert*-butyl ester of trimer **35** under basic conditions required very long reaction times. The application of an aqueous phosphoric acid solution however removed the ester within 4 h to give N-terminal trimer **36**.

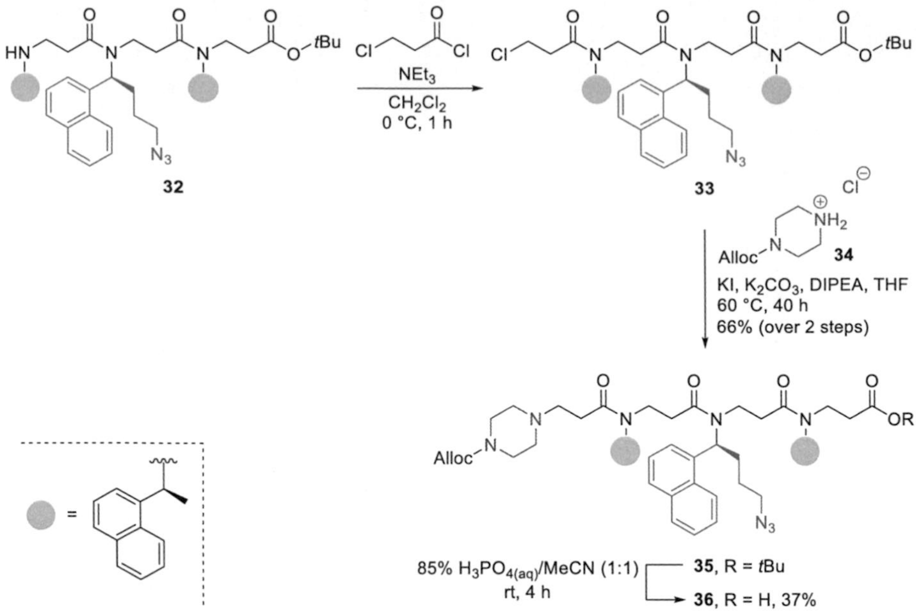

Scheme 8. N-terminal trimer **36** combining piperazine-modification and azide functionalization. Alloc = Allyloxycarbonyl.

Peptide coupling of C-terminal trimer **32** with N-terminal trimer **36** in the presence of coupling reagent HATU yielded peptoid hexamer **37** containing two azide side chains (Scheme 9). Azide containing hexamers have the advantage to readily be functionalized with various alkynes *via* click chemistry. In order to further increase the polarity of the β-peptoid **37**, free hydroxy groups were supposed to be incorporated into the oligomer. For this purpose, β-peptoid hexamer **37** was subjected to a CuAAC reaction with propynol ethoxylate **38**. Subsequent deprotection of the piperazine moiety, purification by preparative reversed-phase high-pressure liquid chromatography (RP-HPLC) with trifluoroacetic acid (TFA) containing buffers and lyophilization yielded hexamer **39** as a TFA-salt in 40% over two steps.

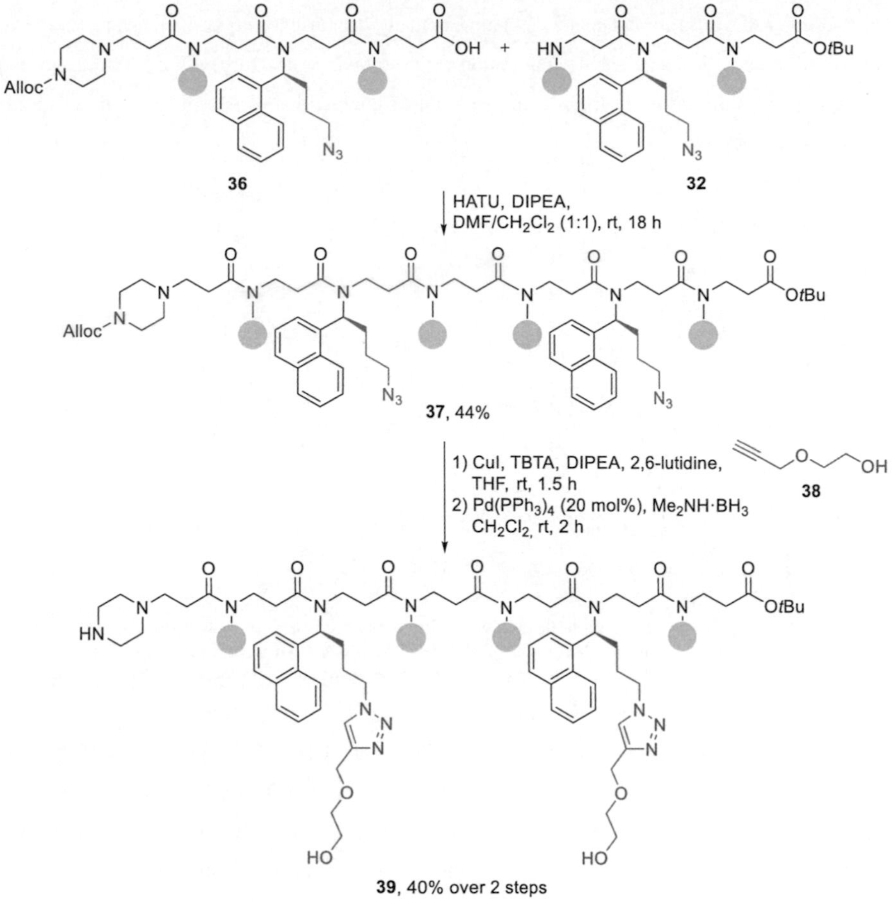

Scheme 9. Synthesis of hexamer **39** combining piperazine-modification and functionalization with an ethylene glycol ether bearing a free hydroxy group. TBTA = Tris(benzyltriazolylmethyl)amine.

Hexamer **39** containing an ethylene glycol ether bearing a free hydroxy group was investigated concerning its behavior in CD spectroscopy. Measurement in acetonitrile at room temperature revealed a spectrum with a trace bearing a maximum at around 216 nm, a minimum at around 225 nm and another positive signal at around 232 nm (Figure 5). This is in agreement with the characteristic peaks related to helical β-peptoids as described in the literature (see Introduction). Heating to 60 °C lead to degradation of the helix as indicated by a strong decrease of the signal intensity at 232 nm. However, after cooling the dissolved hexamer back to room temperature, the structure refolded itself as depicted in the trace that ressembles the CD trace of the first measurement at 25 °C. When hexamer **39** was dissolved in different mixtures of methanol in acetonitrile (20 – 80%), the signal at 232 nm gradually decreased with an increasing amount of methanol (Figure 5). This gradual denaturation of the secondary structure with increasing addition of protic solvent might result from hydrogen bonding interactions of the solvent with the amide bonds of the backbone.

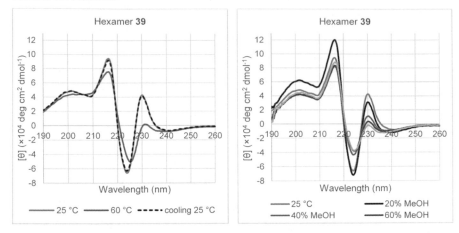

Figure 5. CD spectra of **39** in acetonitrile (MeCN) at different temperatures and with varying amounts of MeOH. Spectra were measured in MeCN (48–71 μM) at 25 °C if not stated otherwise. The mean residue molar ellipticity [θ] is normalized with regards to the number of residues and the concentration of the peptoids.

With the level of polarity obtained for hexamer **39**, it was for the first time possible to measure CD spectra of β-peptoid hexamers in solutions of up to 60% PBS buffer (phosphate-buffered saline) in acetonitrile (Figure 6). Upon addition of PBS buffer to solutions of oligomer **39** in MeCN, a decrease of folding was observed with increasing amount of buffer, as indicated by the decreasing signal at 232 nm. Nevertheless, a small positive signal at 232 nm was obtained even with 60% PBS buffer in acetonitrile. Presumably protic solvents enable hydrogen bonding between the backbone amides and the solvent molecules and therefore disturb the helical folding. To conclude, the folding of hexamer **39** is strongly dependent on the temperature as well as the solvent.

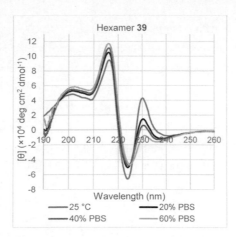

Figure 6. CD spectra of **39** in MeCN with varying amounts of PBS buffer. Spectra were measured in MeCN (48–71 μM) at 25 °C if not stated otherwise. The mean residue molar ellipticity [θ] is normalized with regards to the number of residues and the concentration of the peptoids.

3.2   Metal Complexes with β-Peptoids

Oligomers that upon coordination with metal ions fold into specific secondary structures are called metallofoldamers.[35] This coordination can either template helical structure or nucleate its formation.[36-37] It is known that α-peptoids can complex metal ions enabling applications like metal-ion sensing or catalysis.[38-46] Therefore, in this project it was intended to design β-peptoids with the ability of metal-ion binding in order to build tertiary structure-like formations. By making use of the bis-azide-containing helical β-peptoid **37** described in chapter 3.1.2, a copper-catalyzed click reaction with ethynylpyridine (**41**) was performed (Scheme 10).

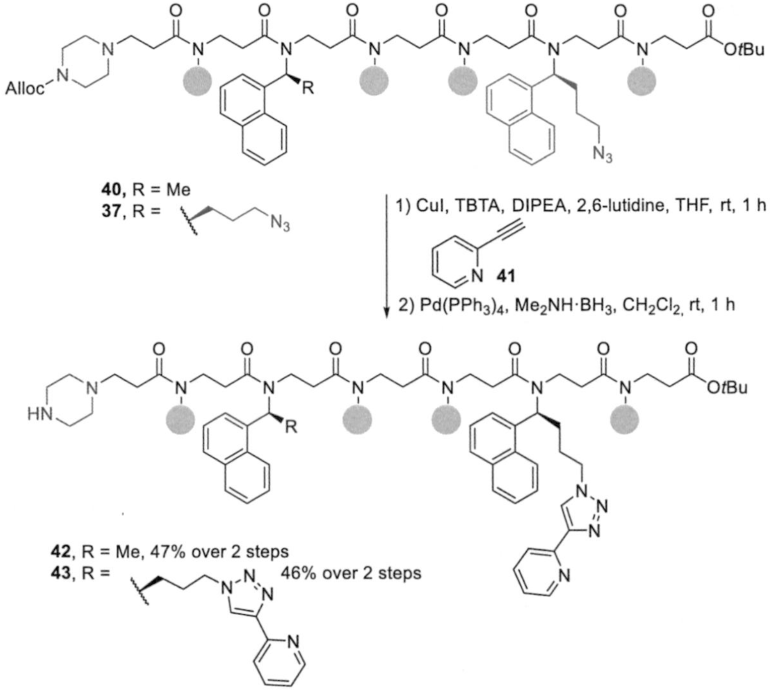

Scheme 10. Copper-catalyzed click reaction of (bis-)azide-containing helical β-peptoid **40** or **37** with ethynylpyridine (**41**) to afford hexamers **42** and **43** containing metal-binding groups.

Via triazole formation multidentate metal-binding 2-(1*H*-1,2,3-triazol-4-yl)pyridine ligands were introduced into the oligomers. As it was envisioned to investigate intramolecular as well as intermolecular peptoid complexes, a hexamer containing only one azide functionality was synthesized. A trimer made of only *N*-(*S*)-1-(1-naphthyl)ethyl building blocks **IWE-155** was provided by Isabelle Wellhöfer and subjected to peptide coupling with azide-containing trimer **32** to give hexamer **40**. Reaction with ethynylpyridine (**41**) yielded product **42** containing one 2-(1*H*-

1,2,3-triazol-4-yl)pyridine ligand (Scheme 10). After deprotection of the Alloc group at the piperazine moiety, purification by preparative RP-HPLC and lyophilization, both hexamers **42** and **43** were obtained in good yield.

The secondary structure of the above described hexamers **42** and **43** was investigated *via* CD measurements in acetonitrile. With the occurrence of a minimum at around 223 nm and a maximum at around 230 nm in the CD spectrum, helical folding was indicated (Figure 7, Figure 8). The investigation of the behaviour of hexamers **42** and **43** equipped with metal-binding groups at elevated temperatures revealed that in both oligomers the secondary structures were denatured upon heating to 60 °C, however the molecules were capable of refolding when the temperature was decreased to 25 °C. The secondary structure of hexamer **42** containing one metal-binding group gradually denatured with an increasing amount of methanol in acetonitrile.

Unpublished studies of the group of C. A. Olsen showed pH dependency of the folding for *β*-peptoids comprising basic or acidic side chains. In response to pH changes, *β*-peptoids containing side chains with varying protonation state can switch between a folded and an unfolded state. As the final oligomer **43** was obtained as a TFA-salt, the side chain is expected to be protonated. For 2-(1*H*-1,2,3-triazol-4-yl)pyridine ligand containing hexamer **43** indeed basic conditions were preferred. It is assumed that in acidic environments protonated helices interact with the solvent and therefore the folding is disrupted. As seen before, the degree of folding of hexamers **42** and **43** was dependent on the temperatures as well as the solvent.

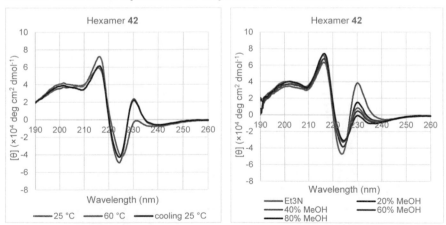

Figure 7. CD measurements of mono-pyridyl ligand containing helix **42** in MeCN in the presence of Et₃N at different temperatures (left) and CD spectra of **42** in MeCN with varying amounts of MeOH (right). Spectra were measured in MeCN (48–71 µM) at 25 °C if not stated otherwise. The mean residue molar ellipticity [θ] is normalized with regards to the number of residues and the concentration of the peptoids.

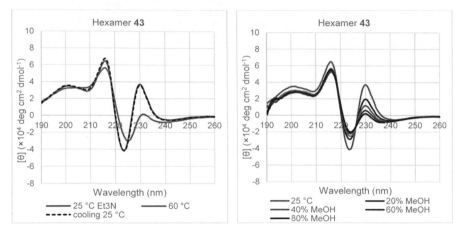

Figure 8. CD measurements of bis-pyridyl ligand containing helix **43** in MeCN in the presence of Et₃N at different temperatures (left) and CD spectra of **43** in MeCN with varying amounts of MeOH (right). Spectra were measured in MeCN (48–71 μM) at 25 °C if not stated otherwise. The mean residue molar ellipticity [θ] is normalized with regards to the number of residues and the concentration of the peptoids.

A comparison of the folding behaviour of oligomers **42** and **43** with and without the addition of triethylamine (Et₃N) emphasized the preference for basic conditions (Figure 9, Figure 10). The degree of helicity of both hexamers **42** and **43** in solutions of 20% PBS in acetonitrile was reduced in about the same way as in the hexamer **39** containing free hydroxy groups.

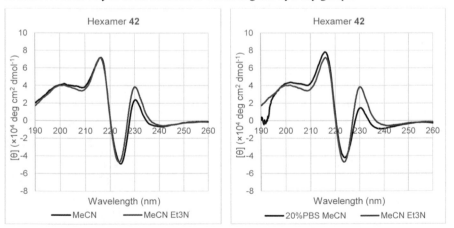

Figure 9. CD measurements of hexamer **42** with comparison of the traces with and without the addition of Et₃N (20 equiv.) (left) as well as with 20% PBS buffer (right). Spectra were measured in MeCN (48–62 μM) at 25 °C if not stated otherwise. The mean residue molar ellipticity [θ] is normalized with regards to the number of residues and the concentration of the peptoids.

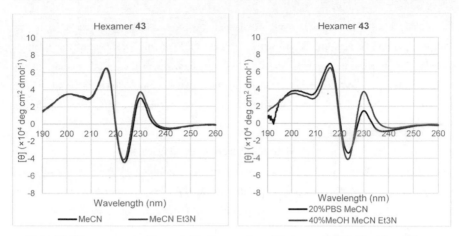

Figure 10. CD measurements of hexamer **43** with comparison of the traces with and without the addition of Et₃N (20 equiv.) (left) as well as with 20% PBS buffer (right). Spectra were measured in MeCN (48–62 μM) at 25 °C if not stated otherwise. The mean residue molar ellipticity [θ] is normalized with regards to the number of residues and the concentration of the peptoids.

After the investigation of the folding behavior of hexamers **42** and **43**, their ability to complexate metal-ions was supposed to be studied.

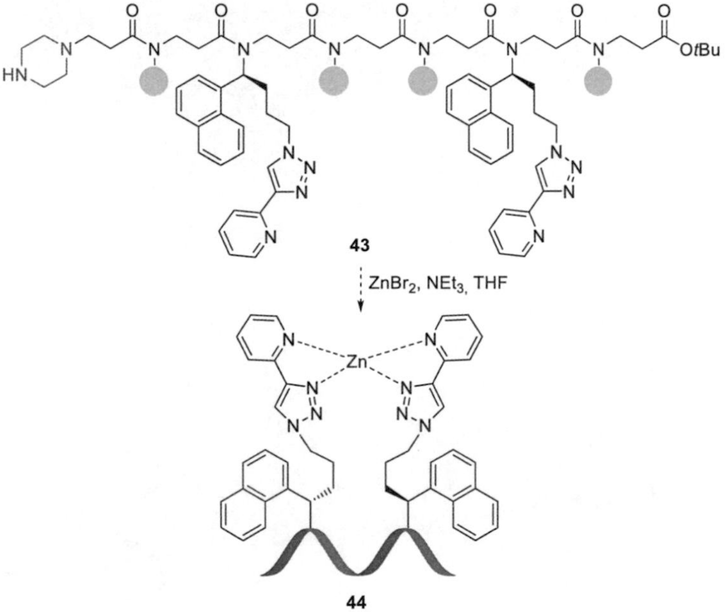

Scheme 11. Attempted synthesis of an intramolecular complex **44** of zinc(II) and hexamer **43**.

Intramolecular complexation of hexamer **43** containing two metal-binding groups was attempted with the addition of zinc bromide in tetrahydrofuran (THF) in the presence of triethylamine (Scheme 11). Electrospray ionization (ESI) mass measurements of the resulting crude showed protonated as well as sodiated hexamer **43** but only traces of the complexed hexamer **44**. In order to generate intramolecular metallofoldamers, it would be necessary to perform further experiments.

In addition, it was tried to form of an intermolecular complex of two helices connected *via* a copper ion (Scheme 12). Therefore, hexamer **42** containing one metal-binding group was stirred with copper(II)acetate in THF in the presence of triethylamine. ESI mass measurements of the resulting crude product showed one hexamer in complexation with a copper ion, instead of the desired product **45**.

As both attempts for the complexation of hexamers with metal-ions were only performed once with residual 2 mg of product, further studies are required to investigate the complexation behaviour of β-peptoid metallofoldamers in more detail.

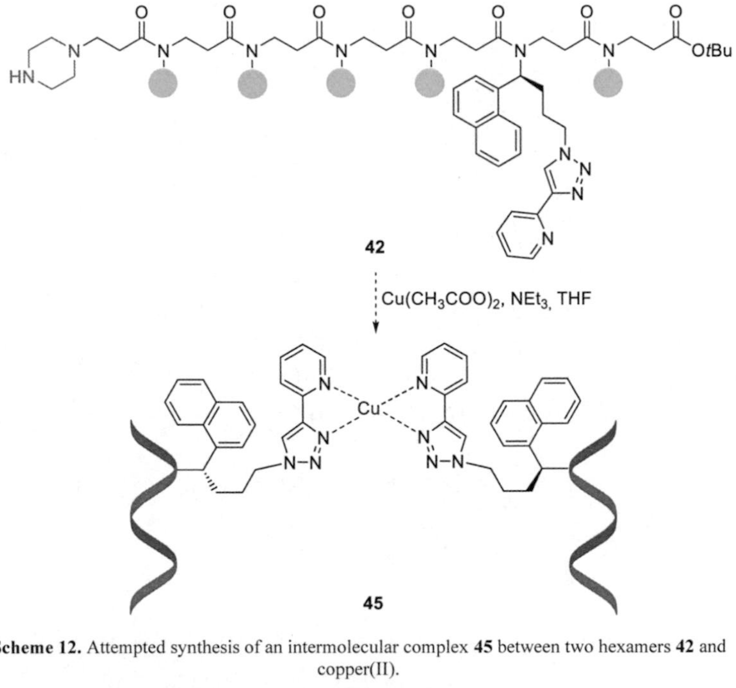

Scheme 12. Attempted synthesis of an intermolecular complex **45** between two hexamers **42** and copper(II).

4   Summary and Outlook

The overall aim of the functionalization of β-peptoids is the development of foldamers with a desired function, such as catalysis or polyvalent display. In this project important steps towards the applicability of β-peptoids were made by well-defined positioning of side chain functionalities.

A precursor for the synthesis of a β-peptoid monomer containing a side chain with hydroxy functionality as well as a chiral azide building block **27** were successfully prepared by making use of Ellman's auxiliary **11**. With the incorporation of the azide building block **27** into peptoids, a series of hexamers containing different functionalities was designed *via* late-stage functionalization with copper-catalyzed click chemistry. The attachment of other functional groups *via* various alkynes in order to implement different side chains on one hexamer remains to be investigated in future studies.

The late-stage functionalization of hexamers with ethylene glycol ethers **39** bearing a free hydroxy group *via* CuAAC gave rise to substantially increased polarity of the oligomers which enabled the investigation of their folding propensity even in aqueous solutions. Furthermore, the handling of the oligomers, including synthesis, purification and analysis was improved significantly.

The design of hexamers containing ligands for metal chelation gave helical β-peptoids **42** and **43** that make ideal precursors for metallofoldamers. Moreover, pH dependency was shown for secondary structure formation with the addition of base to the solution of hexamer **43**. In order to be able to control the folding for special applications, this might be a convenient strategy.

5 Experimental Part[2]

5.1 General Remarks

5.1.1 Preparative Work

The starting materials, solvents and reagents were purchased from abcr, Acros, Alfa Aesar, Carbolution, ChemPUR, Fluka, Iris, Merck, Riedel-de Haën, TCI, Thermo Fisher Scientific, Sigma Aldrich and used without further purification. Liquid column chromatography was performed either manually on silica gel (particle size $40 - 63$ µm) or with a Büchi Pure C-810 flash chromatography device.

All reactions containing air- and/or moisture-sensitive compounds were performed under nitrogen or argon atmosphere using oven-dried glassware applying Schlenk-techniques. Liquids were added *via* steel cannulas and solids were added directly in powdered shape. Reactions were accomplished at room temperature, if nothing else is mentioned. For low reaction temperatures flat dewars with ice/water or isopropanol/dry ice mixtures were used. The solvents were removed at 40 °C with a rotary evaporator under reduced pressure. For solvent mixtures each solvent was measured volumetrically. If not stated otherwise, saturated, aqueous solutions of inorganic salts were used. Celite® for filtrations was purchased from Alfa Aesar (Celite® 545, treated with Na_2CO_3).

5.1.2 Solvents and Reagents

All chemicals and solvents were of analytical grade and used without further purification. All reactions under nitrogen atmosphere were performed in dry solvents. Dichloromethane, *N,N*-dimethylformamide (DMF), and tetrahydrofuran (THF) were retrieved from a solvent purification system.

5.1.3 Analytics and Equipment

5.1.3.1 Nuclear Magnetic Resonance (NMR)

^{1}H NMR and ^{13}C NMR spectra were recorded at 298 K with a cryogenically cooled probe at 600 MHz and 151 MHz, respectively. Chemical shifts are reported in ppm, relative to deuterated solvents as internal standard (δ_H = CDCl$_3$ 7.26 ppm, CD$_3$OD 3.31 ppm, CD$_3$CN 1.94 ppm, C$_6$D$_6$ 7.16 ppm; δ_C = CDCl$_3$ 77.16 ppm, CD$_3$OD 49.00 ppm, CD$_3$CN 1.32 ppm, C$_6$D$_6$ 128.06 ppm).

---

[2] Excerpts of this chapter are published in
I. Wellhöfer, J. Beck, K. Frydenvang, S. Bräse, C. A. Olsen, *J. Org. Chem.* **2020**, *85*, 16, 10466–10478 as well as in the PhD thesis of Isabelle Wellhöfer (University of Copenhagen, DK) "Functionalization of Foldamers with Peptidomimetic Backbone Architectures".

Coupling constants (J) are reported in Hertz [Hz]. Multiplicities of NMR signals are reported as follows: s, singlet; br, broad singlet; d, doublet; t, triplet; q, quartet; m, multiplet. Assignments of peak identities are based on 2D NMR experiments (COSY, HSQC, HMBC). The following abbreviations are used for assignments: CH_{ar}, aromatic proton/carbon; C_q, quaternary carbon.

5.1.3.2 Mass Spectrometry

Ultra high-performance liquid chromatography (UPLC)-mass spectrometry (MS) analyses were performed on an HPLC (high-performance liquid chromatography) system equipped with a C18 column (50 mm × 2.1 mm, 1.7 μm, 100 Å) using a gradient of eluent I (0.1% HCOOH in water) and eluent II (0.1% HCOOH in acetonitrile) rising linearly from 0% to 95% of eluent II during t = 0.00 – 5.00 min at a flow rate of 0.6 mL/min.

High-resolution mass spectra (HRMS) were recorded using a quadrupole time-of-flight (TOF) mass spectrometer equipped with an electrospray (ESI) source. Alternatively, they were recorded on an UHPLC equipped with a diode array detector and coupled to a QTOF mass spectrometer operated in positive electrospray or by either matrix-assisted laser desorption/ionization (MALDI) or ESI.

5.1.3.3 Analytical HPLC

Analytical reversed-phase HPLC was performed on a system equipped with a C8 column (250 × 4.6 mm, 5 μm, 100 Å) and a diode-array UV detector, using a gradient of eluent III (water-acetonitrile-TFA, 95:5:0.1) and eluent IV (0.1% TFA in acetonitrile) rising linearly from 5% to 95% of eluent IV during t = 5 – 35 min (gradient A), from 50% to 95% of eluent IV during t = 5 – 40 min (gradient B), or from 50% to 95% of eluent IV during t = 5 – 25 min (gradient C) with a flow rate of 1.0 mL/min.

5.1.3.4 Preparative HPLC

Preparative HPLC purification was performed on a system equipped with a C8 column (250 × 21.2 mm, 5 μm, 100 Å), a diode-array UV detector, and an evaporative light-scattering detector (ELSD) at a flow rate of 20 mL/min. The applied gradients using eluent III (water-acetonitrile-TFA, 95:5:0.1) and IV (0.1% TFA in acetonitrile) are specified for each individual compound. Fractions containing the target compound were identified using UPLC-MS and analytical HPLC. Selected fractions were pooled and lyophilized.

5.1.3.5 Thin Layer Chromatography (TLC)

All reactions were monitored by thin layer chromatography (TLC) using silica gel coated aluminum plates (Merck, silica gel 60, F_{254}). The detection was performed with UV light (254 nm) and/or dipped into a solution of Seebach reagent (2.5% phosphor molybdic acid, 1.0% Cerium(IV) sulfate tetrahydrate and 6.0% sulfuric acid in H_2O, dipping solution) and heated with a heat gun.

5.1.3.6 Circular Dichroism Spectroscopy (CD)

Spectra were acquired with a JASCO J-1500 spectrophotometer equipped with a water-circulating bath and a nitrogen gas flowmeter with sensor. Measurements were carried out in 1 mm quartz cuvettes, and compound solutions in acetonitrile were prepared using the dry weight of the lyophilized material followed by dilution to give the desired concentrations. Concentrations were verified using a NanoDrop to measure the absorbance at $\lambda = 280$ nm ($A_{280} = \varepsilon \cdot c \cdot l$). The CD data were obtained at 298 K with a bandwidth of 1.00 nm, a scanning speed of 20 nm/min, three accumulations, and a data integration time of 4 s. The measurements were performed in triplicates. Spectra were recorded in millidegree units (m°), corrected for solvent contributions and normalized to mean residue ellipticity (θ) = $100 \cdot$ m°$/l \cdot c \cdot$n, with c being the concentration in mM, l being the path length (0.1 cm), and n being the number of peptoid amide bonds.

5.2   Syntheses and Characterizations

5.2.1   General Procedures (GPs)

GP A: *Aza*-Michael addition. The respective amine (1.00 – 2.00 equiv.) was added to a solution of the acrylated β-peptoid (0.1 M, 1.00 equiv.) in MeOH. The reaction mixture was stirred at 50 °C for 18 h. After completion, the solvent was evaporated. In case of the dimeric and trimeric peptoids, the residue was redissolved in EtOAc and washed with 2 M aq. HCl and sat. aq. NaHCO$_3$. The organic phase was dried over MgSO$_4$, filtered and concentrated *in vacuo*. The residue was purified by column chromatography (2% MeOH, 0.25% NH$_3$ in CH$_2$Cl$_2$).

GP B: Acryloylation reaction. Under nitrogen atmosphere, the peptoid (1.00 equiv.) and Et$_3$N (1.20 equiv.) were dissolved in anhydrous THF (0.05 M) and the mixture was cooled to 0 °C. Acryloyl chloride (1.40 equiv.) was added and the solution was stirred for 1 h at this temperature. The reaction mixture was filtered, and the filter cake was washed with cold EtOAc. The combined filtrates were concentrated *in vacuo* to give the crude acrylamide, which was used without further purification.

GP C: *N*-terminal acylation. To a solution of the peptoid (1.00 equiv.) in anhydrous THF (0.02 M) at −10 °C, Et$_3$N (1.40 equiv.) and 3-chloropropionyl chloride (1.50 equiv.) were added. The reaction mixture was stirred for 1 h at −10 °C. After completion, the solution was diluted with EtOAc and washed with water. The organic phase was dried over MgSO$_4$ and the solvent was removed *in vacuo* to give the crude product, which was used without further purification.

GP D: Piperazine coupling. The peptoid trimer (1.00 equiv.) was dissolved in THF (0.05 M). K$_2$CO$_3$ (5.00 equiv.), KI (0.1 equiv.), DIPEA (3.00 equiv.) and the protected piperazine·HCl salt (3.00 equiv.) were added and the reaction mixture was stirred at 60 °C for 48 h. After completion, the mixture was diluted with EtOAc (15 mL) and washed with 0.1 M aq. HCl (10 mL) and sat. aq. NaHCO$_3$ (10 mL). The organic phase was dried over MgSO$_4$ and the solvent was removed *in vacuo*. The crude residue was purified by column chromatography.

GP E: *C*-terminal deprotection. 1) The peptoid trimer (1.00 equiv., 5 mM) was dissolved in MeCN–H$_3$PO$_4$ (aq., 80%) 1:1 and stirred for 4 h at room temperature. After reaction completion, the solution was diluted with water and extracted with EtOAc. The combined organic phases were washed with sat. aq. NaHCO$_3$, dried over MgSO$_4$ and concentrated *in vacuo*. The product was used without further purification.

2) A solution of the peptoid (6.00 mM, 1.00 equiv.) in 1 M aq. LiOH–MeOH 1:1 was stirred at room temperature for 48 h. After dilution with water, the solution was acidified (pH 1) with aq.

conc. HCl. This caused precipitation and the suspension was extracted with EtOAc. The combined organic layers were dried over MgSO$_4$, filtered, and concentrated. The resulting crude product was purified by column chromatography (1 → 2% MeOH, 0.25% AcOH in CH$_2$Cl$_2$).

GP F: Amide bond formation. The N-terminal peptoid trimer (1.00 equiv.) and the C-terminal peptoid trimer (1.00 equiv.) were placed in a vial and dissolved in anhydrous DMF–CH$_2$Cl$_2$ 1:1 (0.1 M). Subsequently, DIPEA (3.00 equiv.) and HATU (1.20 equiv.) were added and the reaction was left on a shaker at room temperature for 18 – 48 h. After completion, the solution was concentrated using a nitrogen stream and either subjected to Alloc-deprotection or purification by preparative HPLC.

5.2.2   Synthesis of Building Blocks

(*S*)-2-Methyl-*N*-(naphthalen-1-ylmethylene)propane-2-sulfinimine (12)

To a 0.5 M solution of Ti(OEt)$_4$ (5.18 mL, 5.65 g, 24.8 mmol, 1.00 equiv.) in anhydrous THF under nitrogen atmosphere were added 1-naphthaldehyde (**10**) (3.70 mL, 4.25 g, 27.2 mmol, 1.10 equiv.) and (*S*)-*tert*-butanesulfinamide (**11**) (3.00 g, 24.8 mmol, 1.00 equiv.). The reaction mixture was stirred at room temperature for 18 h. It was then poured into a flask with brine (100 mL) upon rapid stirring. The suspension was filtered through a pad of Celite, and the filter cake was washed well with EtOAc. The filtrate was washed with brine (2 × 100 mL), and the aqueous layer was extracted with EtOAc (1 × 50 mL). The combined organic phases were dried over MgSO$_4$ and concentrated *in vacuo*. The residue was purified using column chromatography (short flash column as product is labile on silica, 0 → 1.5% MeOH in CH$_2$Cl$_2$) to give sulfinimine **12** as a yellow oil that solidified upon storage at 4 °C (4.95 g, 77%).

– R_f (CH$_2$Cl$_2$) = 0.18. – ^{1}H NMR (600 MHz, CD$_3$CN): δ = 1.28 (s, 9H, C(C*H*$_3$)$_3$), 7.61–7.66 (m, 2H, C*H*$_{ar}$), 7.70 (ddd, J = 8.5, 6.8, 1.4 Hz, 1H, C*H*$_{ar}$), 8.02 (ddt, J = 8.1, 1.3, 0.6 Hz, 1H, C*H*$_{ar}$), 8.08–8.14 (m, 1H, C*H*$_{ar}$), 9.05 (dq, J = 8.6, 0.9 Hz, 1H, C*H*$_{ar}$), 9.09 (s, 1H, N=C*H*) ppm. – ^{13}C NMR (151 MHz, CD$_3$CN): δ = 22.3 (3 × C*C*H$_3$), 57.7 (*C*CH$_3$), 124.9 (*C*H$_{ar}$), 126.0 (*C*H$_{ar}$), 127.2 (*C*H$_{ar}$), 128.8 (*C*H$_{ar}$), 129.5 (*C*H$_{ar}$), 130.0 (naphthyl-*C*$_q$), 131.5 (naphthyl-*C*$_q$), 132.9 (*C*H$_{ar}$), 133.9 (*C*H$_{ar}$), 134.5 (naphthyl-*C*$_q$), 163.2 (N=*C*H) ppm. – UPLC-MS *m/z*: calcd. for [M+H]$^+$, C$_{15}$H$_{18}$NOS$^+$, 260.11; found 260.14.

(*R*)-2-Methyl-*N*-((*S*)-1-(naphthalen-1-yl)allyl)propane-2-sulfinamide (16)

Under nitrogen atmosphere, a solution of the aldimine **12** (500 mg, 1.93 mmol, 1.00 equiv.) in anhydrous toluene (50.0 mL) at –20 °C was treated with dropwise addition of vinylmagnesium bromide **15** (3.86 mL, 506 mg, 3.85 mmol, 2.00 equiv.). The reaction mixture was stirred at –20 °C for 1 h, slowly heated up to reach room temperature and after 3 h quenched by the addition of sat. aq. NH$_4$Cl (10 mL). The mixture was diluted with EtOAc (20 mL) and the organic phase was washed with water (50 mL). The aqueous phase was extracted with EtOAc (2 × 30 mL). The combined organic phases were dried over MgSO$_4$ and the solvent was removed *in vacuo*. The residue was purified by automated column chromatography (20 → 80% EtOAc in heptane) yielding the product **16** as a clear oil (430 mg, 77%).

– R_f (heptane–EtOAc, 1:2) = 0.37. – ^{1}H NMR (600 MHz, CD$_3$CN): δ = 1.13 (s, 9H, C(C*H*$_3$)$_3$), 4.32 (d, *J* = 5.8 Hz, 1H, N*H*), 5.22 (dt, *J* = 10.3, 1.3 Hz, 1H, C=C*H*$_{cis}$), 5.31 (dt, *J* = 17.1, 1.4 Hz, 1H, C=C*H*$_{trans}$), 5.69 (td, *J* = 5.8, 1.5 Hz, 1H, NHC*H*), 6.30 (ddd, *J* = 17.1, 10.3, 6.2 Hz, 1H, C*H*=CH$_2$), 7.46–7.56 (m, 3H, C*H*$_{ar}$), 7.59–7.64 (m, 1H, C*H*$_{ar}$), 7.84 (dd, *J* = 8.3, 1.1 Hz, 1H, C*H*$_{ar}$), 7.89–7.93 (m, 1H, C*H*$_{ar}$), 8.23 (dt, *J* = 8.0, 1.0 Hz, 1H, C*H*$_{ar}$) ppm. – ^{13}C NMR (151 MHz, CD$_3$CN): δ = 23.0 (3 × CC*H*$_3$), 56.4 (*C*CH$_3$), 59.8 (NH*C*H), 117.0 (*C*H=CH$_2$), 125.0 (*C*H$_{ar}$), 126.4 (*C*H$_{ar}$), 126.7 (*C*H$_{ar}$), 126.9 (*C*H$_{ar}$), 127.0 (*C*H$_{ar}$), 129.2 (*C*H$_{ar}$), 129.7 (*C*H$_{ar}$), 131.8 (naphthyl-*C*$_q$), 135.0 (naphthyl-*C*$_q$), 137.8 (naphthyl-*C*$_q$), 140.3 (*C*H=CH$_2$) ppm. – UPLC-MS *m/z*: calcd. for [M+H]$^+$, C$_{17}$H$_{22}$NOS$^+$, 288.14; found 288.14. Analytical data is in accordance with previously published literature.[47]

(*R*)-*N*-((*S*)-3-Hydroxy-1-(naphthalen-1-yl)propyl)-2-methylpropane-2-sulfinamide (17)

9-BBN monomer (0.5 M in THF, 0.70 mL, 42.5 mg, 350 µmol, 2.00 equiv.) was added dropwise to a solution of the alkene **16** (50.0 mg, 170 µmol, 1.00 equiv.) in anhydrous THF (1.70 ml) at 0 °C under nitrogen atmosphere. After addition, the mixture was stirred at 0 °C for 30 min, before it was allowed to warm to room temperature and subsequently stirred for further 60 min. NaOH (3 M, 0.07 mL) and 30% H$_2$O$_2$ (0.07 mL) were added. The resulting mixture was heated under reflux for 2 h. After reaction completion, the solution was concentrated *in vacuo*. The residue was partitioned between EtOAc (3 mL) and water (3 mL). The separated aqueous layer was extracted with EtOAc (2 × 3 mL). The combined organic extracts were washed with brine (2 × 3 mL), dried over Na$_2$SO$_4$, filtered, and evaporated *in vacuo*. The residue was purified by automated column chromatography (0 → 6% MeOH, 0.25% NH$_3$ in CH$_2$Cl$_2$), yielding product **17** as a white solid (35.5 mg, 67%).

– R_f (6% MeOH, 0.25% NH$_3$ in CH$_2$Cl$_2$) = 0.11. – ^{1}H NMR (600 MHz, CD$_3$OD): δ = 1.19 (s, 9H, C(CH_3)$_3$), 2.12–2.19 (m, 1H, CHCH$_2$CH$_2$), 2.19–2.27 (m, 1H, CHCH_2CH$_2$), 3.68–3.74 (m, 1H, CH$_2$CH$_2$OH), 3.75–3.81 (m, 1H, CH$_2$CH$_2$OH), 5.32–5.40 (m, 1H, NHCH), 7.46–7.51 (m, 2H, CH_{ar}), 7.53 (ddd, J = 8.5, 6.8, 1.5 Hz, 1H, CH_{ar}), 7.63 (dd, J = 7.3, 1.2 Hz, 1H, CH_{ar}), 7.80 (dd, J = 8.2, 1.0 Hz, 1H, CH_{ar}), 7.85–7.94 (m, 1H, CH_{ar}), 8.26 (d, J = 8.5 Hz, 1H, CH_{ar}). Missing signals (2H, OH, NH) ppm. – ^{13}C NMR (151 MHz, CD$_3$OD): δ = 23.1 (3 × CCH_3), 40.8 (CHCH$_2$CH$_2$), 56.2 (C(CH$_3$)$_3$), 56.8 (NHCH), 60.9 (CH$_2$CH$_2$OH), 124.4 (CH_{ar}), 126.1 (CH_{ar}), 126.3 (CH_{ar}), 126.6 (CH_{ar}), 127.1 (CH_{ar}), 129.1 (CH_{ar}), 130.0 (CH_{ar}), 132.2 (naphthyl-C_q), 135.5 (naphthyl-C_q), 139.5 (naphthyl-C_q) ppm. – HRMS (ESI-TOF) m/z: calcd. for [M+H]$^+$, C$_{17}$H$_{24}$NO$_2$S$^+$, 306.1522; found 306.1528.

3-(Imidazole-1-sulfonyl)-1-methyl-3H-imidazol-1-ium triflate (24)

To a solution of N,N'-sulfuryldiimidazole (496 mg, 2.50 mmol, 1.00 equiv.) in CH$_2$Cl$_2$ (5.00 mL) at 0 °C, methyl triflate (0.25 mL, 369 mg, 2.25 mmol, 0.90 equiv.) was added dropwise over 15 min. After 2 h at 0 °C, the solid was filtered and dried under high vacuum to give triflate salt **24** as a white solid (837 mg, quant.).

– ^{1}H NMR (400 MHz, D$_2$O): δ = 4.04 (s, 3H, CH_3), 7.31(s, 1H, CH_{ar}), 7.75 (s, 1H, CH_{ar}), 7.83 (s, 1H, CH_{ar}), 8.20 (s, 1H, CH_{ar}), 8.54 (s, 1H, CH_{ar}), 9.96 (s, 1H, CH_{ar}) ppm. – ^{13}C NMR (101 MHz, D$_2$O): δ = 37.1 (CH_3), 118.8 (CH_{ar}), 120.6 (CH_{ar}), 126.0 (CH_{ar}), 131.8 (CH_{ar}), 138.5 (2 × CH_{ar}) ppm. – Analytical data is in accordance with previously published literature.[33]

Imidazole-1-sulfonyl azide (25)

Triflate salt **24** (822 mg, 2.27 mmol, 1.00 equiv.) was dissolved in H$_2$O (2.70 mL) at 0 °C, and then the equal volume of EtOAc (2.70 mL) was added. After stirring for 0.5 h, NaN$_3$ (177 mg, 2.72 mmol, 1.20 equiv.) was added in portions and the reaction mixture was stirred at 0 °C for 1 h. The aqueous phase was extracted with EtOAc (2 × 2 mL) and the combined organic phases were dried over Na$_2$SO$_4$. The crude **25** was used for the diazo transfer reaction directly without further purification, the yield was not determined. – Analytical data is in accordance with previously published literature.[33]

(*S*)-*N*-((*S*)-4-Azido-1-(naphthalen-1-yl)butyl)-2-methylpropane-2-sulfinamide (26)

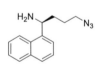

To the *in situ* generated imidazole-1-sulfonyl azide (**25**) in an EtOAc solution (6.00 mL), were sequentially added (*S*)-*N*-((*S*)-4-amino-1-(naphthalen-1-yl)butyl)-2-methylpropane-2-sulfinamide (**22**) (753 mg, 2.36 mmol, 1.20 equiv.) in MeOH (7.60 mL), K_2CO_3 (490 mg, 3.55 mmol, 1.80 equiv.), and anhydrous $CuSO_4$ (3.77 mg, 23.6 µmol, cat.). The mixture was stirred at room temperature for 16 h. The crude residue was purified by column chromatography (0 → 10% MeOH, 0.25% NH_3 in CH_2Cl_2) to give sulfinamide **26** as a yellow oil (561 mg, 83%).

– R_f (*n*-heptane/EtOAc, 2:1) = 0.28. – ^1H NMR (600 MHz, CD_3CN): δ = 1.14 (s, 9H, C(CH_3)$_3$), 1.56–1.64 (m, 1H, CH$_2$CH_2CH$_2$), 1.69–1.76 (m, 1H, CH$_2$CH_2CH$_2$), 2.07–1.98 (m, 2H, CHCH_2CH$_2$), 3.30 (td, *J* = 6.8, 2.3 Hz, 2H, CH$_2$CH_2N), 4.39 (d, *J* = 6.9 Hz, 1H, NH), 5.14 (q, *J* = 6.9 Hz, 1H, NCH), 7.54–7.49 (m, 2H, CH_{ar}), 7.57 (ddd, *J* = 8.5, 6.8, 1.5 Hz, 1H, CH_{ar}), 7.63–7.66 (m, 1H, CH_{ar}), 7.83–7.85 (m, 1H, CH_{ar}), 7.95–7.90 (m, 1H, CH_{ar}), 8.21 (d, *J* = 8.5 Hz, 1H, CH_{ar}) ppm. – ^{13}C NMR (151 MHz, CD_3CN): δ = 23.0 (3 × CCH_3), 26.5 (CH$_2$$CH_2CH_2$), 35.2 (CH$CH_2CH_2$), 51.9 (CH$_2$$CH_2$N), 55.7 ($C$(CH$_3$)$_3$), 56.6 (N$C$H), 124.1 ($CH_{ar}$), 125.4 ($CH_{ar}$), 126.4 ($CH_{ar}$), 126.7 ($CH_{ar}$), 127.2 ($CH_{ar}$), 128.8 ($CH_{ar}$), 129.9 ($CH_{ar}$), 131.7 (naphthyl-$C_q$), 134.9 (naphthyl-$C_q$), 140.0 (naphthyl-$C_q$) ppm. – UPLC-MS: t_R = 1.85 min, *m/z*: calcd. for [M+H]$^+$, $C_{18}H_{25}N_4OS^+$, 345.17; found 345.12.

(*S*)-4-Azido-1-(naphthalen-1-yl)butan-1-amine (27)

The sulfinamide **26** (561 mg, 1.63 mmol, 1.00 equiv.) was dissolved in 2 M HCl in diethyl ether–MeOH 1:1 (12 mL) and stirred at room temperature for 1 h. Then the mixture was concentrated *in vacuo* and purified by flash chromatography (0 → 10% MeOH, 0.25% NH_3 in CH_2Cl_2), yielding product **27** as a light yellow oil (337 mg, 86%).

– R_f (5% MeOH, 0.25% NH_3 in CH_2Cl_2) = 0.31. – ^1H NMR (600 MHz, CD_3OD): δ = 1.57 (ddtd, *J* = 13.7, 10.5, 6.8, 5.5 Hz, 1H, CH$_2$CH_2CH$_2$), 1.67 (ddtd, *J* = 13.7, 10.6, 6.8, 5.0 Hz, 1H, CH$_2$CH_2CH$_2$), 1.92–1.80 (m, 1H, CHCH_2CH$_2$), 1.97 (ddt, *J* = 13.6, 10.6, 5.9 Hz, 1H, CHCH_2CH$_2$), 3.26 (t, *J* = 6.8 Hz, 2H, CH$_2$CH_2N), 4.78 (t, *J* = 6.7 Hz, 1H, NCH), 7.51–7.45 (m, 2H, CH_{ar}), 7.53 (ddd, *J* = 8.4, 6.7, 1.4 Hz, 1H, CH_{ar}), 7.61 (dd, *J* = 7.2, 1.1 Hz, 1H, CH_{ar}), 7.77 (d, *J* = 8.1 Hz, 1H, CH_{ar}), 7.87 (dd, *J* = 8.2, 1.4 Hz, 1H, CH_{ar}), 8.14 (d, *J* = 8.5 Hz, 1H, CH_{ar}). Missing signals (2H, NH_2) ppm. – ^{13}C NMR (151 MHz, CD_3OD): δ = 27.0 (CH$_2$$CH_2CH_2$), 37.0 (CH$CH_2CH_2$), 51.1 (CH$_2$$CH_2$N), 52.4 (N$C$H), 123.6 ($CH_{ar}$), 123.7 ($CH_{ar}$), 126.6 ($CH_{ar}$), 126.6 ($CH_{ar}$), 127.2 ($CH_{ar}$),

128.5 (CH_{ar}), 130.0 (CH_{ar}), 132.3 (naphthyl-C_q), 135.4 (naphthyl-C_q), 142.5 (naphthyl-C_q) ppm. – HRMS (MALDI-TOF) m/z: calcd. for [M + H]$^+$C$_{14}$H$_{17}$N$_4^+$ 241.1448, found 241.1447.

5.2.3 Synthesis of Trimers and Hexamers

β-Peptoid dimer 29

The peptoid dimer was synthesized from the peptoid monomer **28** (9.54 g, 31.9 mmol, 1.00 equiv.) according to GP B. The crude acryloylated peptoid (744 mg, 2.10 mmol, 1.50 equiv.) was then reacted with the amine **27** (337 mg, 1.40 mmol, 1.00 equiv.) according to GP A. The crude was purified by automated column chromatography (0.5 → 4% MeOH, 0.25% NH$_3$ in CH$_2$Cl$_2$), yielding the product **29** as a light yellow foam (400 mg, 48%).

– R_f (3% MeOH, 0.25% NH$_3$ in CH$_2$Cl$_2$) = 0.26. – UPLC-MS: t$_R$ = 1.98 min, m/z: calcd. for [M+H]$^+$, C$_{36}$H$_{44}$N$_5$O$_3^+$, 594.34; found 594.34.

β-Peptoid trimer 32

The peptoid trimer was synthesized from **29** (150 mg, 250 μmol, 1.00 equiv.) according to GP B, followed by GP A using (*S*)-1-(1-naphthyl)ethylamine **31** (0.16 mL, 170 mg, 1.00 mmol, 4.00 equiv.). The crude was purified by automated column chromatography (0.5 → 4% MeOH, 0.25% NH$_3$ in CH$_2$Cl$_2$) yielding product **32** as a yellow foam (90.0 mg, 44%).

– R_f (3% MeOH, 0.25% NH$_3$ in CH$_2$Cl$_2$) = 0.30. – HRMS (MALDI-TOF) m/z: calcd. for [M + H]$^+$ C$_{51}$H$_{59}$N$_6$O$_4^+$ 819.4592, found 819.4593.

β-Peptoid trimer 36

The peptoid trimer was synthesized from **33** (200 mg, 240 μmol, 1.00 equiv.) according to GP C followed by GP D using Alloc-piperazine·HCl salt **34** (149 mg, 720 μmol, 3.00 equiv.). The peptoid trimer **35** was obtained as a white foam (166 mg, 66%) after automated column chromatography (0 → 10% MeOH, 0.25% NH$_3$ in CH$_2$Cl$_2$). Trimer **35** (166 mg, 159 μmol, 1.00 equiv.) was then subjected to GP E1. The peptoid trimer **36** was obtained as a light yellow foam after column chromatography (58.2 mg, 24% over 2 steps).

$-R_f = 0.28$ (5% MeOH, 0.25% NH$_3$ in CH$_2$Cl$_2$). $-$ HRMS (MALDI-TOF) m/z: calcd. for [M + H]$^+$ C$_{58}$H$_{67}$N$_8$O$_7^+$ 987.5127, found 987.5112.

β-Peptoid hexamer 37

The peptoid hexamer **37** was synthesized according to GP F using peptoid trimers **32** (48.1 mg, 59.0 µmol, 1.00 equiv.) and **36** (58.0 mg, 59.0 µmol, 1.00 equiv.). The residual crude product was purified by preparative HPLC (gradient of eluent IV rising linearly from 50 → 95% in eluent III over 25 min, t_R = 29.1 – 30.3 min). The peptoid hexamer **37** was obtained as a white solid (46.0 mg, 44%).

$-$ HRMS (MALDI-TOF) m/z: calcd. for [M + H]$^+$ C$_{109}$H$_{123}$N$_{14}$O$_{10}^+$ 1787.9540, found 1787.9543. Purity according to analytical HPLC (gradient C): 93%.

β-Peptoid hexamer 39

In a vial, propynol ethoxylate (**38**) (2.4 µL, 2.46 mg, 24.6 µmol, 2.00 equiv.) and β-peptoid **37** (22.0 mg, 12.3 µmol, 1.00 equiv.) were dissolved in dry THF (250 µL, 0.05 M). 2.6-Lutidine (2.9 µL, 2.64 mg, 25.0 µmol, 2.00 equiv.) and DIPEA (4.3 µL, 3.18 mg, 24.6 µmol, 2.00 equiv.), CuI (0.469 mg, 2.50 µmol, 20 mol%) and TBTA (1.31 mg, 2.50 µmol, 20 mol%) were added. The reaction was stirred at room temperature for 1 h. After completion, the reaction mixture was filtered, and the solvent was removed in *vacuo* to give the crude Alloc-protected hexamer. The crude (24.0 mg, 12.1 µmol, 1.00 equiv.) was dissolved in 0.30 mL anhydrous CH$_2$Cl$_2$ and

Me$_2$NH$_2$·BH$_3$ (14.2 mg, 241 µmol, 20.0 equiv.) was added. Then, Pd(PPh$_3$)$_4$ (2.79 mg, 2.40 µmol, 20 mol%) was added and the mixture was stirred for 90 min. The reaction mixture was evaporated to dryness and the crude was purified by preparative HPLC (gradient of eluent IV rising linearly from 30 → 95% in eluent III over 25 min, t$_R$ = 22.3 – 23.2 min). The peptoid hexamer **39** was obtained as a white solid (9.20 mg, 40% over 2 steps).

– HRMS (MALDI-TOF) *m/z*: calcd. for [M + H]$^+$ C$_{115}$H$_{135}$N$_{14}$O$_{12}$$^+$ 1904.0377, found 1904.0396. Purity according to analytical HPLC (gradient B): 91%.

β-Peptoid hexamer 40

The peptoid hexamer **40** was synthesized according to GP F using peptoid trimers **32** (29.6 mg, 36.0 µmol, 1.00 equiv.) and **IWE-155** (33.2 mg, 36.0 µmol, 1.00 equiv.). The residual crude product was purified by preparative HPLC (gradient of eluent IV rising linearly from 50 → 95% in eluent III over 25 min, t$_R$ = 28.7 – 29.6 min). The peptoid hexamer **40** was obtained as a white solid (23.9 mg, 39%).

– HRMS (MALDI-TOF) *m/z*: calcd. for [M + H]$^+$ C$_{107}$H$_{120}$N$_{11}$O$_{10}$$^+$ 1718.9213; found 1718.9220. Purity according to analytical HPLC (gradient A): 91%.

β-Peptoid hexamer 42

In a vial, ethynylpyridine (**41**) (2.5 µL, 2.55 mg, 25.5 µmol, 2.00 equiv.) and *β*-peptoid **40** (21.9 mg, 12.7 µmol, 1.00 equiv.) were dissolved in dry THF (257 µL, 0.05 M). 2,6-Lutidine (3.00 µL, 2.73 mg, 25.0 µmol, 2.00 equiv.) and DIPEA (4.40 µL, 3.29 mg, 25.5 µmol, 2.00 equiv.), CuI (490 µg, 2.50 µmol, 20 mol%) and TBTA (1.35 mg, 2.50 µmol, 20 mol%) were added. The reaction was stirred at room temperature for 4 h. After completion, the reaction mixture was filtered, and the solvent was removed in *vacuo* to give the crude Alloc-protected hexamer. The crude hexamer (23.0 mg, 12.6 μmol, 1.00 equiv.) was dissolved in 315 μL anhydrous CH_2Cl_2 and $Me_2NH_2 \cdot BH_3$ (14.9 mg, 252 μmol, 20.0 equiv.) was added. Then, $Pd(PPh_3)_4$ (2.92 mg, 2.50 μmol, 20 mol%) was added and the mixture was stirred for 60 min. The reaction mixture was evaporated to dryness, redissolved in DMF, and after filtration the crude was purified by preparative HPLC (gradient of eluent IV rising linearly from 50 → 95% in eluent III over 25 min, t_R = 16.3 – 17.5 min). The peptoid hexamer **42** was obtained as a light yellow solid (10.4 mg, 47%).

– HRMS (MALDI-TOF) *m/z*: calcd. for $[M + H]^+$ $C_{110}H_{121}N_{12}O_8^+$ 1737.9424; found 1737.9432. Purity according to analytical HPLC (gradient B): 95%.

β-Peptoid hexamer 43

In a vial, ethynylpyridine (**41**) (2.5 μL, 2.54 mg, 24.6 μmol, 2.00 equiv.) and β-peptoid **37** (22.0 mg, 12.3 μmol, 1.00 equiv.) were dissolved in dry THF (250 μL, 0.05 M). 2,6-Lutidine (2.9 μL, 2.64 mg, 25.0 μmol, 2.00 equiv.) and DIPEA (4.3 μL, 3.18 mg, 24.6 μmol, 2.00 equiv.), CuI (0.469 mg, 2.50 μmol, 20 mol%) and TBTA (1.31 mg, 2.50 μmol, 20 mol%) were added. The reaction was stirred at room temperature for 1 h. After completion, the reaction mixture was filtered, and the solvent was removed in *vacuo* to give the crude Alloc-protected hexamer. The crude hexamer (24.0 mg, 12.0 μmol, 1.00 equiv.) was dissolved in 0.30 mL anhydrous CH_2Cl_2 and $Me_2NH_2 \cdot BH_3$ (14.2 mg, 241 μmol, 20.0 equiv.) was added. Then, $Pd(PPh_3)_4$ (2.78 mg, 2.40 μmol, 20 mol%) was added and the mixture was stirred for 90 min. The reaction mixture was evaporated to dryness, redissolved in DMF, and after filtration the crude was purified by preparative HPLC (gradient of eluent IV rising linearly from 30 → 95% in eluent III over 25 min, t_R = 20.9 min– 22.1 min). The peptoid hexamer **43** was obtained as a light yellow solid (10.7 mg, 47%).

– HRMS (MALDI-TOF) *m/z*: calcd. for $[M + H]^+$ $C_{120}H_{128}N_{16}O_8^+$ 1910.0173, found 1910.0198. Purity according to analytical HPLC (gradient B): 94%.

6 List of Abbreviations

A

Alloc .... allyloxycarbonyl

aq .... aqueous

B

9-BBN.................9-borabicyclo[3.3.1]nonane

Bu .... butyl

C

CD.. circular dichroism

COSY .... correlation spectroscopy

CuAAC.................................... Cu(I)-catalyzed

azide–alkyne cycloaddition

D

DIPEA ....N,N-diisopropylethylamine

DMF .... dimethylformamide

E

ELSD ..... evaporative light-scattering detector

ESI ....electrospray ionization

Et.. ethyl

H

HATU ....hexafluorophosphate

azabenzotriazole tetramethyl uranium

HMBC ....heteronuclear multiple-bond

correlation spectroscopy

HPLC....................... high-performance liquid

chromatography

HRMS.......high-resolution mass spectrometry

HSQC ....heteronuclear single-quantum

correlation spectroscopy

M

MALDI ....matrix-assisted laser

desorption/ionization

Me.. methyl

MeCN ....acetonitrile

N

Nap..naphthyl

NMR .... nuclear magnetic resonance

Ns1npe .... N-(S)-1-(1-naphthyl)ethyl

P

PBS .... phosphate-buffered saline

R

RP-HPLC... reversed-phase high-performance

liquid chromatography

rt...room temperature

T

TBDMS .... $tert$-butyldimethylsilyl

TBTA..........tris(benzyltriazolylmethyl)amine

TFA.................................... trifluoroacetic acid

THF..tetrahydrofuran

TLC...................... thin layer chromatography

TOF... time of flight

U

UPLC............... ultra high-performance liquid

chromatography

7   References

[1]   R. N. Zuckermann, T. Kodadek, *Curr. Opin. Mol. Ther.* **2009**, *11*, 299–307.
[2]   R. Gopalakrishnan, A. I. Frolov, L. Knerr, W. J. Drury III, E. Valeur, *J. Med. Chem.* **2016**, *59*, 9599–9621.
[3]   Z. C. Girvin, M. K. Andrews, X. Liu, S. H. Gellman, *Science* **2019**, *366*, 1528–1531.
[4]   A. Battigelli, J. H. Kim, D. C. Dehigaspitiya, C. Proulx, E. J. Robertson, D. J. Murray, B. Rad, K. Kirshenbaum, R. N. Zuckermann, *ACS nano* **2018**, *12*, 2455–2465.
[5]   E. J. Robertson, A. Battigelli, C. Proulx, R. V. Mannige, T. K. Haxton, L. Yun, S. Whitelam, R. N. Zuckermann, *Acc. Chem. Res.* **2016**, *49*, 379–389.
[6]   S. H. Gellman, *Acc. Chem. Res.* **1998**, *31*, 173–180.
[7]   A. D. Bautista, C. J. Craig, E. A. Harker, A. Schepartz, *Curr. Opin. Chem. Biol.* **2007**, *11*, 685–692.
[8]   C. M. Goodman, S. Choi, S. Shandler, W. F. DeGrado, *Nat. Chem. Biol.* **2007**, *3*, 252–262.
[9]   C. A. Olsen, *ChemBioChem* **2010**, *11*, 152–160.
[10] D. Seebach, A. K. Beck, D. J. Bierbaum, *Chem. Biodiv.* **2004**, *1*, 1111–1239.
[11] D. Seebach, J. Gardiner, *Acc. Chem. Res.* **2008**, *41*, 1366–1375.
[12] Y.-D. Wu, S. Gellman, *Acc. Chem. Res.* **2008**, *41*, 1231–1232.
[13] B. Yoo, K. Kirshenbaum, *Curr. Opin. Chem. Biol.* **2008**, *12*, 714–721.
[14] K. Kirshenbaum, A. E. Barron, R. A. Goldsmith, P. Armand, E. K. Bradley, K. T. Truong, K. A. Dill, F. E. Cohen, R. N. Zuckermann, *Proc. Nat. Acad. Sci.* **1998**, *95*, 4303–4308.
[15] C. W. Wu, T. J. Sanborn, R. N. Zuckermann, A. E. Barron, *J. Am. Chem. Soc.* **2001**, *123*, 2958–2963.
[16] C. W. Wu, T. J. Sanborn, K. Huang, R. N. Zuckermann, A. E. Barron, *J. Am. Chem. Soc.* **2001**, *123*, 6778–6784.
[17] C. W. Wu, K. Kirshenbaum, T. J. Sanborn, J. A. Patch, K. Huang, K. A. Dill, R. N. Zuckermann, A. E. Barron, *J. Am. Chem. Soc.* **2003**, *125*, 13525–13530.
[18] B. C. Gorske, B. L. Bastian, G. D. Geske, H. E. Blackwell, *J. Am. Chem. Soc.* **2007**, *129*, 8928–8929.
[19] B. C. Hamper, S. A. Kolodziej, A. M. Scates, R. G. Smith, E. Cortez, *J. Org. Chem.* **1998**, *63*, 708–718.
[20] S. Mukherjee, G. Zhou, C. Michel, V. A. Voelz, *J. Phys. Chem. B* **2015**, *119*, 15407–15417.
[21] J. Fernandez-Santin, J. Aymamí, A. Rodríguez-Galán, S. Munoz-Guerra, J. Subirana, *Nature* **1984**, *311*, 53–54.
[22] P. Armand, K. Kirshenbaum, R. A. Goldsmith, S. Farr-Jones, A. E. Barron, K. T. Truong, K. A. Dill, D. F. Mierke, F. E. Cohen, R. N. Zuckermann, *Proc. Nat. Acad. Sci.* **1998**, *95*, 4309–4314.
[23] M. L. DeRider, S. J. Wilkens, M. J. Waddell, L. E. Bretscher, F. Weinhold, R. T. Raines, J. L. Markley, *J. Am. Chem. Soc.* **2002**, *124*, 2497–2505.
[24] M. P. Hinderaker, R. T. Raines, *Prot. Sci.* **2003**, *12*, 1188–1194.
[25] A. Choudhary, D. Gandla, G. R. Krow, R. T. Raines, *J. Am. Chem. Soc.* **2009**, *131*, 7244–7246.
[26] J. S. Laursen, J. Engel-Andreasen, P. Fristrup, P. Harris, C. A. Olsen, *J. Am. Chem. Soc.* **2013**, *135*, 2835–2844.
[27] J. S. Laursen, J. Engel-Andreasen, C. A. Olsen, *Acc. Chem. Res.* **2015**, *48*, 2696–2704.
[28] J. S. Laursen, P. Harris, P. Fristrup, C. A. Olsen, *Nat. Commun.* **2015**, *6*, 7013.
[29] I. Wellhöfer, K. Frydenvang, S. Kotesova, A. M. Christiansen, J. S. Laursen, C. A. Olsen, *J. Org. Chem.* **2019**, *84*, 3762–3779.
[30] T. Hjelmgaard, S. Faure, C. Caumes, E. De Santis, A. A. Edwards, C. Taillefumier, *Org. Lett.* **2009**, *11*, 4100–4103.

[31]   V. V. Rostovtsev, L. G. Green, V. V. Fokin, K. B. Sharpless, *Angew. Chem. Int. Ed.* **2002**, *41*, 2596–2599.

[32]   C. W. Tornøe, C. Christensen, M. Meldal, *J. Org. Chem.* **2002**, *67*, 3057–3064.

[33]   H. Ye, R. Liu, D. Li, Y. Liu, H. Yuan, W. Guo, L. Zhou, X. Cao, H. Tian, J. Shen, *Org. Lett.* **2013**, *15*, 18–21.

[34]   C. M. Darapaneni, P. J. Kaniraj, G. Maayan, *Org. Biomol. Chem.* **2018**, *16*, 1480–1488.

[35]   G. Maayan, *Eur. J. Org. Chem.* **2009**, *2009*, 5699–5710.

[36]   F. Zhang, S. Bai, G. P. Yap, V. Tarwade, J. M. Fox, *J. Am. Chem. Soc.* **2005**, *127*, 10590–10599.

[37]   M. Baskin, H. Zhu, Z.-W. Qu, J. H. Chill, S. Grimme, G. Maayan, *Chem. Sci.* **2019**, *10*, 620–632.

[38]   B.-C. Lee, T. K. Chu, K. A. Dill, R. N. Zuckermann, *J. Am. Chem. Soc.* **2008**, *130*, 8847–8855.

[39]   G. Maayan, M. D. Ward, K. Kirshenbaum, *Chem. Commun.* **2009**, 56–58.

[40]   T. Ghosh, P. Ghosh, G. Maayan, *ACS Catal.* **2018**, *8*, 10631–10640.

[41]   D. C. Mohan, A. Sadhukha, G. Maayan, *J. Catal.* **2017**, *355*, 139–144.

[42]   M. Baskin, G. Maayan, *Chem. Sci.* **2016**, *7*, 2809–2820.

[43]   M. C. Pirrung, K. Park, L. N. Tumey, *J. Comb. Chem.* **2002**, *4*, 329–344.

[44]   G. Maayan, M. D. Ward, K. Kirshenbaum, *Proc. Nat. Acad. Sci.* **2009**, *106*, 13679–13684.

[45]   K. J. Prathap, G. Maayan, *Chem. Commun.* **2015**, *51*, 11096–11099.

[46]   N. Maulucci, I. Izzo, G. Bifulco, A. Aliberti, C. De Cola, D. Comegna, C. Gaeta, A. Napolitano, C. Pizza, C. Tedesco, *Chem. Commun.* **2008**, 3927–3929.

[47]   X. Feng, Y. Wang, B. Wei, J. Yang, H. Du, *Org. Lett.* **2011**, *13*, 3300–3303.

Curriculum Vitae

Personal information

Name	Janina Beck
Date/Place of Birth	13.08.1990 in Mannheim
Address	Veilchenstr. 26, 76131 Karlsruhe
E-Mail	janina_beck@yahoo.de

Education

11/2016 – 04/2020 **Ph. D.** (magna cum laude), Institute of Organic Chemistry at the Karlsruhe Institute of Technology (KIT), group of Prof. Dr. Stefan Bräse

10/2014 – 09/2016 **M. Sc.** (1.0), Chemical Biology at the Karlsruhe Institute of Technology (KIT), focus on Chemical Biology and Organic Chemistry
Master Thesis: *Studies on the Tetracyclic Core Structure of Beticolin 0*

10/2011 – 09/2014 **B. Sc.** (1.7), Chemical Biology at the Karlsruhe Institute of Technology (KIT), focus on Biochemistry and Chemical Biology
Bachelor Thesis: *Enzymatic Solid Phase Synthesis of Short Hyaluronic Acids with a Magnetic Bioreactor*

Experience/Internships

08/2019 – 12/2019 **Research Internship**, Department of Drug Design and Pharmacology, University of Copenhagen, DK, group of Prof. Dr. Christian A. Olsen

03/2016 – 02/2015 **Research Assistant**, ScreenFect GmbH, Eggenstein-Leopoldshafen

Awards

07/2017– 04/2020 Carl-Zeiss Ph. D. scholarship

08/2019 – 12/2019 Karlsruhe House of Young Scientists (KHYS) Research Travel Grant

Conferences/Posters

01/2018 Global Young Scientists Summit, NTU, Singapore
Poster: *Studies Towards the Total Synthesis of Beticolin 0*

02/2018 Irseer Naturstofftage (natural product conference), Irsee, Germany

Publications

A versatile Diels-Alder Approach to Functionalized Hydroanthraquinones. J. Beck, O. Fuhr, M. Nieger, S. Bräse, *submitted*

Increasing the Functional Group Diversity in Helical β-Peptoids: Achievement of Solvent- and pH-Dependent Folding. I. Wellhöfer, J. Beck, K. Frydenvang, S. Bräse, C. A. Olsen, *J. Org. Chem.* **2020**, *85*, 16, 10466–10478, DOI: 10.1021/acs.joc.0c00780

Danksagung – Acknowledgements

An dieser Stelle möchte ich mich vom ganzen Herzen bei allen bedanken, die mich während meiner Promotion auf die eine oder andere Weise unterstützt haben und so zum Gelingen dieser Arbeit beigetragen haben.

Mein besonderer Dank gilt meinem Doktorvater Prof. Dr. Stefan Bräse für die Möglichkeit in seiner Arbeitsgruppe zu forschen und das mir anvertraute spannende und anspruchsvolle Promotionsthema. Vielen Dank für die fachliche sowie menschliche Betreuung und Unterstützung, die mir überlassene Freiheit, sowie die Möglichkeit wertvolle Erfahrungen im Rahmen von Konferenzen und während eines Auslandsaufenthaltes zu sammeln.

Herrn Prof. Dr. Joachim Podlech danke ich für die freundliche Übernahme des Korreferats.

Für die finanzielle Förderung dieser Arbeit möchte ich der Carl-Zeiss-Stiftung danken.

I want to thank Prof. Dr. Christian Adam Olsen from the University of Copenhagen for the opportunity to do research in his group and explore the chemistry of foldamers. A special thanks goes to the whole working group for making this a unique experience and to Isabelle Wellhöfer for the great teamwork and the good times in beautiful Copenhagen. Der Auslandsaufenthalt wurde vom Karlsruhe House of Young Scientists (KHYS) gefördert, auch dafür möchte ich danken.

Ein großer Dank gilt Dr. Martin Nieger und Dr. Olaf Fuhr für zahlreiche Kristallstrukturanalysen.

Für Ihren Einsatz jeglicher Art möchte ich mich bei Dr. Norbert Foitzik, Angelika Mösle, Danny Wagner, Lara Hirsch, Rieke Schulte, Dr. Andreas Rapp, Pia Lang, Tanja Ohmer-Scherrer, Lennart Oberle, Richard von Budberg, Karolin Kohnle und Despina Savvidou-Kourmpidou herzlich bedanken. Vielen Dank auch an das ComPlat-Team!

Danke an Christiane Lampert, Selin Samur, Janine Bolz, Dr. Ilona Wehl und Dr. Christin Bednarek für die stets freundliche, schnelle und kompetente Unterstützung bei organisatorischen Fragen.

Bei Ahmad Qais Parsa, Rieke Schulte, Nasrin Salaeh-arae und Rafael Sterzik möchte ich mich für die fleißige Unterstützung im Labor, gute Ideen und unermüdliche Motivation bedanken. Qais, es war eine Bereicherung mit Dir zusammen zu arbeiten, ich bin sehr stolz auf Dich!

Der gesamten Arbeitsgruppe Bräse möchte ich für Anregungen, Hilfsbereitschaft und Feedback und vor allem für eine unvergessliche Zeit danken. Ich durfte tolle Menschen kennenlernen und besondere Freundschaften schließen. Mein besonderer Dank gilt dabei Dr. Yuling Hu und Dr. Robin Bär für die beste Zeit in 306, Dr. Alexander Braun für spaßige Erlebnisse inner- und außerhalb des Labors, Dr. Florian Mohr für die schönste Frisur, Dr. Stefan Marschner für das tollste Karlsruher Oktoberfest, Dr. Vanessa Koch und Jasmin Busch für kulinarische Erlebnisse, meinen Lieblings-Labornachbarn Susanne Kirchner und Dr. Johannes Karcher für jede Menge Spaß und gute Gespräche, Dr. Larissa Geiger und Dr. Christina Retich für gemeinsames Auspowern und Dr. Marco Mende dafür, dass Du so bist wie Du bist.

Dr. Stefan Marschner, Dr. Robin Bär und Dr. Vanessa Koch danke ich für die gewissenhafte Korrektur dieser Arbeit. Danke, dass Ihr Euch so viel Zeit dafür genommen habt!

Danke auch an meine lieben Freunde, die mein Leben nun schon seit so vielen Jahren begleiten und jegliche Höhen und Tiefen miterlebt und mitgetragen haben. Danke für Eure treue Freundschaft, auch wenn ich mich manchmal etwas rar gemacht habe.

Von Herzen danke ich meinen Eltern Daniela und Karl, meinem Bruder Julian sowie meinen Großeltern Anita und Otto für Ihren Glauben an mich, die immerwährende Unterstützung und die Möglichkeit, das tun zu können was ich mir wünsche. Mein Dank gilt ebenso meinen Großeltern Elsa und Karl, die diesen Moment leider nicht mehr miterleben dürfen.

Tobi, ich bin so dankbar dafür Dich zu haben! Danke für Deinen klugen Rat, Deine Geduld, Zuversicht, ein immer offenes Ohr und bedingungsloses Vertrauen in mich.

Meine Gedanken sind auch bei Patricia, die so viel für die Möglichkeit zu promovieren gegeben hätte. Danke, dass Du mir gezeigt hast was unermüdliche Zuversicht, Lebensfreude und Stärke bedeuten.